e's
nomy Series

Springer
London
Berlin
Heidelberg
New York
Hong Kong
Milan
Paris
Tokyo

Other Titles in this Series

Telescopes and Techniques
Chris Kitchin

The Art and Science of CCD Astronomy
David Ratledge (Ed.)

The Observer's Year
Patrick Moore

Seeing Stars
Chris Kitchin and Robert W. Forrest

Photo-guide to the Constellations
Chris Kitchin

The Sun in Eclipse
Michael Maunder and Patrick Moore

Software and Data for Practical Astronomers
David Ratledge

Amateur Telescope Making
Stephen F. Tonkin

Observing Meteors, Comets, Supernovae and other Transient Phenomena
Neil Bone

Astronomical Equipment for Amateurs
Martin Mobberley

Transit: When Planets Cross the Sun
Michael Maunder and Patrick Moore

Practical Astrophotography
Jeffrey R. Charles

Observing the Moon
Peter T. Wlasuk

Deep-Sky Observing
Steven R. Coe

AstroFAQs
Stephen F. Tonkin

The Deep-Sky Observer's Year
Grant Privett and Paul Parsons

Field Guide to the Deep Sky Objects
Mike Inglis

Choosing and Using a Schmidt–Cassegrain Telescope
Rod Mollise

Astronomy with Small Telescopes
Stephen F. Tonkin (Ed.)

Solar Observing Techniques
Chris Kitchin

Observing the Planets
Peter T. Wlasuk

Light Pollution
Bob Mizon

Using the Meade ETX
Mike Weasner

Practical Amateur Spectroscopy
Stephen F. Tonkin (Ed.)

More Small Astronomical Observatories
Patrick Moore (Ed.)

The Practical Astronomer's Deep-sky Companion

Jess K. Gilmour

With 841 Illustrations

Springer

British Library Cataloguing in Publication Data
Gilmour, Jess K.
The practical astronomer's deep-sky companion. – (Patrick Moore's practical astronomy series)
1. Astronomy – Observers' manuals 2. Stars
I. Title
522
ISBN 1852334746

Library of Congress Cataloging-in-Publication Data
Gilmour, Jess K., 1963–
The practical astronomer's deep-sky companion/Jess K. Gilmour.
p. cm. – (Patrick Moore's practical astronomy series, ISSN 1617-7185)
Includes index.
ISBN 1-85233-474-6 (alk. paper)
1. Astronomy – Observers' manuals. I. Title. II. Series.
QB64.G46 2002
522–dc21 2002070838

Patrick Moore's Practical Astronomy Series ISSN 1617-7185
ISBN 1-85233-474-6 Springer-Verlag London Berlin Heidelberg
a member of BertelsmannSpringer Science+Business Media GmbH
http://www.springer.co.uk

Printed in Singapore

Typeset by EXPO Holdings, Malaysia
Printed and bound by Kyodo Printing Co. (S'pore) Pte. Ltd., Singapore
58/3830-543210 Printed on acid-free paper SPIN 10833992

This book is dedicated to my wife Roxie,
whose patience and understanding has
helped me be the best person I can be.

To my children Nathan and Rowan, whose
very presence makes me strive for perfection.

Last I wish to thank all my observing buddies.
You know who you are. We have spent many
a clear, star-filled night solving the world's problems
and observing all that the night sky has to offer.

Dreams do come true!

Preface

As an amateur astronomer with years of experience, I marvel at the joy experienced by a beginner who successfully hunts down their first deep-space object in a telescope. No matter what age or skill level, "nailing" a previously unobserved object through the eyepiece, both instantly defines their love of the hobby and gives a feeling of scientific accomplishment no matter how well known to others the object may be.

With the advancement in computer-guided telescopes and automatic object centering, the amateur astronomy hobby has experienced tremendous and unprecedented growth. First timers are attending public observing sessions or summer star parties with low-cost computer-controlled telescopes, and are instantly rewarded with views of celestial objects with strange names and numbers. But: what to look at? Can I see it through my telescope? For the seasoned observer the problem is different: "I've seen that object a thousand times, can anybody show me something new?" Astrophotographers, novice and seasoned, often wonder about capturing new objects on film or CCD, but first spend hours poring through star charts familiarizing themselves with the star field, selecting guide star, etc.

The contents of this book combines, in a clear and concise manner, information that will assist beginner, novice, intermediate and advanced amateur astronomy hobbyists. The objects are all visible in medium-to-large-aperture telescopes and provide a wide selection of objects to observe or photograph.

Contents

Introduction

Ancient societies and religions have associated the battle between Good and Evil with visual observations of the rising and setting Sun, so many cultures began to mark the approach or passing of the winter and summer solstice with remarkable accuracy and elaborate ceremonies. It is believed that these ceremonies came from an ancient fear that the failing light would not return unless some sort of sacred vigil observing the rebirth of the Sun's ascent into the sky was practiced.

Widely disparate cultures built tombs, temples, cairnes and sacred observatories, aligned to mark the solstices and equinoxes. There is even archeological evidence that medieval Catholic churches were built not just as houses of worship, but also as solar observatories. The church, reinforcing the ancient ties between religious celebration and the seasonal changes, needed astronomy to predict the date of Easter – which even now is held on the Monday following the full moon following the Spring equinox – so observatories were built into cathedrals and churches. Usually a small hole in the roof admitted a beam of sunlight, which would trace a path along the floor known as the meridian line. Monks meticulously marked the Sun's movement along this line, noting the coming and going of the seasons.

As time passed and man's understanding evolved, we attempted to understand the shifting mosaic of stars. Soon scientists began to grind glass and fashion the first telescopes to magnify views of the night sky, only to be confronted with even more intriguing mysteries of fuzzy, blurred, greenish patches of – what? The night sky seemed to be a continuously more complex mystery.

Since the nineteenth century, astronomy has been a science practiced at night from remote locations with big, expensive, imposing instruments. For the media and many people the difficulty of the resulting science meant that astronomy was largely ignored. Advances and discoveries that profoundly affected our understanding of the universe and its origins went unnoticed.

The launch of the Hubble Space Telescope changed everything. One of NASA's riskiest launches, and astronomy's greatest research endeavor, suddenly took center-stage. The problem for the media, however, was being disinterested resulted in being unprepared. Scrambling for news, television media departments began to ask questions. What was astronomy all about? Who was doing it? Where were they doing it? Why were they doing it? Globally – and especially in the United States – interest in astronomy soared. Fortunately for the media but unfortunately for NASA, the science stayed in the public eye because of flawed optics in the telescope. The media loves nothing better than a problem or disaster, and this was a mistake in the order of billions of dollars! Coverage of the discovery of the problem, the fix, and the repair mission kept Hubble, NASA and astronomy under public and media scrutiny for many years. It wasn't long after the repair mission that the Hubble team began to use the telescope to its full potential, releasing photographs of billowing structures of gas and star-forming regions, planetary nebulae, galaxies, and even spectacular images of the planets. And the images and discoveries kept on coming ...

There have always been those of us to whom the night sky is a constant companion – not to be ignored but rather, embraced – its secrets revealed through the eyepiece of a telescope or captured on film with a camera. Beyond our galaxy there are others, nebulae within our own galactic spiral arms, super nova remnants – the illuminated dust and gas of long-ago catastrophic and explosive deaths of massive stars. It is to these distant objects that the interested are almost invariably drawn. For many, the pursuit of practical astronomy began at an early age when they received a small telescope as a Christmas present.

And just as the professional astronomers' tools have changed, so too have those available to the hobby astronomer. Now for a thousand dollars or so, it is possible to buy a reasonable-sized backyard, portable telescope that is computer controlled. No longer do we amateur observers need to get frustrated trying to locate unknown objects in unknown star fields; we can just set the scope up and have the computer guide us through a tour of the universe automatically. Instant viewing gratification, along with a continuous stream of new astronomical discoveries, has resulted in unprecedented growth in the hobby of astronomy as a pass-time.

I have been observing and photographing the night sky for twenty-five years. I began my pursuit of the hobby when I was ten years old, using a six-inch Newtonian reflector telescope, but now I too have become one of those computer-controlled-telecope users and have lost the art of locating objects in the celestial sphere, armed only with star maps and setting circles. The ability to locate almost any object in the sky provides unlimited observing possibilities. But what objects should I look at first? Should I use a list of planetary nebulae? What about galaxies? (Galaxies would be a challenge and probably take the rest of my life to complete.) Settling on the familiar, I compiled a list of objects that I had meticulously observed over the years, hoping that it would permit me the time to research and compile a list of new objects, not as yet observed. It didn't.

I wanted to compile my new list quickly. I started by creating a spreadsheet of northern hemisphere and zodiacal constellations observable from my various viewing locations throughout the Province of Ontario, Canada, which range from latitude 46 to 48 degrees North. I then researched, constellation-by-constellation, the position of thousands of objects. Objects selected, were those that could best be observed with popular commercially available telescopes (4–12 inch, that's 100–300 mm). As the list took shape and grew in size, further selection criteria were introduced to trim the number of objects. The new selection criteria became (i) visual presentation, (ii) photographic splendor, (iii) obscurity, (iv) observing challenge, and (v) sky track.

This book represents that final list. I hope that you find it as useful as I have. The objects selected range from common, beginner level naked eye objects to those that will challenge even the most seasoned observer or astrophotographer.

Book Format

Just as the original list was, the book is divided into the separate constellations visible throughout the year from the northern hemisphere. A few constellations were deliberately omitted because they did not contain any objects that met the selection criteria. The next challenge was how to represent the data in a way that would appeal to all amateur astronomers, from beginners to the truly dedicated. A combination of maps, photos and data tables seemed to be the best solution to the problem of presenting the information in a format that would not be overwhelming or impossible to read in the dark.

Each section begins with a map of the entire constellation, followed by pertinent data such as: latitudes from which it is visible; number of degrees of sky the constellation covers; sky track; when it crosses the meridian; correct pronunciation. Following the constellation page, begins the presentation of the information for each selected object in that particular constellation. Here the reader will find localized star charts for each object, an accompanying photograph, and a data frame containing the pertinent data to locate each object in the sky.

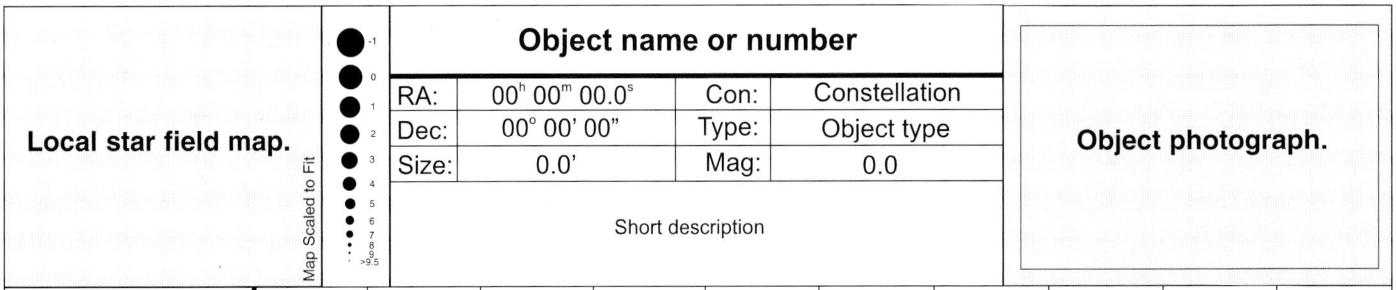

Telescope Aperture:	4" f/5	4" f/9	6" f/7	6" f/9	8" f/6.3	8" f/10	10" f/6.3	10" f/10	12" f/6.3	12" f/10
FOV(35mm film):	2.7° x 4.1°	1.50° x 2.26°	1.29° x 1.93°	1.0° x 1.50°	1.07° x 1.61°	0.68° x 1.02°	0.86° x 1.29°	0.54° x 0.81°	0.72° x 1.07°	0.45° x 0.68°

RA:	given Right Ascension coordinates in hh:mm:ss for Epoch 2000.
Dec:	given Declination coordinates in dd:mm:ss for Epoch 2000.
Size:	the data is represented as either a number (00.0') in arcminute's or as a set of numbers with a multiplier between denoting the major and minor axes of the object in arcminute's.
Con:	the constellation within which the object is located.
Type:	identifies the object type. Categories include Nebula (emission and reflection), Clusters (globular and open), Planetary Nebula, Galaxy (spiral, barred-spiral, irregular, etc.), Super Nova Remnant and Dark Nebula.
Mag:	records the visual magnitude of the object.
Object Notes:	short narrative describing visual notes of the object to aid in locating and observing.

Each localized star field is marked in four-minute intervals of right ascension and by one-degree increments in declination. North is "up" (ascending Dec). Some fields are noted as having been "scaled to fit", as they are so big they could not be placed within the allowed frame. Scaling is arbitrary, and was done to maintain the dimensions within the frame with offsets included. Photographs were fitted into their respective 4 × 3 format frames by stretching, rotating, and scaling their attributes to maintain the symmetry of the page and the data's position within the entire frame.

Below the object data frame is located the telescope frame. This frame is the same for all objects and is an aid to quickly determine if the object will fit on a 35 mm negative through various sizes and focal length telescopes. The Field of View (FOV) is represented in arc minutes, major axis first. Comparing these values to the object's major and minor axes will quickly determine if the object will fit a 35 mm negative.

By combining the information contained in both frames and performing simple calculations those interested in photography or CCD imaging will be able to frame the object, calculate its size on the image plane, and calculate exposure times for different speeds of film.

Relating FOV to CCD Imaging

"Will it fit?" Is a frequently asked question. Beginners in both film and CCD imaging experience the same problem. Therefore understanding the FOV calculation is essential to successful imaging! FOV is directly related to the size of the imaging chip and the focal length of the telescope used. A good rule to remember is "the longer the focal length of the telescope, the smaller the section of sky imaged." Another way to look at things: "the longer the focal length, the more magnified the image will be."

To calculate the Field of View, three things need to be defined:

1. the dimensions of the CCD chip that your particular camera uses, measured in millimeters;
2. the focal length of the telescope being used, measured in millimeters;
3. the size of the object you want to photograph, measured in degrees or arc-minutes.

The calculations are as follows:

$$\text{Arcmin (FOV)} = (S \times 3438) / f$$
$$\text{Deg (FOV)} = (S \times 57.3) / f$$

Where:

S = dimension of one side of the CCD chip or film negative in millimeters;
f = focal length of the telescope measured in millimeters.

Working through an example we see the following:

Telescope:	12″ SCT operating at f/6.3
Object size:	M96, which measures 7.5′ × 5.2′
CCD size:	KAF-0401 chip, whose size is 6.89 × 4.59 mm.

To convert inches to millimeters multiply the inch measurement by 25.4 mm:

Diameter in mm: $12 \times 25.4 = 304.8$ mm
Focal length: $304.8 \times 6.3 = 1920.24$ mm
Chip dimensions:

$S_x = 6.89$ mm
$S_y = 4.59$ mm

Therefore:

$$\text{Arcmin}_x = (6.89 \times 3438) / 1920.24 = 12.34'$$

$$\text{Arcmin}_y = (4.59 \times 3438) / 1920.24 = 8.22'$$

The calculation shows that m96 will fit on the KAF-0401 chip if imaged through a 12-inch SCT operating at a focal ratio of 6.3. The major and minor axes are $12.34' \times 8.22'$ compared to the measurements of M96 which are $7.5' \times 5.2'$. It is clear that the image will fit, but will leave little room for framing any of the objects associated star field.

The same calculation works equally well for 35 mm film negatives. Using the above formula and the measurements of 36×24 mm for the film, the FOV major and minor axes work out to $64.45' \times 42.97'$. The image formed on the film not only fits nicely, but will provide room to include a larger star field and create a more visually pleasing photograph.

As I've explained, this book is the culmination of a personal project to compile a list of challenging new objects to observe and photograph. As the idea for this book came into focus I decided early that it would have no precisely defined target audience, but contains a cross-section of information required to track down and observe objects whether you are a beginner or advanced observer.

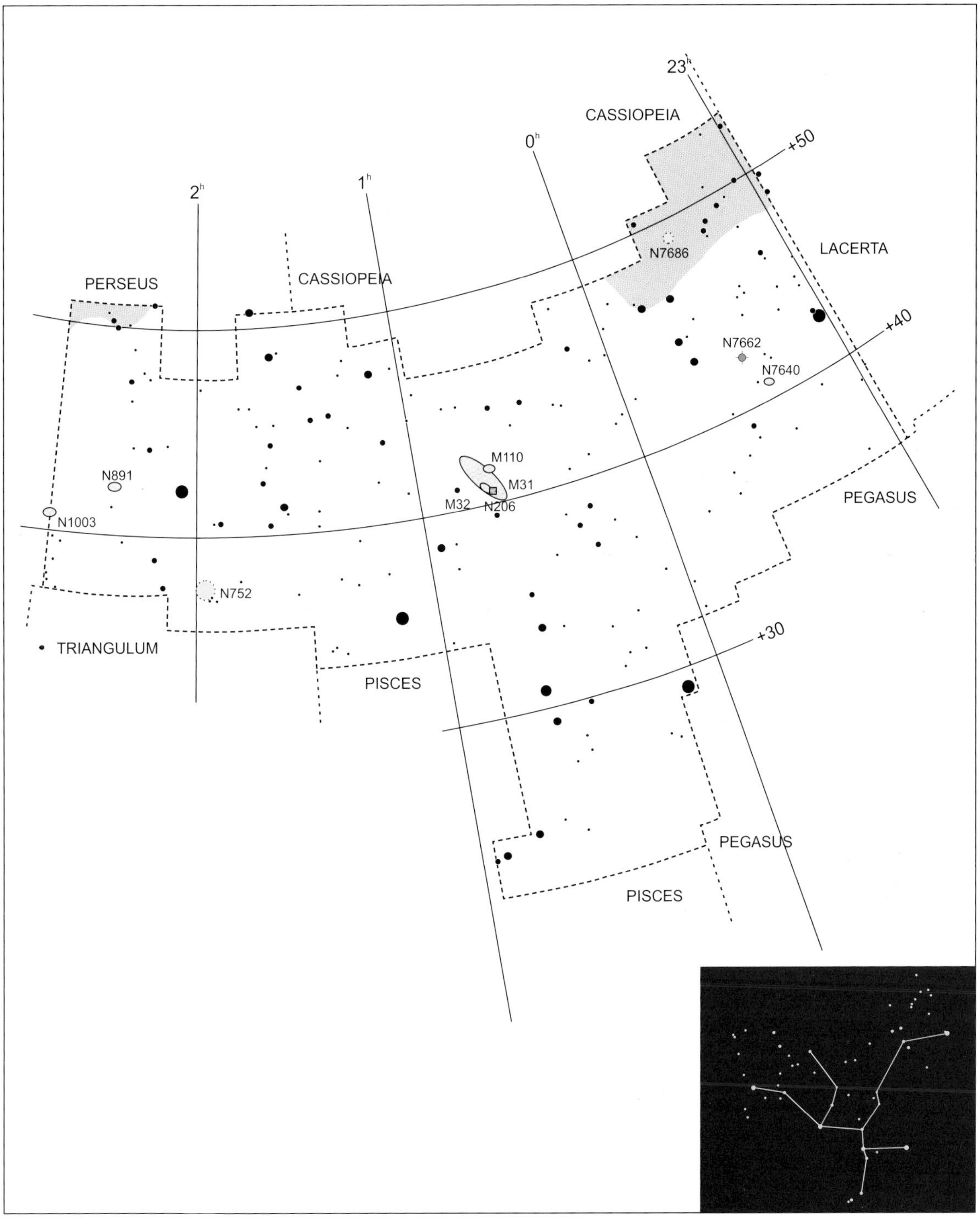

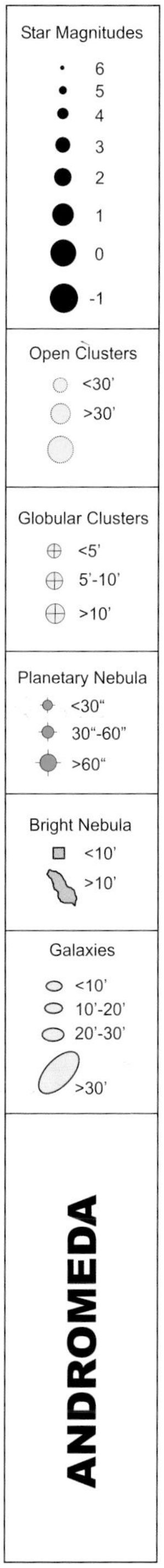

Constellation Facts:

Andromeda; (an-DROM-eh-da)

Andromeda, the Chained Lady;
rises in the northeastern sky, passes overhead,
and moves northwest.
Andromeda's head is represented by the star
Alpheratz which also marks the northeastern corner
of the Great Square of Pegasus.
The constellation covers 722 square degrees.

Andromeda is visible from 90° N to 37° S. Partially visible from 37° S to 90° S.

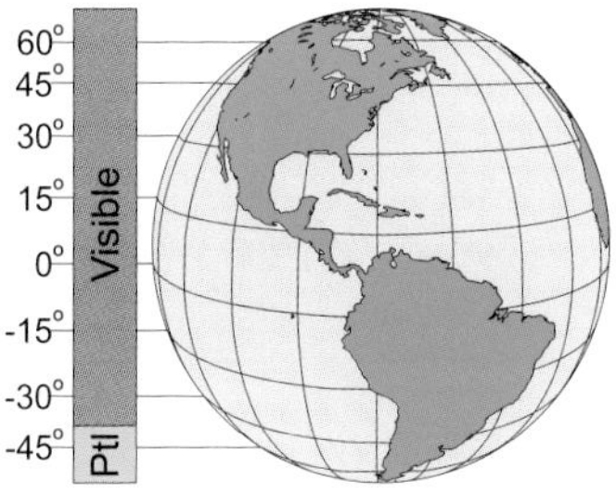

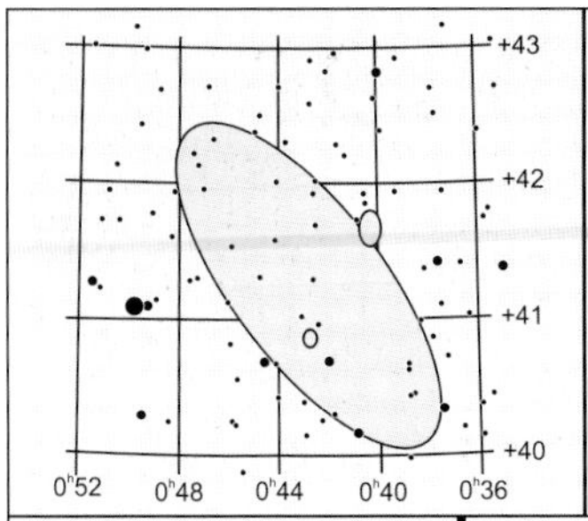

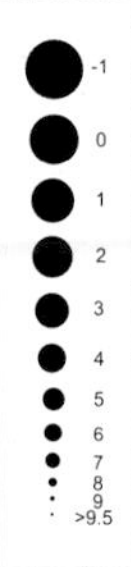

M31 (NGC 224)

RA:	$00^h 42^m 45.7^s$	Con:	Andromeda
Dec:	+41° 16' 20"	Type:	Spiral Galaxy
Size:	188.8' x 61.5'	Mag:	4.0

NGC 224 is the brightest galaxy in the northern sky. The galaxy is elongated with a dusty bright core. It is the largest galaxy in the Local Group and lies 2.4 million light years away.

Telescope Aperture:	4" f/5	4" f/9	6" f/7	6" f/9	8" f/6.3	8" f/10	10" f/6.3	10" f/10	12" f/6.3	12" f/10
FOV(35mm film):	2.7° x 4.1°	1.50° x 2.26°	1.29° x 1.93°	1.0° x 1.50°	1.07° x 1.61°	0.68° x 1.02°	0.86° x 1.29°	0.54° x 0.81°	0.72° x 1.07°	0.45° x 0.68°

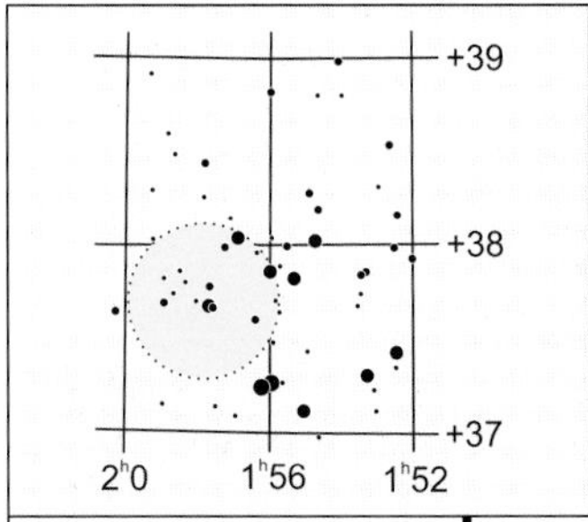

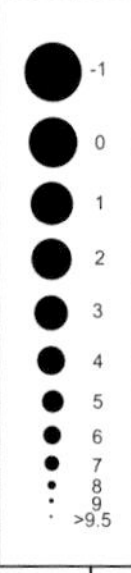

NGC 752

RA:	$01^h 57^m 49.8^s$	Con:	Andromeda
Dec:	+37° 41' 13"	Type:	Open Cluster
Size:	50.0'	Mag:	5.7

NGC752 is a large open cluster containing over 70 stars.

Telescope Aperture:	4" f/5	4" f/9	6" f/7	6" f/9	8" f/6.3	8" f/10	10" f/6.3	10" f/10	12" f/6.3	12" f/10
FOV(35mm film):	2.7° x 4.1°	1.50° x 2.26°	1.29° x 1.93°	1.0° x 1.50°	1.07° x 1.61°	0.68° x 1.02°	0.86° x 1.29°	0.54° x 0.81°	0.72° x 1.07°	0.45° x 0.68°

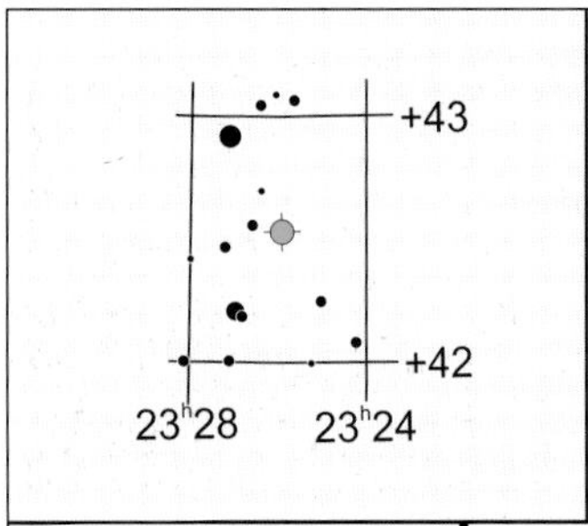

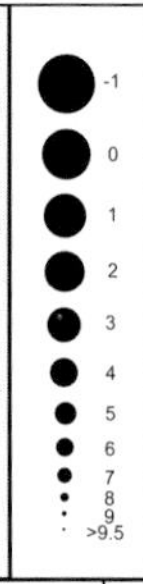

NGC7662

RA:	$23^h 25^m 55.0^s$	Con:	Andromeda
Dec:	+42° 33' 08"	Type:	Planetary
Size:	32" x 28"	Mag:	9.0

NGC7662 is a double-ringed planetary nebula with a bright, well defined ring of gas wrapped in a much larger, dimmer and hazier envelope.

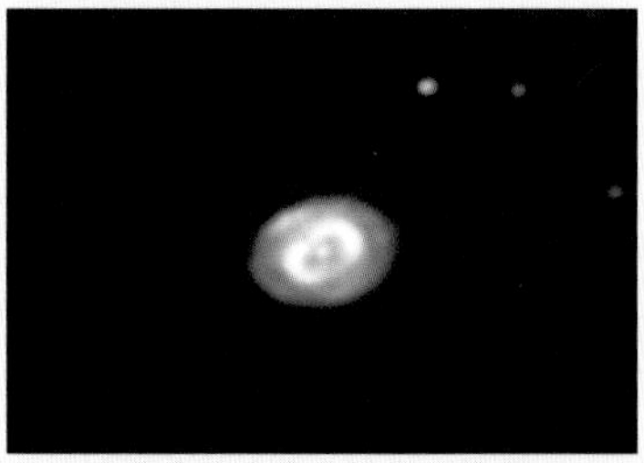

Telescope Aperture:	4" f/5	4" f/9	6" f/7	6" f/9	8" f/6.3	8" f/10	10" f/6.3	10" f/10	12" f/6.3	12" f/10
FOV(35mm film):	2.7° x 4.1°	1.50° x 2.26°	1.29° x 1.93°	1.0° x 1.50°	1.07° x 1.61°	0.68° x 1.02°	0.86° x 1.29°	0.54° x 0.81°	0.72° x 1.07°	0.45° x 0.68°

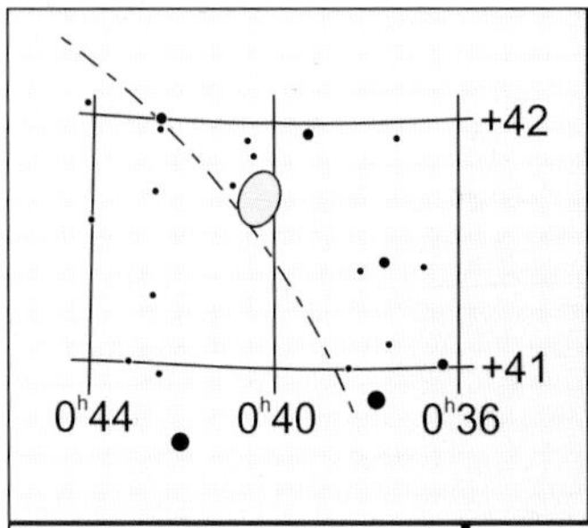

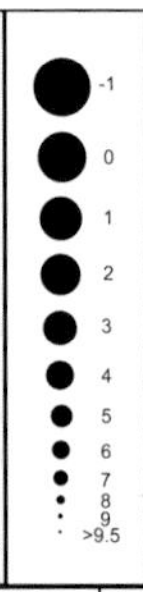

M110 (NGC 205)

RA:	$00^h 40^m 23.8^s$	Con:	Andromeda
Dec:	+41° 41' 22"	Type:	Elliptical Galaxy
Size:	21.9" x 10.8"	Mag:	8.9

NGC 205 is an elliptical satellite galaxy of M31.

Telescope Aperture:	4" f/5	4" f/9	6" f/7	6" f/9	8" f/6.3	8" f/10	10" f/6.3	10" f/10	12" f/6.3	12" f/10
FOV(35mm film):	2.7° x 4.1°	1.50° x 2.26°	1.29° x 1.93°	1.0° x 1.50°	1.07° x 1.61°	0.68° x 1.02°	0.86° x 1.29°	0.54° x 0.81°	0.72° x 1.07°	0.45° x 0.68°

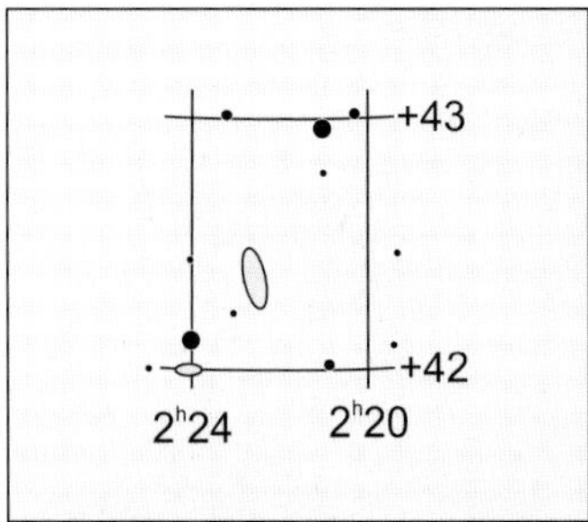

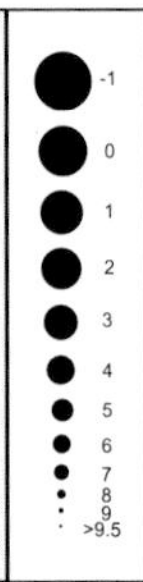

NGC 891

RA:	$02^h 22^m 37.9^s$	Con:	Andromeda
Dec:	+42° 21' 13"	Type:	Edge-on Spiral
Size:	14.0' x 2.0'	Mag:	10.0

NGC 891 is a faint edge-on spiral galaxy with a well defined central dust lane.

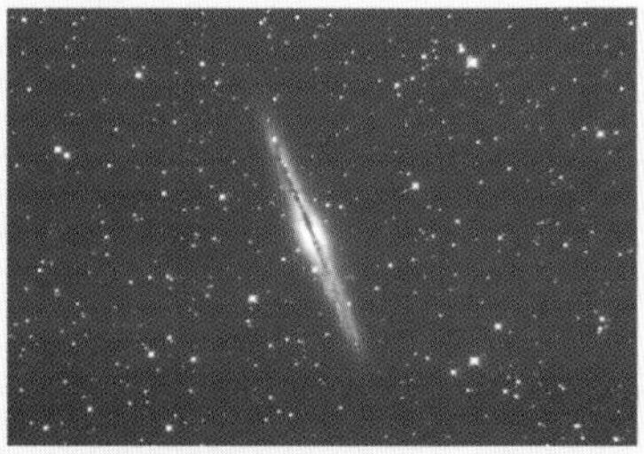

Telescope Aperture:	4" f/5	4" f/9	6" f/7	6" f/9	8" f/6.3	8" f/10	10" f/6.3	10" f/10	12" f/6.3	12" f/10
FOV(35mm film):	2.7° x 4.1°	1.50° x 2.26°	1.29° x 1.93°	1.0° x 1.50°	1.07° x 1.61°	0.68° x 1.02°	0.86° x 1.29°	0.54° x 0.81°	0.72° x 1.07°	0.45° x 0.68°

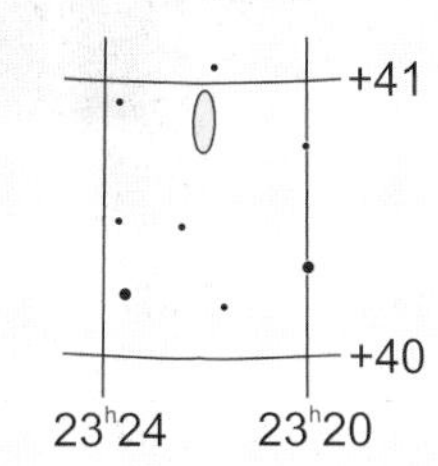

NGC 7640

RA:	$23^h\ 22^m\ 7.2^s$	Con:	Andromeda
Dec:	+40° 51' 07"	Type:	Barred Spiral
Size:	10.0' x 1.5'	Mag:	10.9

NGC 7640 is a dim Sb-type spiral galaxy

Telescope Aperture:	4" f/5	4" f/9	6" f/7	6" f/9	8" f/6.3	8" f/10	10" f/6.3	10" f/10	12" f/6.3	12" f/10
FOV(35mm film):	2.7° x 4.1°	1.50° x 2.26°	1.29° x 1.93°	1.0° x 1.50°	1.07° x 1.61°	0.68° x 1.02°	0.86° x 1.29°	0.54° x 0.81°	0.72° x 1.07°	0.45° x 0.68°

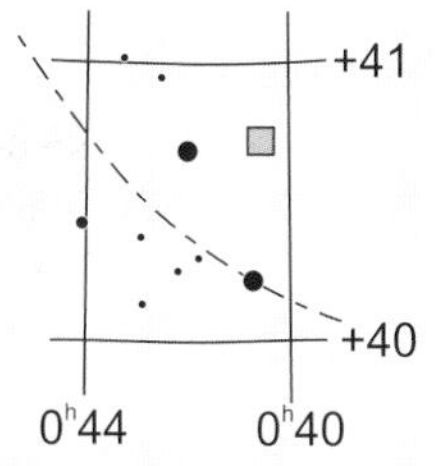

NGC 206

RA:	$00^h\ 40^m\ 37.4^s$	Con:	Andromeda
Dec:	+40° 44' 11"	Type:	Star Cloud
Size:	N/A	Mag:	N/A

NGC 206 is a large star cloud with associated nebulosity located in the southwestern spiral arm of the Andromeda galaxy.

Telescope Aperture:	4" f/5	4" f/9	6" f/7	6" f/9	8" f/6.3	8" f/10	10" f/6.3	10" f/10	12" f/6.3	12" f/10
FOV(35mm film):	2.7° x 4.1°	1.50° x 2.26°	1.29° x 1.93°	1.0° x 1.50°	1.07° x 1.61°	0.68° x 1.02°	0.86° x 1.29°	0.54° x 0.81°	0.72° x 1.07°	0.45° x 0.68°

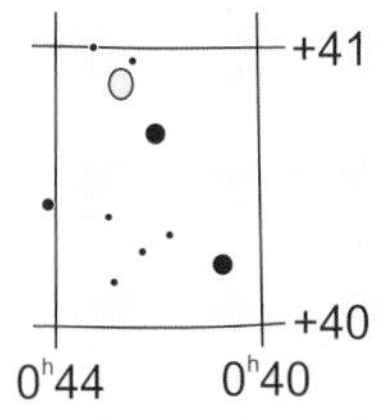

M32 (NGC 221)

RA:	$00^h 42^m\ 43.4^s$	Con:	Andromeda
Dec:	+40° 52' 11"	Type:	Elliptical Galaxy
Size:	8' x 6'	Mag:	8.2

NGC 221 is an elliptical satellite galaxy of M31

Telescope Aperture:	4" f/5	4" f/9	6" f/7	6" f/9	8" f/6.3	8" f/10	10" f/6.3	10" f/10	12" f/6.3	12" f/10
FOV(35mm film):	2.7° x 4.1°	1.50° x 2.26°	1.29° x 1.93°	1.0° x 1.50°	1.07° x 1.61°	0.68° x 1.02°	0.86° x 1.29°	0.54° x 0.81°	0.72° x 1.07°	0.45° x 0.68°

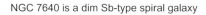

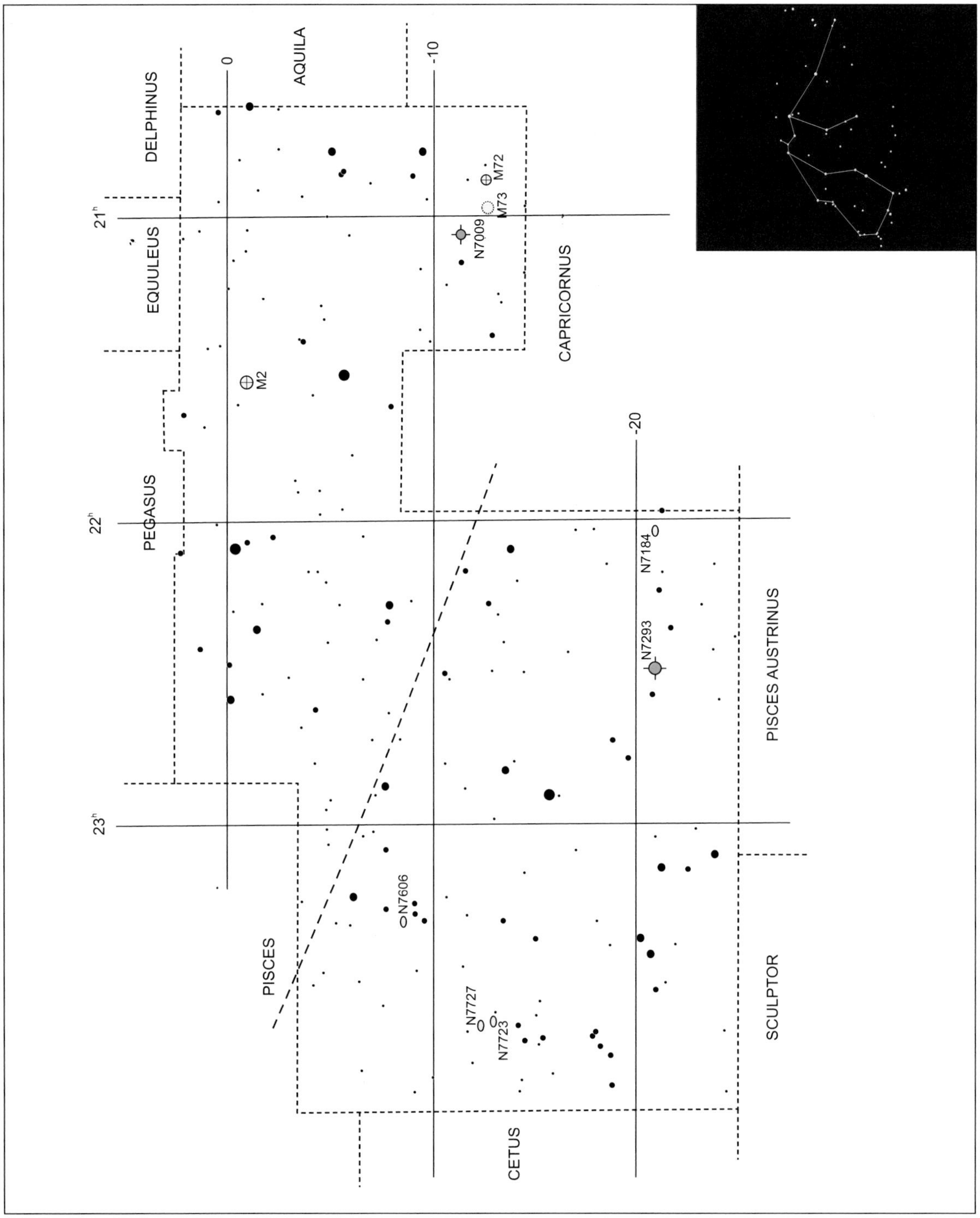

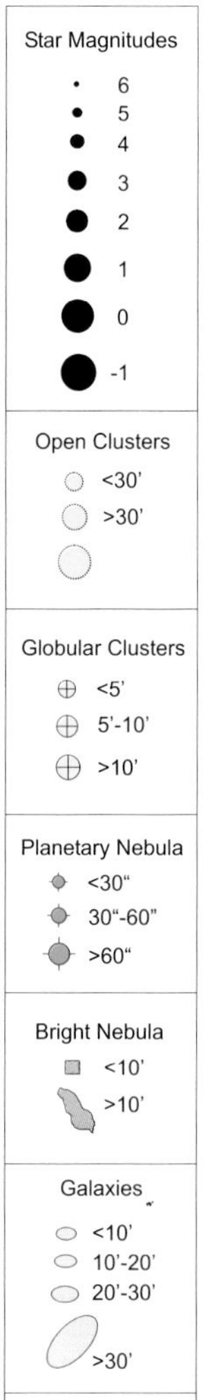

AQUARIUS

Constellation Facts:

Aquarius; (ack-KWAIR-ee-us)

Aquarius, the Water Carrier.
Aquarius rises in the east, crosses the meridian halfway between the horizon and the zenith and sets in the west.
Faint zodiacal constellation, whose principle feature is an asterism known as the Water Jar.
Constellation covers 980 square degrees.

Constellation is visible from 65° N to 86° S. Partially visible from 65° N to 90° N.

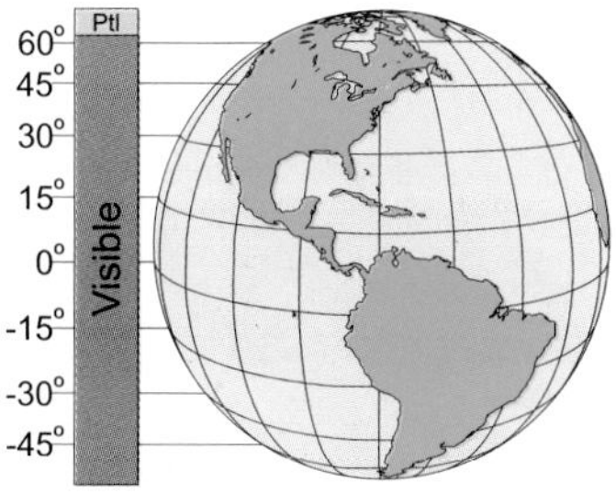

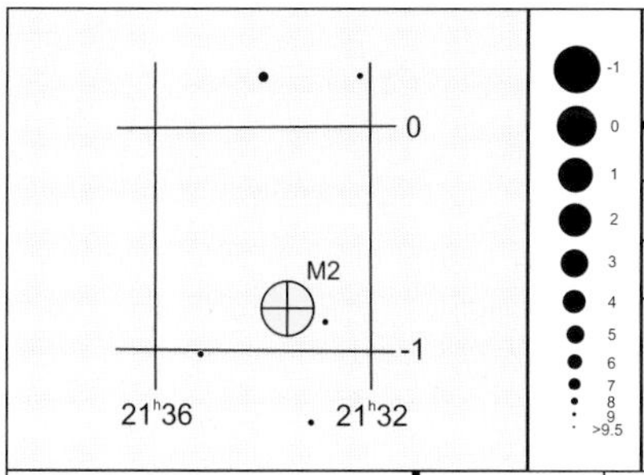

M2 (NGC 7089)

RA:	$21^h\ 33^m\ 34.7^s$	Con:	Aquarius
Dec:	00° 48' 38"	Type:	Globular Cluster
Size:	12.9'	Mag:	6.5

M2 is a bright globular cluster 5° north of the bright double star Beta Aquarii. Globular has a magnitude of 6.5 and covers 12.9' of sky.

Telescope Aperture:	4" f/5	4" f/9	6" f/7	6" f/9	8" f/6.3	8" f/10	10" f/6.3	10" f/10	12" f/6.3	12" f/10
FOV(35mm film):	2.7° x 4.1°	1.50° x 2.26°	1.29° x 1.93°	1.0° x 1.50°	1.07° x 1.61°	0.68° x 1.02°	0.86° x 1.29°	0.54° x 0.81°	0.72° x 1.07°	0.45° x 0.68°

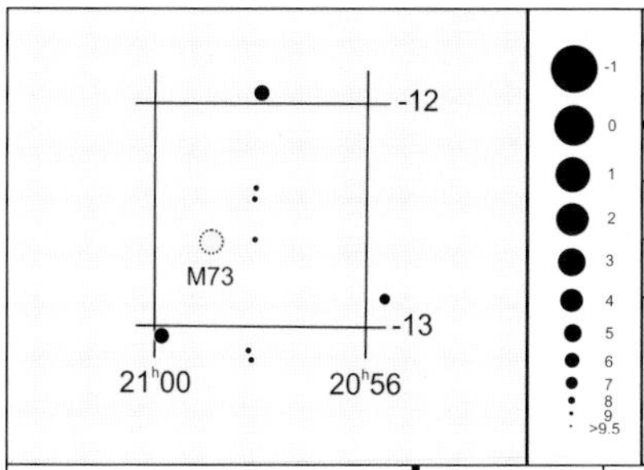

M73 (NGC 6994)

RA:	$20^h\ 59^m\ 5.0^s$	Con:	Aquarius
Dec:	-12° 37' 38"	Type:	Open Cluster
Size:	3.0'	Mag:	9.0

M73 is found 2° southwest of NGC-7009. It is a group of four stars measuring 2.8' across and has a photographic magnitude of 8.9.

Telescope Aperture:	4" f/5	4" f/9	6" f/7	6" f/9	8" f/6.3	8" f/10	10" f/6.3	10" f/10	12" f/6.3	12" f/10
FOV(35mm film):	2.7° x 4.1°	1.50° x 2.26°	1.29° x 1.93°	1.0° x 1.50°	1.07° x 1.61°	0.68° x 1.02°	0.86° x 1.29°	0.54° x 0.81°	0.72° x 1.07°	0.45° x 0.68°

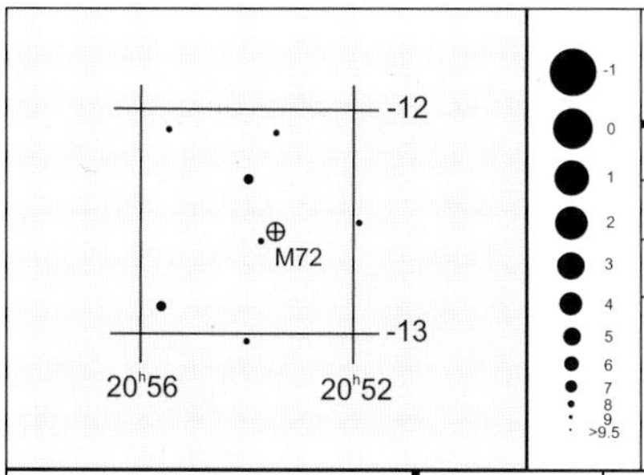

M72 (NGC 6981)

RA:	$20^h\ 53^m\ 35.1^s$	Con:	Aquarius
Dec:	-12° 31' 39"	Type:	Globular Cluster
Size:	5.9'	Mag:	9.4

M72 is located 1.5° west and slightly north of M73. Object is a globular cluster shining at magnitude 9.4 and measures 6' across.

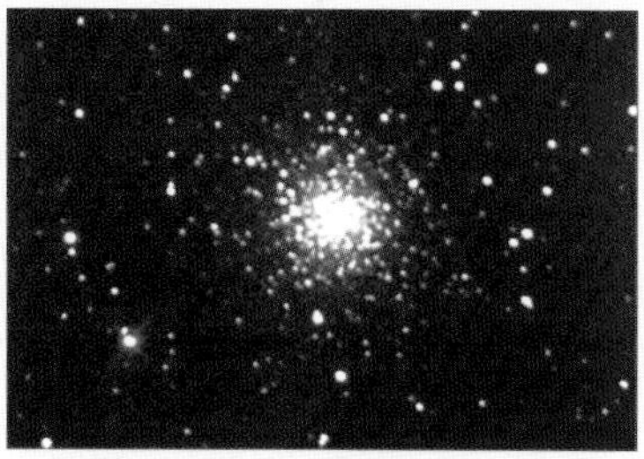

Telescope Aperture:	4" f/5	4" f/9	6" f/7	6" f/9	8" f/6.3	8" f/10	10" f/6.3	10" f/10	12" f/6.3	12" f/10
FOV(35mm film):	2.7° x 4.1°	1.50° x 2.26°	1.29° x 1.93°	1.0° x 1.50°	1.07° x 1.61°	0.68° x 1.02°	0.86° x 1.29°	0.54° x 0.81°	0.72° x 1.07°	0.45° x 0.68°

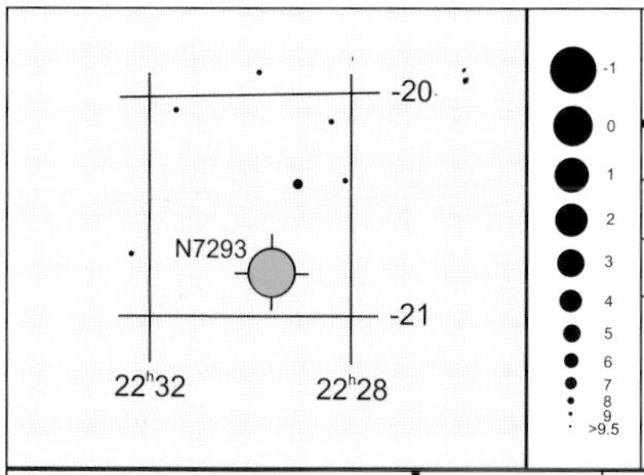

NGC 7293 (Helix Nebula)

RA:	$22^h\ 29^m\ 40.7^s$	Con:	Aquarius
Dec:	-20° 47' 30"	Type:	Planetary Nebula
Size:	12.8'	Mag:	7.3

NGC 7293 is a planetary nebula known as the Helix Nebula. It is the largest and brightest planetary in the night sky. It measures nearly 13' across and shines at photographic magnitude 6.5.

Telescope Aperture:	4" f/5	4" f/9	6" f/7	6" f/9	8" f/6.3	8" f/10	10" f/6.3	10" f/10	12" f/6.3	12" f/10
FOV(35mm film):	2.7° x 4.1°	1.50° x 2.26°	1.29° x 1.93°	1.0° x 1.50°	1.07° x 1.61°	0.68° x 1.02°	0.86° x 1.29°	0.54° x 0.81°	0.72° x 1.07°	0.45° x 0.68°

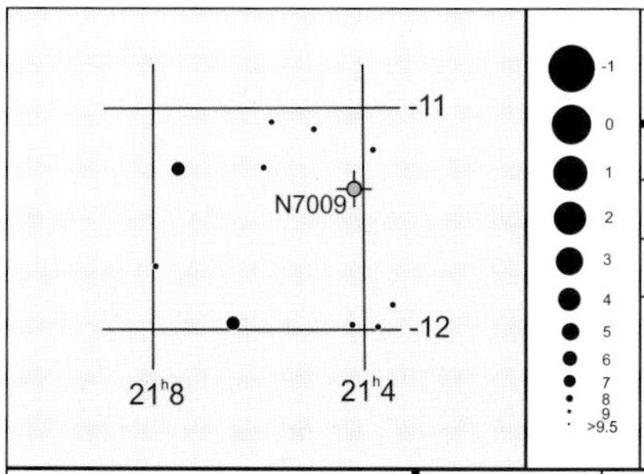

NGC 7009 (Saturn Nebula)

RA:	$21^h\ 04^m\ 17.0^s$	Con:	Aquarius
Dec:	-11° 21' 38"	Type:	Planetary Nebula
Size:	1.7'	Mag:	8.0

NGC 7009 is found in the western edges of the constellation. Object name is derived from the projecting arms of nebulosity extending out either side of the central disk.

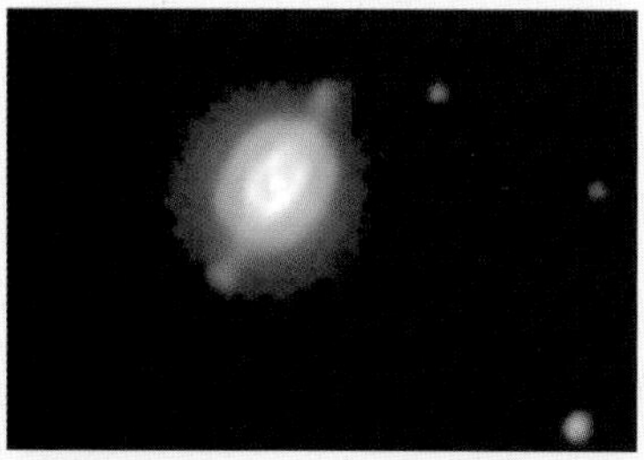

Telescope Aperture:	4" f/5	4" f/9	6" f/7	6" f/9	8" f/6.3	8" f/10	10" f/6.3	10" f/10	12" f/6.3	12" f/10
FOV(35mm film):	2.7° x 4.1°	1.50° x 2.26°	1.29° x 1.93°	1.0° x 1.50°	1.07° x 1.61°	0.68° x 1.02°	0.86° x 1.29°	0.54° x 0.81°	0.72° x 1.07°	0.45° x 0.68°

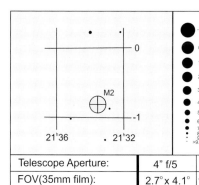

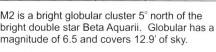

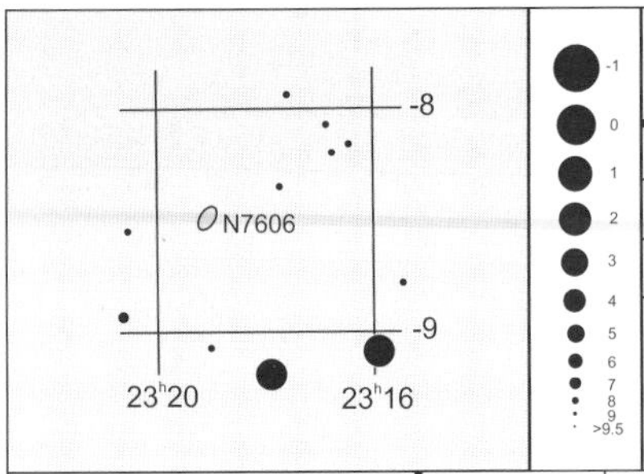

NGC 7606

RA:	$23^h\ 19^m\ 10.2^s$	Con:	Aquarius
Dec:	-08° 28' 31"	Type:	Spiral Galaxy
Size:	4.8' x 1.8'	Mag:	10.8

NGC 7606 is a spiral galaxy measuring 4.8' x 1.8' and shines at magnitude 10.8.

Telescope Aperture:	4" f/5	4" f/9	6" f/7	6" f/9	8" f/6.3	8" f/10	10" f/6.3	10" f/10	12" f/6.3	12" f/10
FOV(35mm film):	2.7° x 4.1°	1.50° x 2.26°	1.29° x 1.93°	1.0° x 1.50°	1.07° x 1.61°	0.68° x 1.02°	0.86° x 1.29°	0.54° x 0.81°	0.72° x 1.07°	0.45° x 0.68°

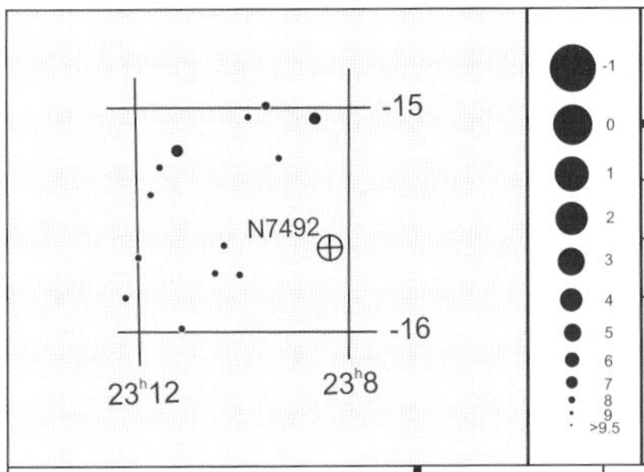

NGC 7492

RA:	$23^h\ 08^m\ 28.3^s$	Con:	Aquarius
Dec:	-15° 36' 30"	Type:	Globular Cluster
Size:	6.2'	Mag:	11.5

NGC 7492 is the third globular cluster found in Aquarius. The object shines at magnitude 11.5 and is 6.2' across.

Telescope Aperture:	4" f/5	4" f/9	6" f/7	6" f/9	8" f/6.3	8" f/10	10" f/6.3	10" f/10	12" f/6.3	12" f/10
FOV(35mm film):	2.7° x 4.1°	1.50° x 2.26°	1.29° x 1.93°	1.0° x 1.50°	1.07° x 1.61°	0.68° x 1.02°	0.86° x 1.29°	0.54° x 0.81°	0.72° x 1.07°	0.45° x 0.68°

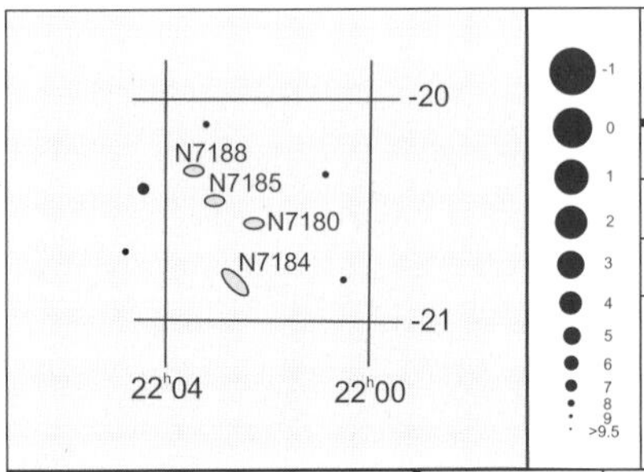

NGC 7184

RA:	$22^h\ 02^m\ 46.9^s$	Con:	Aquarius
Dec:	-20° 48' 32"	Type:	Spiral Galaxy
Size:	5.5' x 1.3'	Mag:	12.0

NGC 7184 is a large Sb-type spiral galaxy shining at magnitude 12.0 and measures 5.5' x 1.3'.

Telescope Aperture:	4" f/5	4" f/9	6" f/7	6" f/9	8" f/6.3	8" f/10	10" f/6.3	10" f/10	12" f/6.3	12" f/10
FOV(35mm film):	2.7° x 4.1°	1.50° x 2.26°	1.29° x 1.93°	1.0° x 1.50°	1.07° x 1.61°	0.68° x 1.02°	0.86° x 1.29°	0.54° x 0.81°	0.72° x 1.07°	0.45° x 0.68°

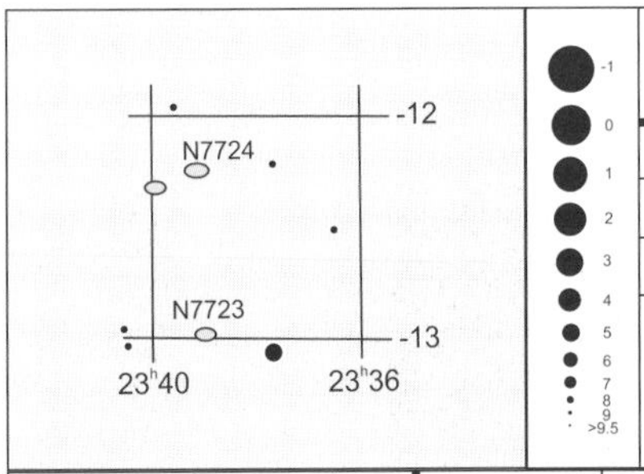

NGC 7723

RA:	$23^h\ 38^m\ 58.1^s$	Con:	Aquarius
Dec:	-12° 57' 30"	Type:	Spiral Galaxy
Size:	3.6' x 2.2'	Mag:	11.1

NGC 7723 is an Sb-type spiral galaxy measuring 3.6' x 2.2' and glows at magnitude 11.1.

Telescope Aperture:	4" f/5	4" f/9	6" f/7	6" f/9	8" f/6.3	8" f/10	10" f/6.3	10" f/10	12" f/6.3	12" f/10
FOV(35mm film):	2.7° x 4.1°	1.50° x 2.26°	1.29° x 1.93°	1.0° x 1.50°	1.07° x 1.61°	0.68° x 1.02°	0.86° x 1.29°	0.54° x 0.81°	0.72° x 1.07°	0.45° x 0.68°

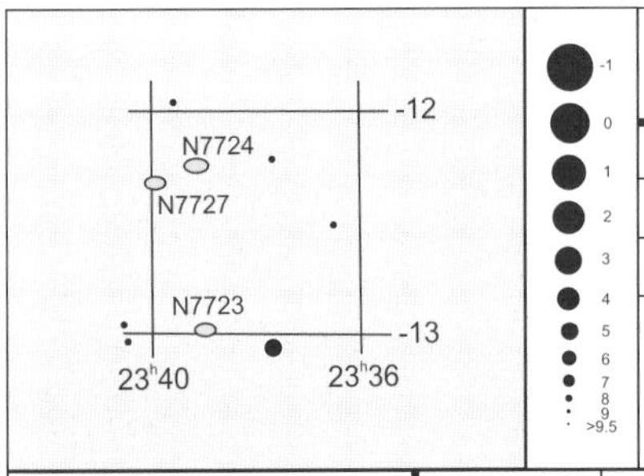

NGC 7727

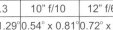

RA:	$23^h\ 39^m\ 58.0^s$	Con:	Aquarius
Dec:	-12° 17' 30"	Type:	Barred Spiral
Size:	4.2' x 3.4'	Mag:	10.7

NGC 7727 is a barred spiral galaxy measuring 4.2' x 3.4' and shines at magnitude 10.7

Telescope Aperture:	4" f/5	4" f/9	6" f/7	6" f/9	8" f/6.3	8" f/10	10" f/6.3	10" f/10	12" f/6.3	12" f/10
FOV(35mm film):	2.7° x 4.1°	1.50° x 2.26°	1.29° x 1.93°	1.0° x 1.50°	1.07° x 1.61°	0.68° x 1.02°	0.86° x 1.29°	0.54° x 0.81°	0.72° x 1.07°	0.45° x 0.68°

Aql

Crosses Prime Meridian: July thru August

SUMMER

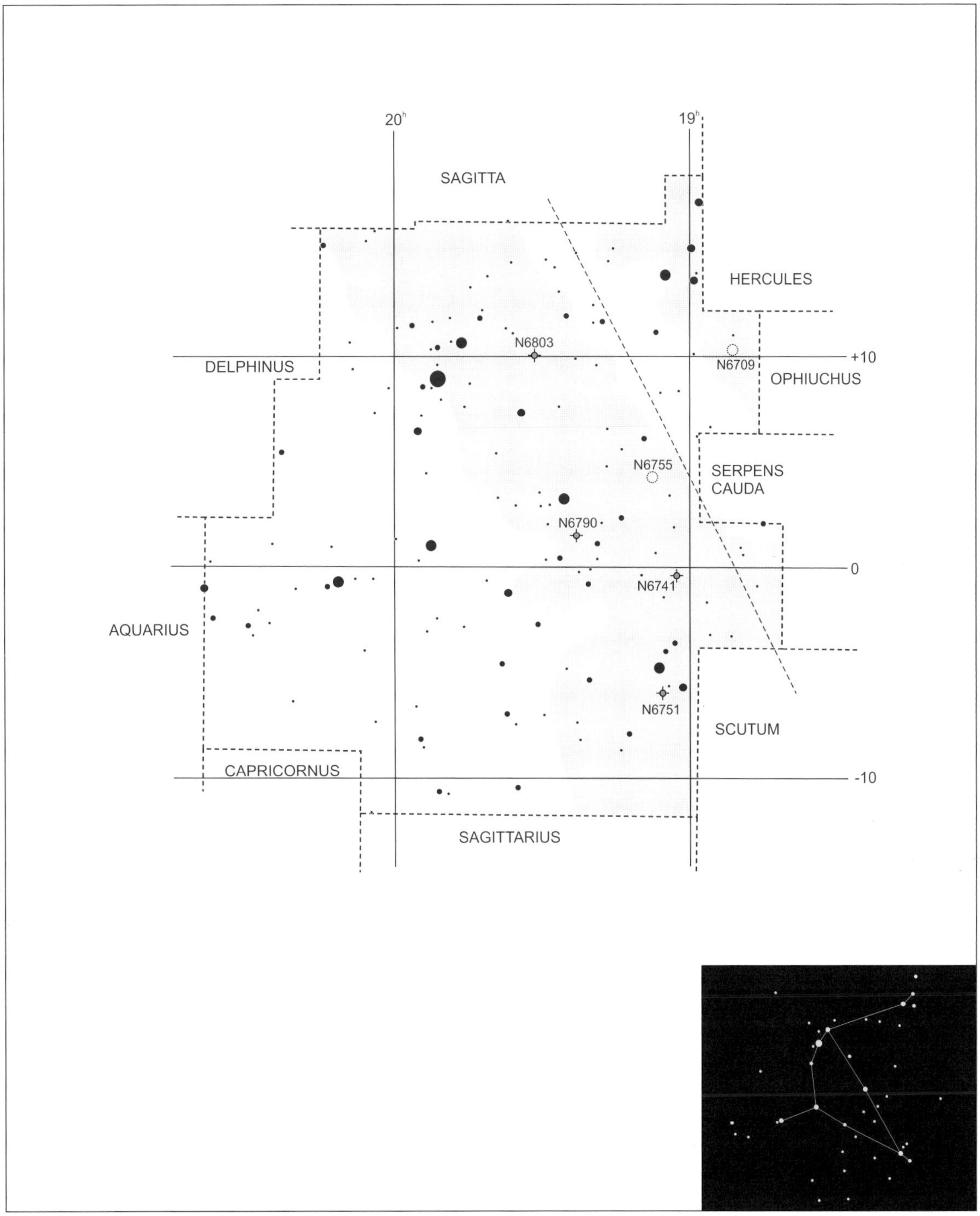

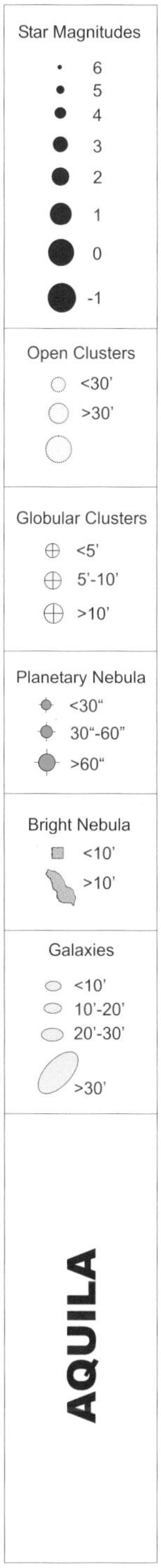

Constellation Facts:

Aquila; (ACK-will-lah)
Aquila, the Eagle.
Constellation lies along the celestial equator. It rises close to the eastern point on the horizon, passes the meridian halfway between the horizon and the zenith and sets directly towards the west.
The Eagle's outline is characterized by its wings outstretched, headed towards the east through the Milky Way.
Aquila covers 652 square degrees.

Constellation is visible from 78° N to 71° S. Partially visible from 78° N to 90° N.

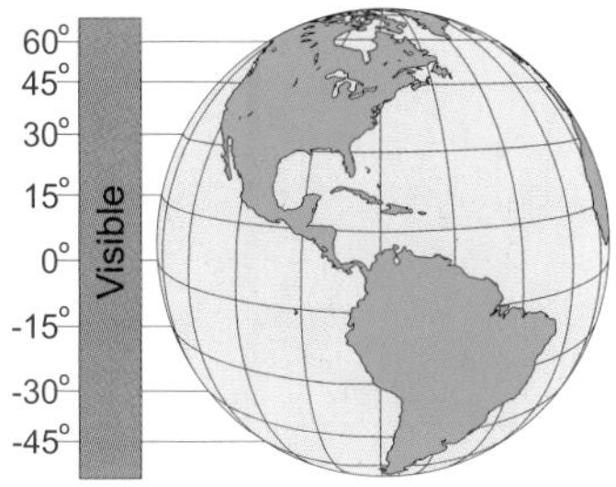

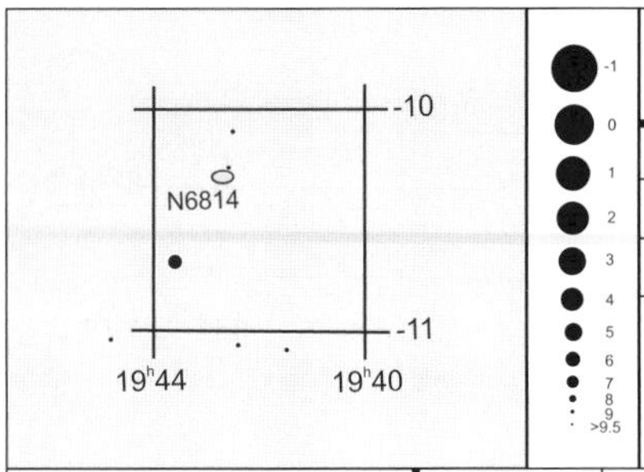

NGC 6814

RA:	19h 42m 47.4s	Con:	Aquila
Dec:	-10° 18' 45"	Type:	Spiral Galaxy
Size:	4.0'	Mag:	11.2

NGC 6814 is a small spiral galaxy with a bright nucleus with visible knotty arms.

Telescope Aperture:	4" f/5	4" f/9	6" f/7	6" f/9	8" f/6.3	8" f/10	10" f/6.3	10" f/10	12" f/6.3	12" f/10
FOV(35mm film):	2.7° x 4.1°	1.50° x 2.26°	1.29° x 1.93°	1.0° x 1.50°	1.07° x 1.61°	0.68° x 1.02°	0.86° x 1.29°	0.54° x 0.81°	0.72° x 1.07°	0.45° x 0.68°

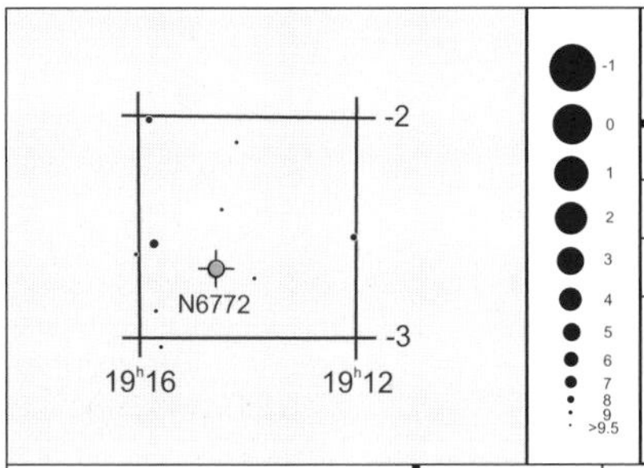

NGC 6772

RA:	19h 14m 41.0s	Con:	Aquila
Dec:	-02° 41' 48"	Type:	Planetary Nebula
Size:	1.0'	Mag:	14.0

NGC 6772 is found in the southern part of the constellation. Object is a faint planetary nebula glowing at magnitude 14.0 and covering a mere 1.0'.

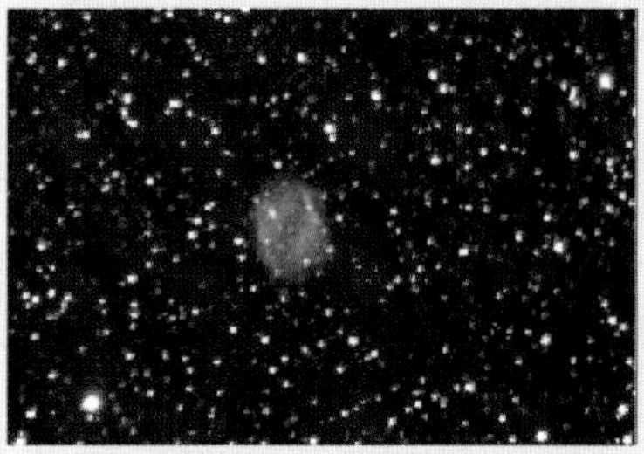

Telescope Aperture:	4" f/5	4" f/9	6" f/7	6" f/9	8" f/6.3	8" f/10	10" f/6.3	10" f/10	12" f/6.3	12" f/10
FOV(35mm film):	2.7° x 4.1°	1.50° x 2.26°	1.29° x 1.93°	1.0° x 1.50°	1.07° x 1.61°	0.68° x 1.02°	0.86° x 1.29°	0.54° x 0.81°	0.72° x 1.07°	0.45° x 0.68°

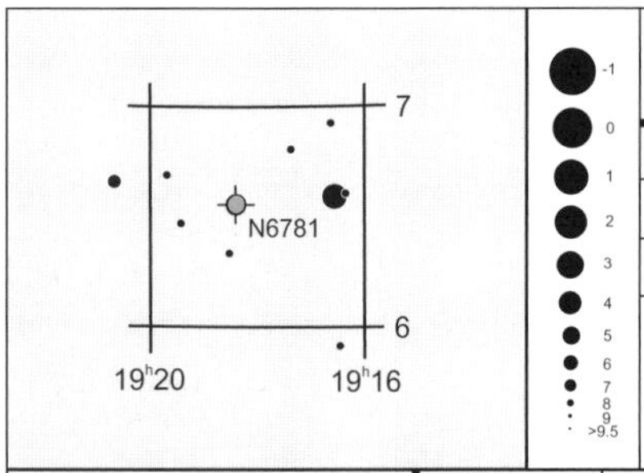

NGC 6781

RA:	19h 18m 28.8s	Con:	Aquila
Dec:	06° 33' 12"	Type:	Planetary Nebula
Size:	1.8'	Mag:	12.0

NGC 6781 displays large ring shaped nebulosity. Object has low surface brightness glowing at magnitude 12.0 and covers 1.8'.

Telescope Aperture:	4" f/5	4" f/9	6" f/7	6" f/9	8" f/6.3	8" f/10	10" f/6.3	10" f/10	12" f/6.3	12" f/10
FOV(35mm film):	2.7° x 4.1°	1.50° x 2.26°	1.29° x 1.93°	1.0° x 1.50°	1.07° x 1.61°	0.68° x 1.02°	0.86° x 1.29°	0.54° x 0.81°	0.72° x 1.07°	0.45° x 0.68°

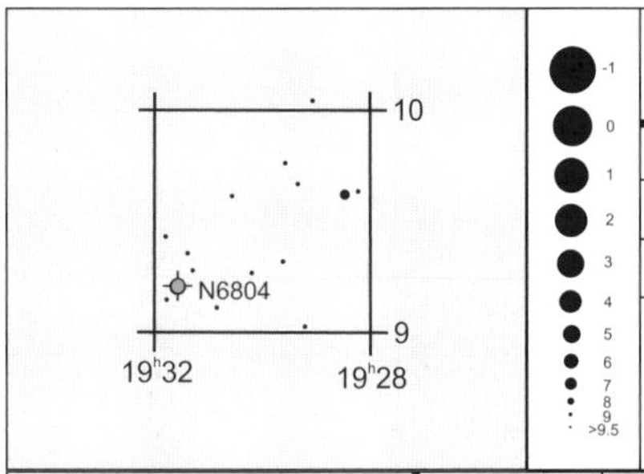

NGC 6804

RA:	19h 31m 40.8s	Con:	Aquila
Dec:	09° 13' 13"	Type:	Planetary Nebula
Size:	1.1'	Mag:	12.0

NGC 6804 is a faint planetary nebula found 3.5° northeast of NGC 6781. Object glows at magnitude 12.0 and covers 1.1' of sky.

Telescope Aperture:	4" f/5	4" f/9	6" f/7	6" f/9	8" f/6.3	8" f/10	10" f/6.3	10" f/10	12" f/6.3	12" f/10
FOV(35mm film):	2.7° x 4.1°	1.50° x 2.26°	1.29° x 1.93°	1.0° x 1.50°	1.07° x 1.61°	0.68° x 1.02°	0.86° x 1.29°	0.54° x 0.81°	0.72° x 1.07°	0.45° x 0.68°

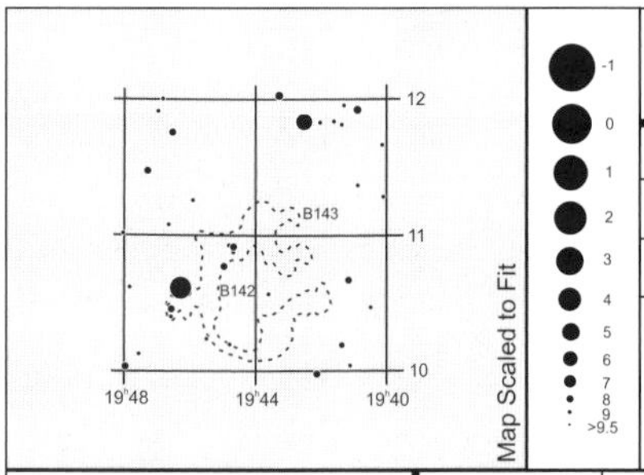

Bernard 142

RA:	19h 40m 42.0s	Con:	Aquila
Dec:	10° 57' 00"	Type:	Dark Nebula
Size:	40.0'	Mag:	

B 142 is a large dark nebula extending east and west. Object combines with B 143 to form a giant "E" shape.

Telescope Aperture:	4" f/5	4" f/9	6" f/7	6" f/9	8" f/6.3	8" f/10	10" f/6.3	10" f/10	12" f/6.3	12" f/10
FOV(35mm film):	2.7° x 4.1°	1.50° x 2.26°	1.29° x 1.93°	1.0° x 1.50°	1.07° x 1.61°	0.68° x 1.02°	0.86° x 1.29°	0.54° x 0.81°	0.72° x 1.07°	0.45° x 0.68°

Ari

Crosses Prime Meridian:
November thru December

AUTUMN

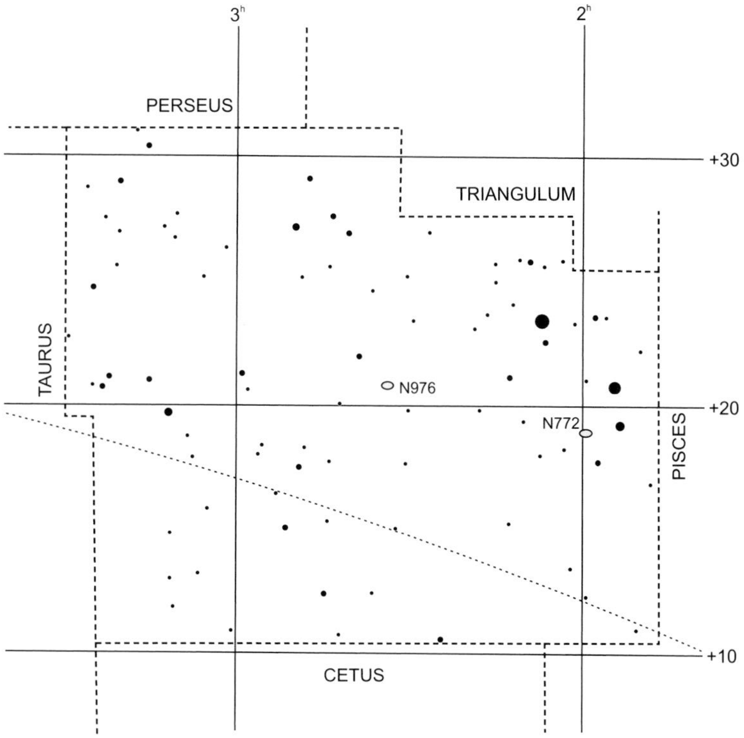

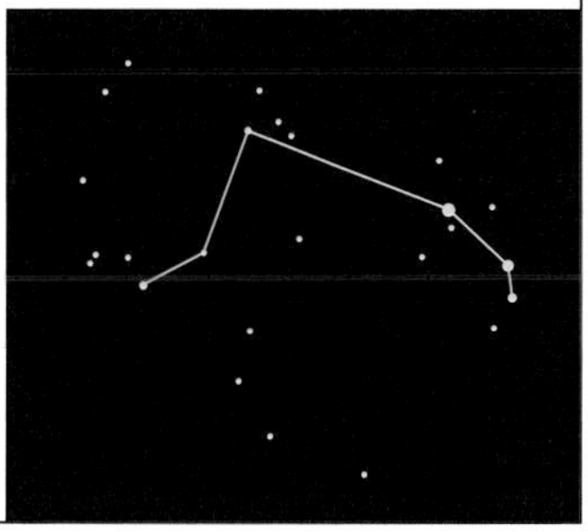

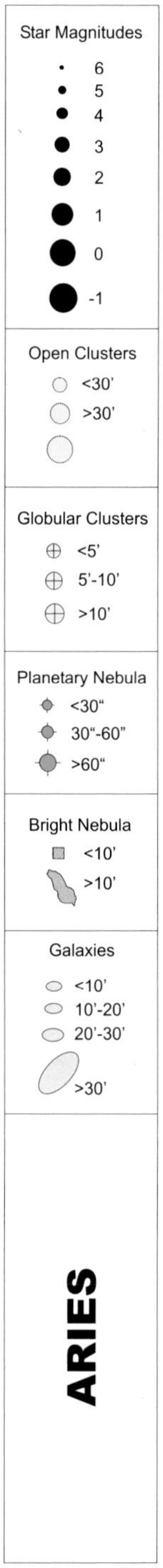

Constellation Facts:

Aries; (AIR-ease)

Aries, the Ram:
tracks through the sky from east to west, crossing the meridian halfway between the horizon and the zenith.

Constellation is visible from 90° N to 58° S. Partially visible from 58° S to 90° S.

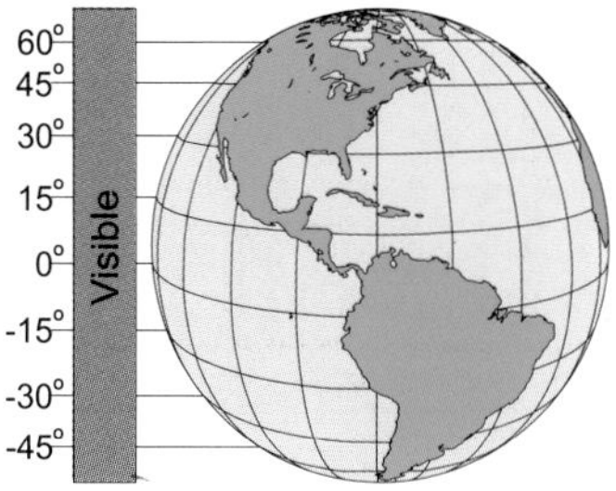

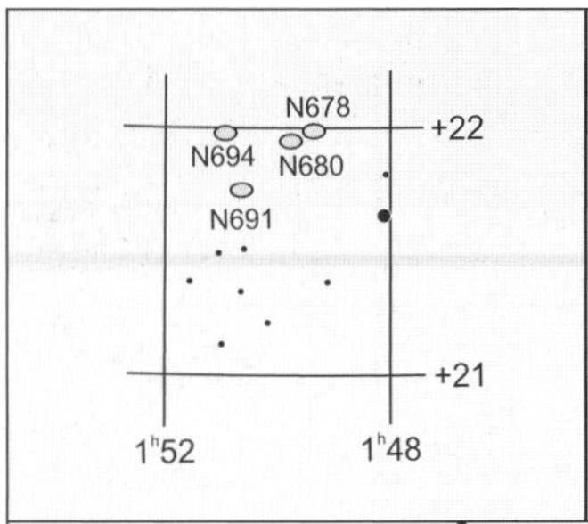

NGC 691

RA:	01h 50m 45.7s	Con:	Aries
Dec:	21° 46' 16"	Type:	Spiral Galaxy
Size:	3.0' x 2.2'	Mag:	12.0

NGC 691 is a spiral galaxy found in the north-western corner of Aries. Galaxy is appears round in the eyepiece. Object shines at magnitude 12.0 and measures 3.0' x 2.2'.

Telescope Aperture:	4" f/5	4" f/9	6" f/7	6" f/9	8" f/6.3	8" f/10	10" f/6.3	10" f/10	12" f/6.3	12" f/10
FOV(35mm film):	2.7° x 4.1°	1.50° x 2.26°	1.29° x 1.93°	1.0° x 1.50°	1.07° x 1.61°	0.68° x 1.02°	0.86° x 1.29°	0.54° x 0.81°	0.72° x 1.07°	0.45° x 0.68°

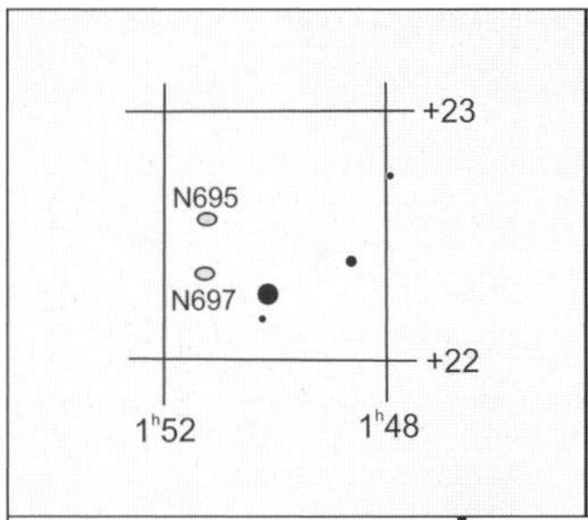

NGC 697

RA:	01h 51m 21.7s	Con:	Aries
Dec:	22° 21' 15"	Type:	Barred Spiral
Size:	4.3' x 1.2'	Mag:	13.0

NGC 697 is a highly inclined barred spiral galaxy found in the northwestern corner of Aries. Object shines at magnitude 13.0 and covers 4.3' x 1.2'.

Telescope Aperture:	4" f/5	4" f/9	6" f/7	6" f/9	8" f/6.3	8" f/10	10" f/6.3	10" f/10	12" f/6.3	12" f/10
FOV(35mm film):	2.7° x 4.1°	1.50° x 2.26°	1.29° x 1.93°	1.0° x 1.50°	1.07° x 1.61°	0.68° x 1.02°	0.86° x 1.29°	0.54° x 0.81°	0.72° x 1.07°	0.45° x 0.68°

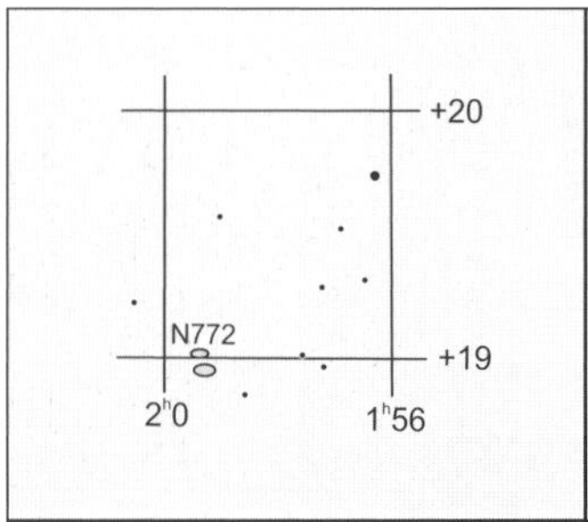

NGC 772

RA:	01h 59m 21.6s	Con:	Aries
Dec:	19° 01' 16"	Type:	Spiral Galaxy
Size:	7.5' x 4.0'	Mag:	10.3

NGC 772 is the brightest galaxy in Aries. It is a spiral galaxy with a bright core. Object shines at magnitude 10.3 and spans 7.5' x 4.0'.

Telescope Aperture:	4" f/5	4" f/9	6" f/7	6" f/9	8" f/6.3	8" f/10	10" f/6.3	10" f/10	12" f/6.3	12" f/10
FOV(35mm film):	2.7° x 4.1°	1.50° x 2.26°	1.29° x 1.93°	1.0° x 1.50°	1.07° x 1.61°	0.68° x 1.02°	0.86° x 1.29°	0.54° x 0.81°	0.72° x 1.07°	0.45° x 0.68°

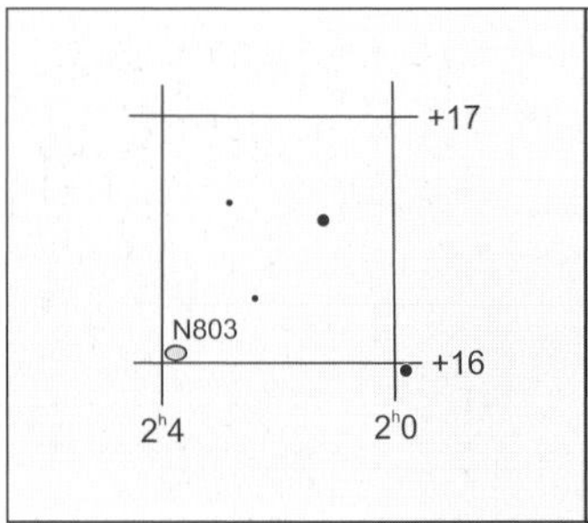

NGC 803

RA:	02h 03m 51.6s	Con:	Aries
Dec:	16° 02' 17"	Type:	Spiral Galaxy
Size:	3.0' x 1.1'	Mag:	12.4

NGC 803 is found midway between NGC 821 and NGC 772. It is an Sb-type spiral galaxy shining at magnitude 12.4 and covers 3.0' x 1.1'.

Telescope Aperture:	4" f/5	4" f/9	6" f/7	6" f/9	8" f/6.3	8" f/10	10" f/6.3	10" f/10	12" f/6.3	12" f/10
FOV(35mm film):	2.7° x 4.1°	1.50° x 2.26°	1.29° x 1.93°	1.0° x 1.50°	1.07° x 1.61°	0.68° x 1.02°	0.86° x 1.29°	0.54° x 0.81°	0.72° x 1.07°	0.45° x 0.68°

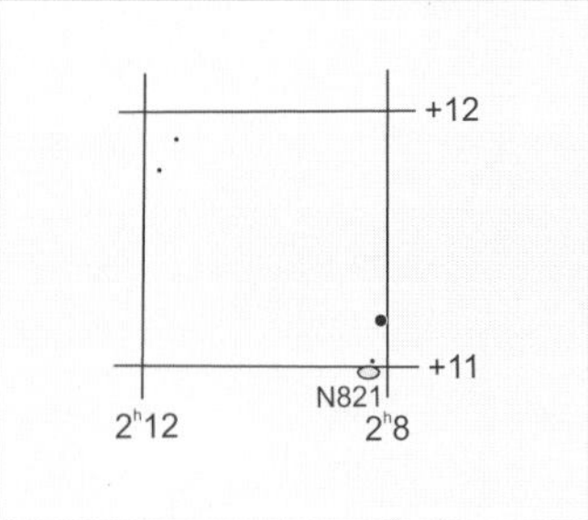

NGC 821

RA:	02h 08m 27.5s	Con:	Aries
Dec:	11° 00' 18"	Type:	Eliptical Galaxy
Size:	3.5' x 2.2'	Mag:	10.8

NGC 821 is located 4° southwest of NGC 877 near the southern border of the constellation. The object shines at magnitude 10.8 and covers 3.5', x 2.2'

Telescope Aperture:	4" f/5	4" f/9	6" f/7	6" f/9	8" f/6.3	8" f/10	10" f/6.3	10" f/10	12" f/6.3	12" f/10
FOV(35mm film):	2.7° x 4.1°	1.50° x 2.26°	1.29° x 1.93°	1.0° x 1.50°	1.07° x 1.61°	0.68° x 1.02°	0.86° x 1.29°	0.54° x 0.81°	0.72° x 1.07°	0.45° x 0.68°

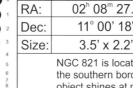

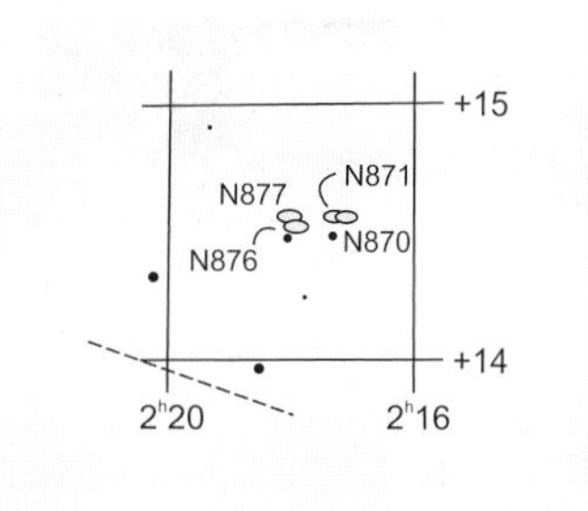

NGC 877

RA:	$02^h\ 18^m\ 3.5^s$	Con:	Aries
Dec:	14° 33' 17"	Type:	Spiral Galaxy
Size:	2.0' x 1.4'	Mag:	11.8

NGC 877 is an Sc-type spiral galaxy, measuring 2.0' x 1.4' shining at magnitude 11.8.

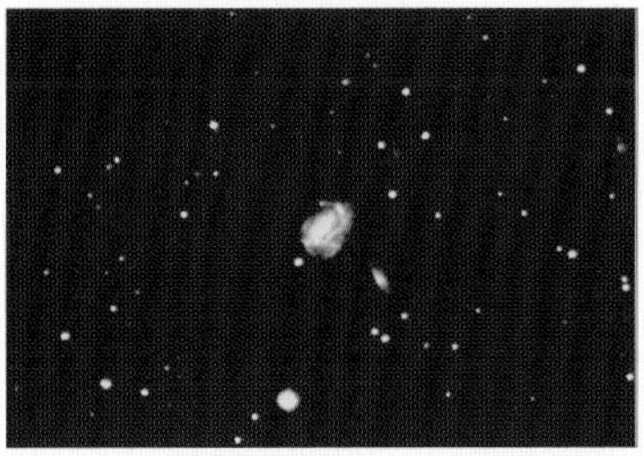

Telescope Aperture:	4" f/5	4" f/9	6" f/7	6" f/9	8" f/6.3	8" f/10	10" f/6.3	10" f/10	12" f/6.3	12" f/10
FOV(35mm film):	2.7° x 4.1°	1.50° x 2.26°	1.29° x 1.93°	1.0° x 1.50°	1.07° x 1.61°	0.68° x 1.02°	0.86° x 1.29°	0.54° x 0.81°	0.72° x 1.07°	0.45° x 0.68°

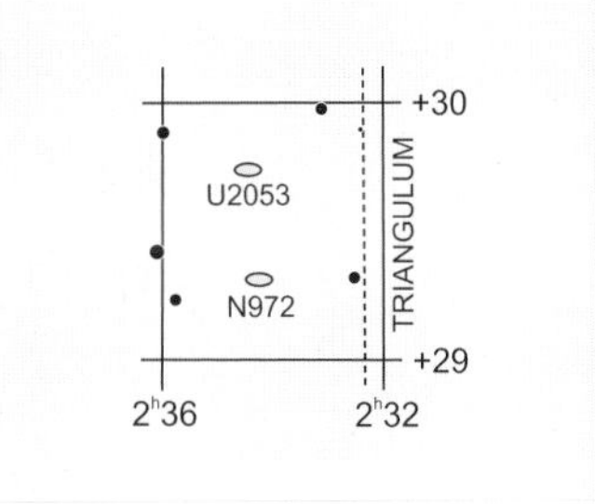

NGC 972

RA:	$02^h\ 34^m\ 15.7^s$	Con:	Aries
Dec:	29° 19' 11"	Type:	Spiral Galaxy
Size:	2.7' x 1.2'	Mag:	11.3

NGC 972 is located in the northern part of the constellation. It is a highly inclined spiral galaxy measuring2.7' x 1.2' and shines at magnitude 11.3.

Telescope Aperture:	4" f/5	4" f/9	6" f/7	6" f/9	8" f/6.3	8" f/10	10" f/6.3	10" f/10	12" f/6.3	12" f/10
FOV(35mm film):	2.7° x 4.1°	1.50° x 2.26°	1.29° x 1.93°	1.0° x 1.50°	1.07° x 1.61°	0.68° x 1.02°	0.86° x 1.29°	0.54° x 0.81°	0.72° x 1.07°	0.45° x 0.68°

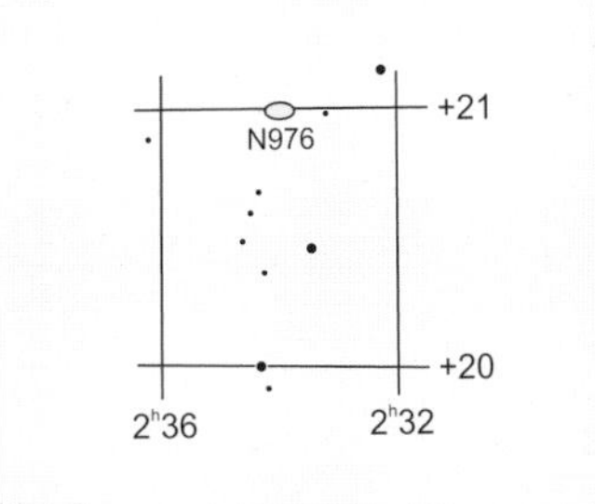

NGC 976

RA:	$02^h\ 34^m\ 3.5^s$	Con:	Aries
Dec:	20° 59' 14"	Type:	Spiral Galaxy
Size:	1.5' x 1.1'	Mag:	12.4

NGC 976 is located in the north-central region of Aries. It's a small spiral galaxy glowing at magnitude 12.4, and measures 1.5' x 1.1'.

Telescope Aperture:	4" f/5	4" f/9	6" f/7	6" f/9	8" f/6.3	8" f/10	10" f/6.3	10" f/10	12" f/6.3	12" f/10
FOV(35mm film):	2.7° x 4.1°	1.50° x 2.26°	1.29° x 1.93°	1.0° x 1.50°	1.07° x 1.61°	0.68° x 1.02°	0.86° x 1.29°	0.54° x 0.81°	0.72° x 1.07°	0.45° x 0.68°

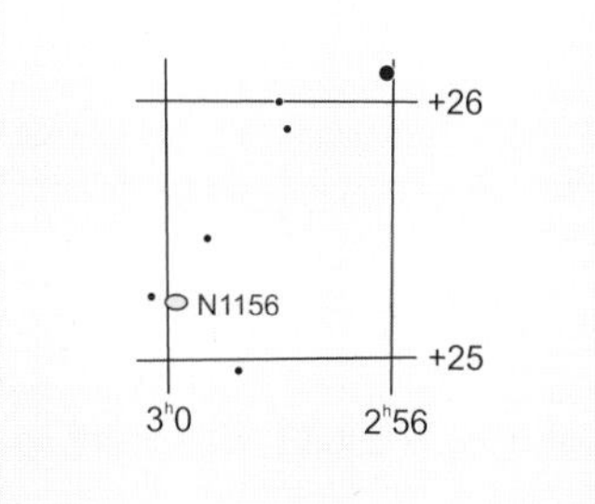

NGC 1156

RA:	$02^h\ 59^m\ 45.5^s$	Con:	Aries
Dec:	25° 14' 11"	Type:	Irregular Galaxy
Size:	3.3' x 2.6'	Mag:	11.7

NGC 1156 is located 7° northeast of NGC 672. The object shines at magnitude 11.7 and is 3.3' x 2.6' in size.

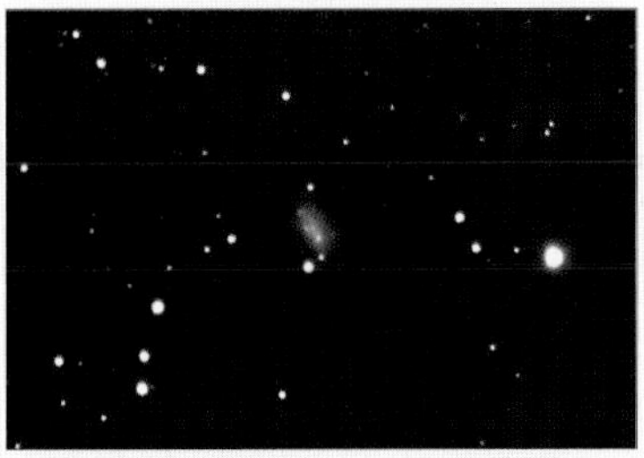

Telescope Aperture:	4" f/5	4" f/9	6" f/7	6" f/9	8" f/6.3	8" f/10	10" f/6.3	10" f/10	12" f/6.3	12" f/10
FOV(35mm film):	2.7° x 4.1°	1.50° x 2.26°	1.29° x 1.93°	1.0° x 1.50°	1.07° x 1.61°	0.68° x 1.02°	0.86° x 1.29°	0.54° x 0.81°	0.72° x 1.07°	0.45° x 0.68°

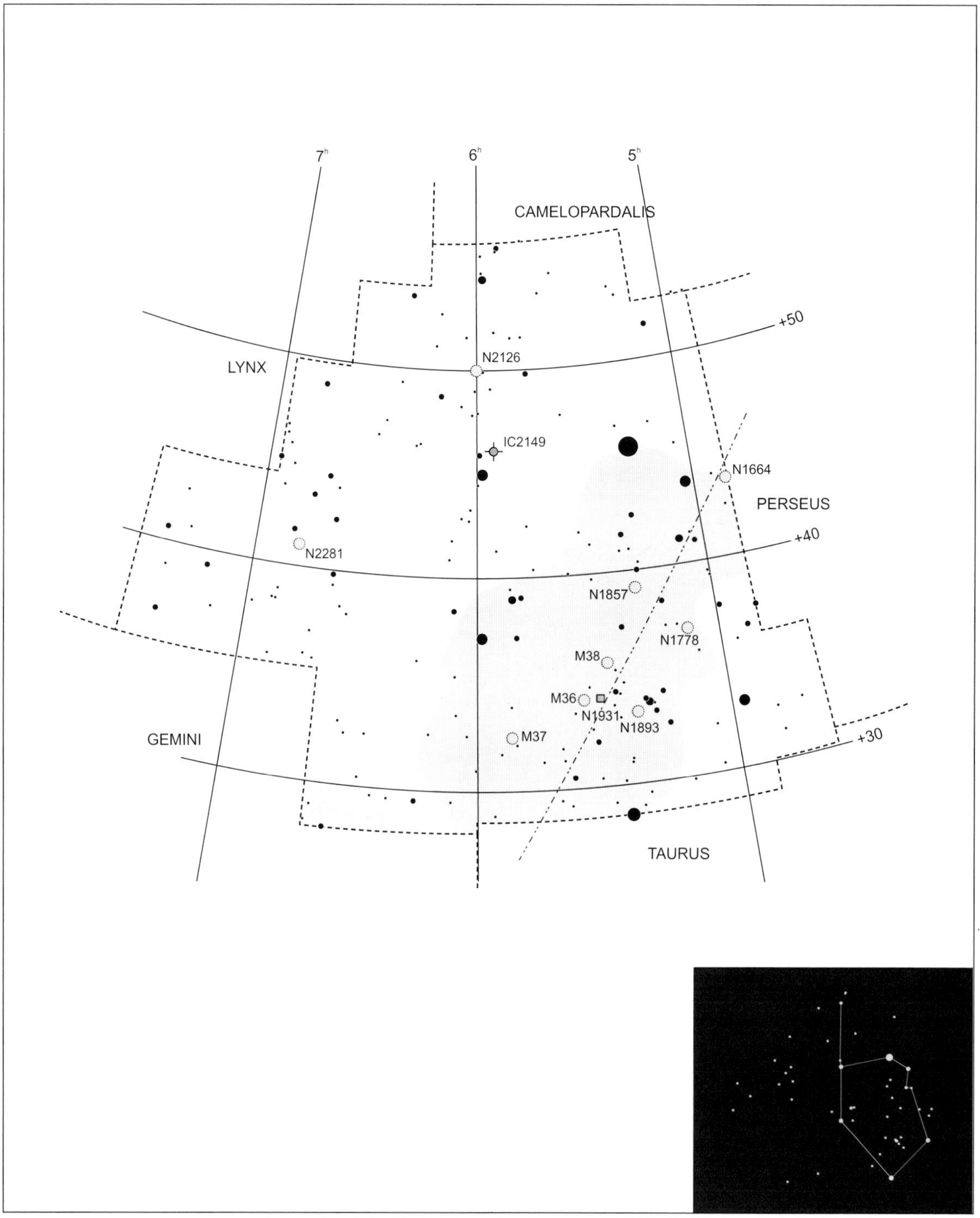

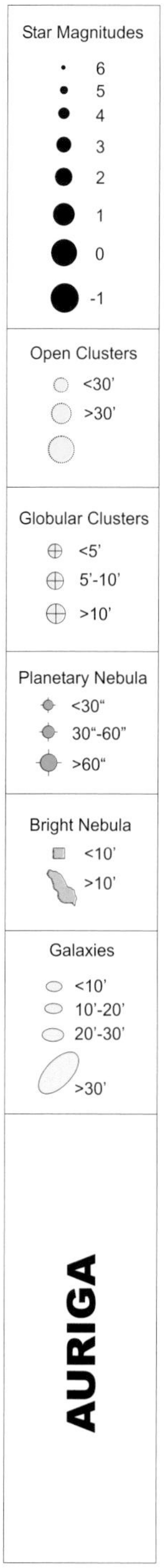

Constellation Facts:

Auriga; (or-EYE-gah)

Auriga, the Charioteer;
rises in the northeastern sky, passes overhead, and moves northwest.
The constellation lies across the centerline of the Milky Way. It contains several open clusters visible in small telescopes. The brightest star in the constellation is Capella, also known as the Great Star.
The constellation covers 657 squares degrees.

Constellation is visible from 90° N to 34° S. Partially visible from 34° S to 90° S.

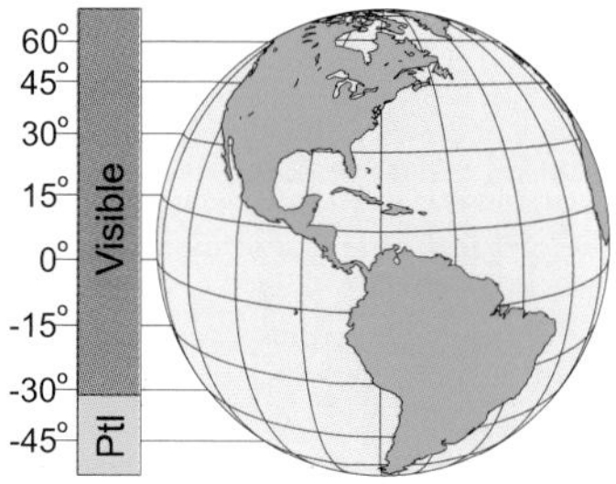

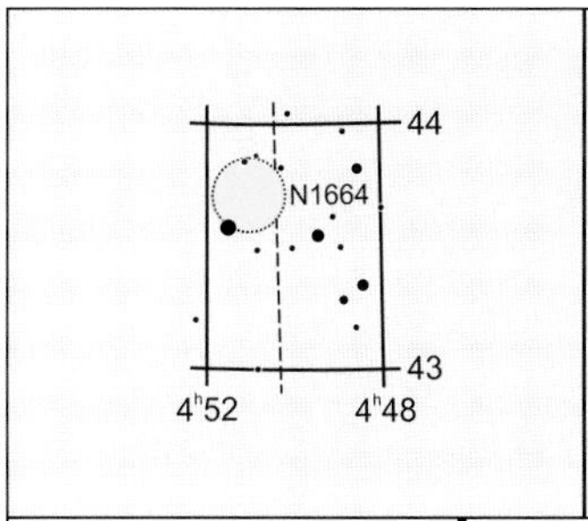

NGC 1664

RA:	$04^h\ 51^m\ 9.7^s$	Con:	Auriga
Dec:	43° 42' 00"	Type:	Open Cluster
Size:	18.0'	Mag:	7.6

NGC 1664 is a rich open cluster found on the border between Perseus and Auriga. Contains 40 stars or magnitude 11.0 and fainter. Cluster measures 18.0' across.

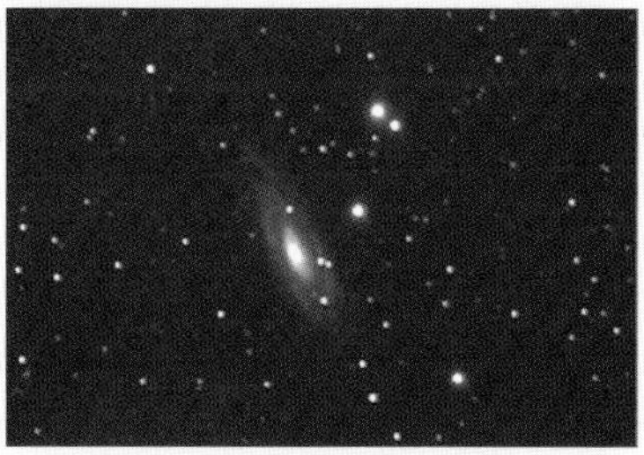

Telescope Aperture:	4" f/5	4" f/9	6" f/7	6" f/9	8" f/6.3	8" f/10	10" f/6.3	10" f/10	12" f/6.3	12" f/10
FOV(35mm film):	2.7° x 4.1°	1.50° x 2.26°	1.29° x 1.93°	1.0° x 1.50°	1.07° x 1.61°	0.68° x 1.02°	0.86° x 1.29°	0.54° x 0.81°	0.72° x 1.07°	0.45° x 0.68°

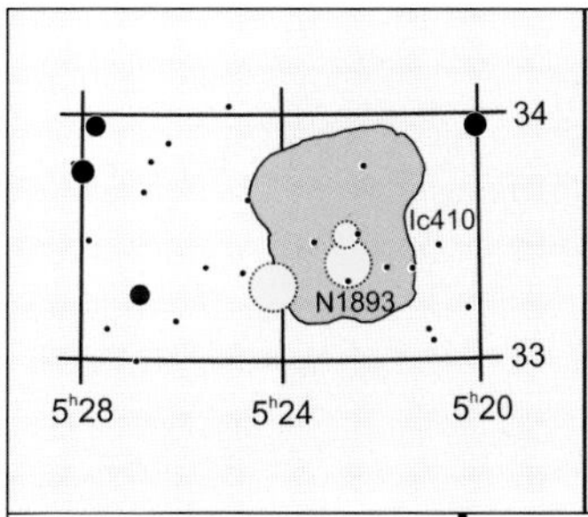

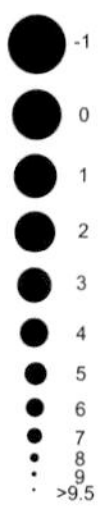

NGC 1893

RA:	$05^h\ 22^m\ 45.3^s$	Con:	Auriga
Dec:	33° 24' 00"	Type:	Open Cluster
Size:	11.0'	Mag:	7.5

NGC 1893 is an open cluster surrounded by nebula. It's an irregular shaped cluster of 20 stars ranging from 9th-12th magnitude. Object spans 11.0'.

Telescope Aperture:	4" f/5	4" f/9	6" f/7	6" f/9	8" f/6.3	8" f/10	10" f/6.3	10" f/10	12" f/6.3	12" f/10
FOV(35mm film):	2.7° x 4.1°	1.50° x 2.26°	1.29° x 1.93°	1.0° x 1.50°	1.07° x 1.61°	0.68° x 1.02°	0.86° x 1.29°	0.54° x 0.81°	0.72° x 1.07°	0.45° x 0.68°

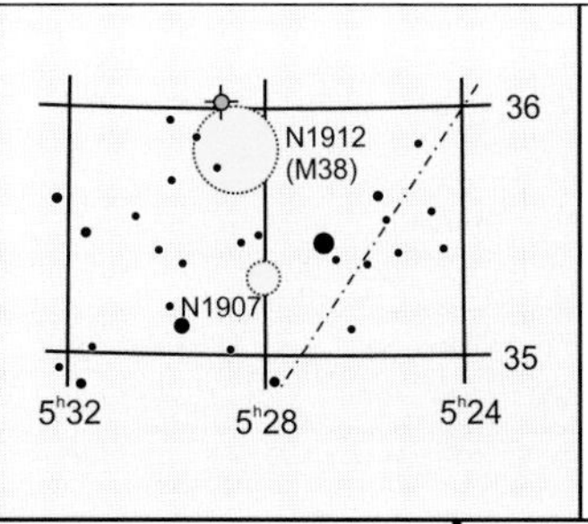

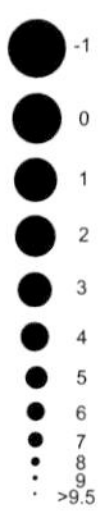

NGC 1907

RA:	$05^h 28^m\ 3.4^s$	Con:	Auriga
Dec:	35° 18' 59"	Type:	Open Cluster
Size:	7.0'	Mag:	8.2

NGC 1907 is found southeast of M38. It's a very small cluster 7.0' in diameter containing about 40 stars of magnitude 10 and fainter.

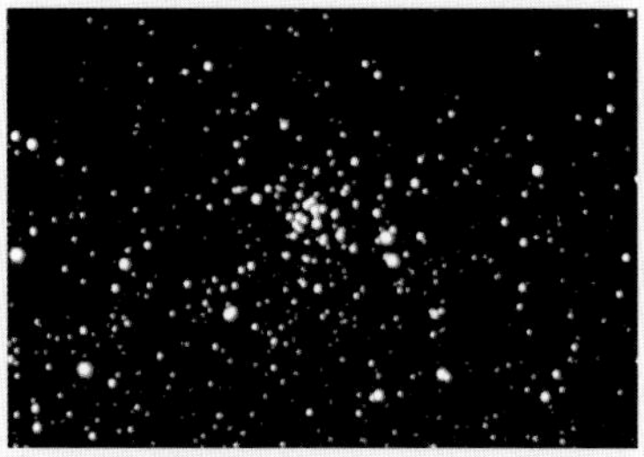

Telescope Aperture:	4" f/5	4" f/9	6" f/7	6" f/9	8" f/6.3	8" f/10	10" f/6.3	10" f/10	12" f/6.3	12" f/10
FOV(35mm film):	2.7° x 4.1°	1.50° x 2.26°	1.29° x 1.93°	1.0° x 1.50°	1.07° x 1.61°	0.68° x 1.02°	0.86° x 1.29°	0.54° x 0.81°	0.72° x 1.07°	0.45° x 0.68°

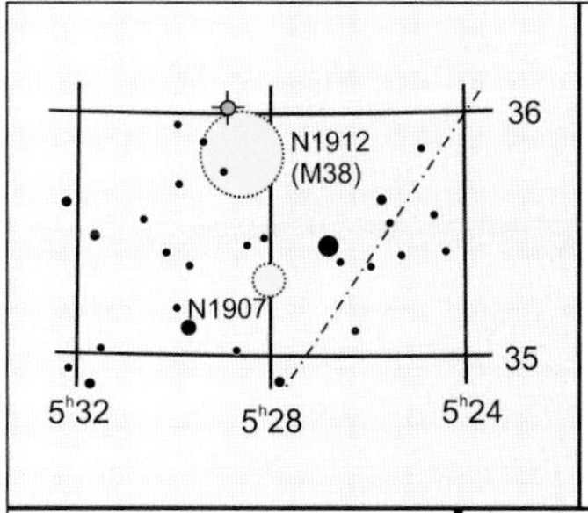

M38 (NGC 1912)

RA:	$05^h 28^m\ 45.4^s$	Con:	Auriga
Dec:	05° 28' 45.4"	Type:	Open Cluster
Size:	21.0'	Mag:	6.4

M38 (NGC 1912) is rich in stars. Observer will find a slight central concentration and outlying detached stars. Object spans 21.0' and shines at magnitude 6.4.

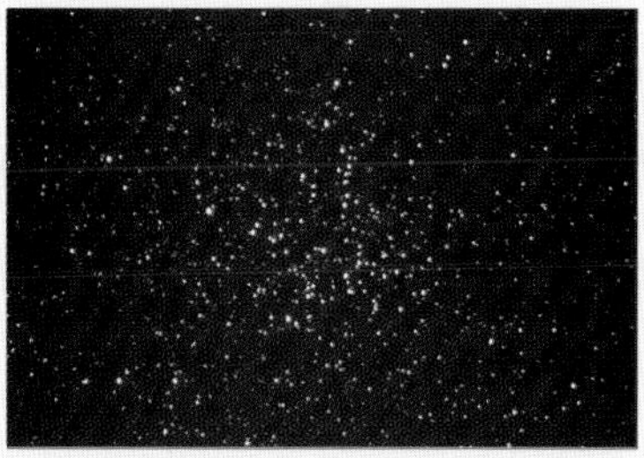

Telescope Aperture:	4" f/5	4" f/9	6" f/7	6" f/9	8" f/6.3	8" f/10	10" f/6.3	10" f/10	12" f/6.3	12" f/10
FOV(35mm film):	2.7° x 4.1°	1.50° x 2.26°	1.29° x 1.93°	1.0° x 1.50°	1.07° x 1.61°	0.68° x 1.02°	0.86° x 1.29°	0.54° x 0.81°	0.72° x 1.07°	0.45° x 0.68°

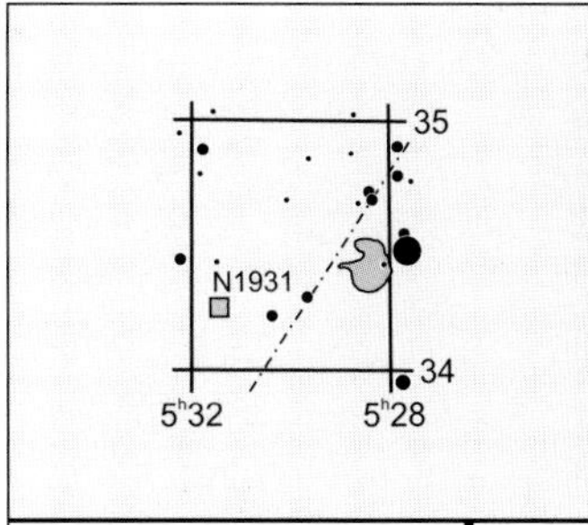

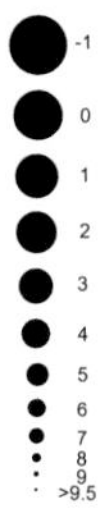

NGC 1931

RA:	$05^h\ 31^m\ 27.3^s$	Con:	Auriga
Dec:	34° 14' 59"	Type:	E + R Nebula
Size:	3.0'	Mag:	11.3

NGC 1931 is a peanut shaped emission nebula involving several clustered stars. Object is small at 3.0' and shines at magnitude 11.3.

Telescope Aperture:	4" f/5	4" f/9	6" f/7	6" f/9	8" f/6.3	8" f/10	10" f/6.3	10" f/10	12" f/6.3	12" f/10
FOV(35mm film):	2.7° x 4.1°	1.50° x 2.26°	1.29° x 1.93°	1.0° x 1.50°	1.07° x 1.61°	0.68° x 1.02°	0.86° x 1.29°	0.54° x 0.81°	0.72° x 1.07°	0.45° x 0.68°

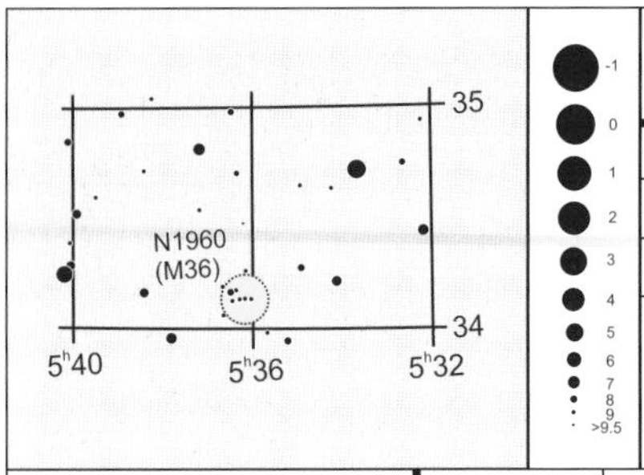

M36 (NGC 1960)

RA:	05^h 36^m 9.3^s	Con:	Auriga
Dec:	34° 07' 59"	Type:	Open Cluster
Size:	12.0'	Mag:	6.0

M36 (NGC 1960) is a rich open cluster with a strong central concentration. Object spans 12.0' and shines at magnitude 6.0.

Telescope Aperture:	4" f/5	4" f/9	6" f/7	6" f/9	8" f/6.3	8" f/10	10" f/6.3	10" f/10	12" f/6.3	12" f/10
FOV(35mm film):	2.7° x 4.1°	1.50° x 2.26°	1.29° x 1.93°	1.0° x 1.50°	1.07° x 1.61°	0.68° x 1.02°	0.86° x 1.29°	0.54° x 0.81°	0.72° x 1.07°	0.45° x 0.68°

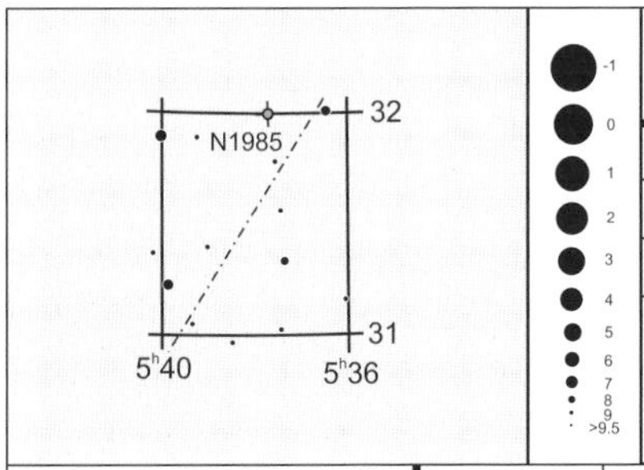

NGC 1985

RA:	05^h 37^m 45.3^s	Con:	Auriga
Dec:	31° 59' 59"	Type:	Bright Nebula
Size:		Mag:	

NGC 1985 is a small emission nebula. Object appears as a round spot of hazy light in large telescopes.

Telescope Aperture:	4" f/5	4" f/9	6" f/7	6" f/9	8" f/6.3	8" f/10	10" f/6.3	10" f/10	12" f/6.3	12" f/10
FOV(35mm film):	2.7° x 4.1°	1.50° x 2.26°	1.29° x 1.93°	1.0° x 1.50°	1.07° x 1.61°	0.68° x 1.02°	0.86° x 1.29°	0.54° x 0.81°	0.72° x 1.07°	0.45° x 0.68°

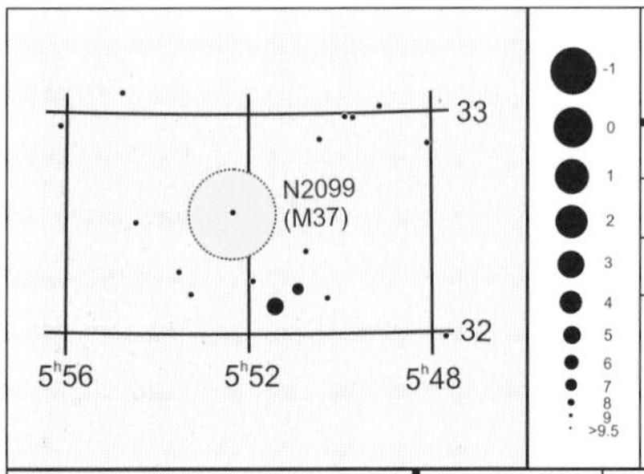

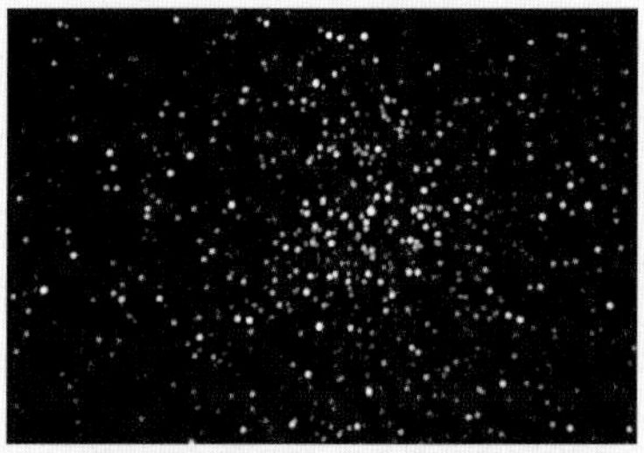

M37 (NGC 2099)

RA:	05^h 52^m 27.2^s	Con:	Auriga
Dec:	32° 32' 57"	Type:	Open Cluster
Size:	24.0'	Mag:	5.6

M37 (NGC 2099) is found just outside the Auriga pentagon asterism, several degrees southeast of M36/M37 area. Object contains about 150 stars.

Telescope Aperture:	4" f/5	4" f/9	6" f/7	6" f/9	8" f/6.3	8" f/10	10" f/6.3	10" f/10	12" f/6.3	12" f/10
FOV(35mm film):	2.7° x 4.1°	1.50° x 2.26°	1.29° x 1.93°	1.0° x 1.50°	1.07° x 1.61°	0.68° x 1.02°	0.86° x 1.29°	0.54° x 0.81°	0.72° x 1.07°	0.45° x 0.68°

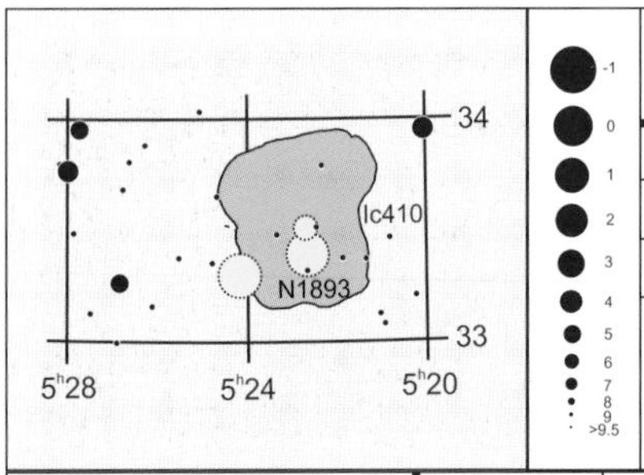

IC 410

RA:	05^h 22^m 39.3^s	Con:	Auriga
Dec:	33° 31' 00"	Type:	Open Cluster
Size:	40.0'	Mag:	

IC 410 is a cluster of stars with a strong central concentration involved with associated nebula. Object spans 40.0' in size.

Telescope Aperture:	4" f/5	4" f/9	6" f/7	6" f/9	8" f/6.3	8" f/10	10" f/6.3	10" f/10	12" f/6.3	12" f/10
FOV(35mm film):	2.7° x 4.1°	1.50° x 2.26°	1.29° x 1.93°	1.0° x 1.50°	1.07° x 1.61°	0.68° x 1.02°	0.86° x 1.29°	0.54° x 0.81°	0.72° x 1.07°	0.45° x 0.68°

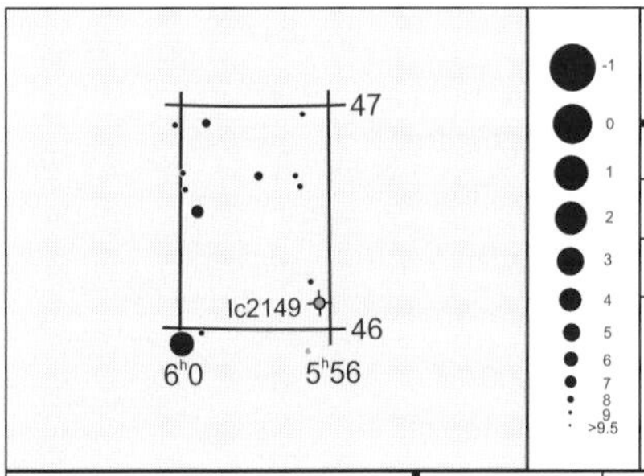

IC 2149

RA:	05^h 56^m 21.5^s	Con:	Auriga
Dec:	46° 06' 56"	Type:	Planetary Nebula
Size:	0.1'	Mag:	11.0

IC 2149 is a bright planetary nebula found 3° north or Beta Aurigae. Object is 0.1' in size and shines at magnitude 11.0.

Telescope Aperture:	4" f/5	4" f/9	6" f/7	6" f/9	8" f/6.3	8" f/10	10" f/6.3	10" f/10	12" f/6.3	12" f/10
FOV(35mm film):	2.7° x 4.1°	1.50° x 2.26°	1.29° x 1.93°	1.0° x 1.50°	1.07° x 1.61°	0.68° x 1.02°	0.86° x 1.29°	0.54° x 0.81°	0.72° x 1.07°	0.45° x 0.68°

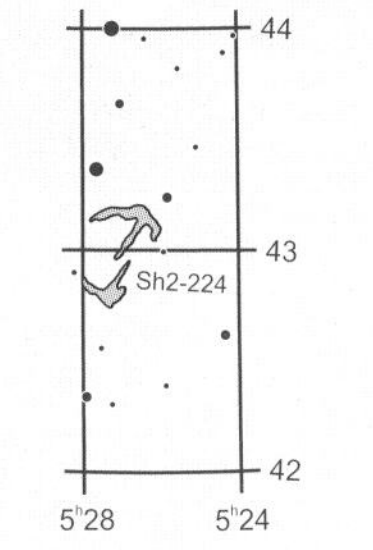

Sh2-224

RA:	05h 36m 9.3s	Con:	Auriga
Dec:	34° 07' 59"	Type:	SN Remnant
Size:	20' x 3.0'	Mag:	

Sh2-224 is a super-nova remnant that forms an incomplete oval ring.

Telescope Aperture:	4" f/5	4" f/9	6" f/7	6" f/9	8" f/6.3	8" f/10	10" f/6.3	10" f/10	12" f/6.3	12" f/10
FOV(35mm film):	2.7° x 4.1°	1.50° x 2.26°	1.29° x 1.93°	1.0° x 1.50°	1.07° x 1.61°	0.68° x 1.02°	0.86° x 1.29°	0.54° x 0.81°	0.72° x 1.07°	0.45° x 0.68°

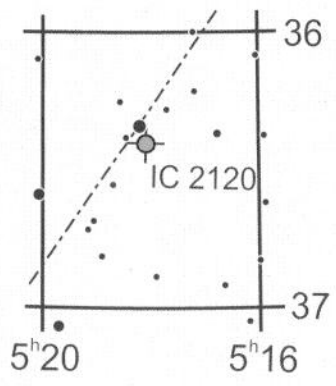

IC 2120

RA:	05h 37m 45.3s	Con:	Auriga
Dec:	31° 59' 59"	Type:	Planetary Nebula
Size:		Mag:	

IC 2120 is a small planetary nebula. Object is hard to locate visually in medium to large scopes.

Telescope Aperture:	4" f/5	4" f/9	6" f/7	6" f/9	8" f/6.3	8" f/10	10" f/6.3	10" f/10	12" f/6.3	12" f/10
FOV(35mm film):	2.7° x 4.1°	1.50° x 2.26°	1.29° x 1.93°	1.0° x 1.50°	1.07° x 1.61°	0.68° x 1.02°	0.86° x 1.29°	0.54° x 0.81°	0.72° x 1.07°	0.45° x 0.68°

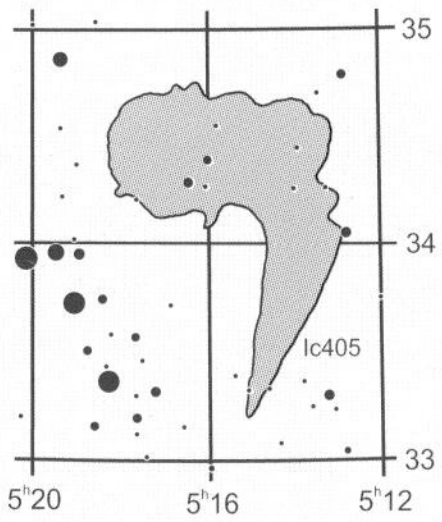

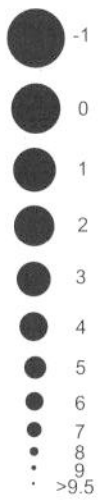

IC 405 (Flaming Star Nebula)

RA:	05h 16m 15.4s	Con:	Auriga
Dec:	34° 16' 00"	Type:	E + R Nebula
Size:	30.0' x 20.0'	Mag:	

IC 405 the "Flaming Star Nebula" is a large emission nebula appearing as a faint, wispy gas cloud. Object contains structures similar in appearance to that of the Veil Nebula in Cygnus.

Telescope Aperture:	4" f/5	4" f/9	6" f/7	6" f/9	8" f/6.3	8" f/10	10" f/6.3	10" f/10	12" f/6.3	12" f/10
FOV(35mm film):	2.7° x 4.1°	1.50° x 2.26°	1.29° x 1.93°	1.0° x 1.50°	1.07° x 1.61°	0.68° x 1.02°	0.86° x 1.29°	0.54° x 0.81°	0.72° x 1.07°	0.45° x 0.68°

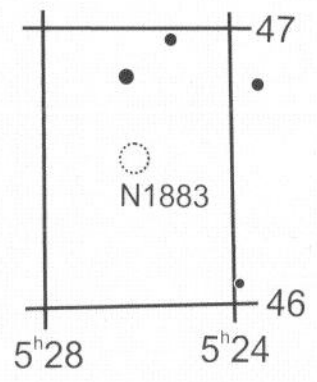

NGC 1883

RA:	05h 22m 39.3s	Con:	Auriga
Dec:	33° 31' 00"	Type:	Open Cluster
Size:	3.0'	Mag:	12.0

NGC 1883 is a nice open cluster moderately rich in stars. Object displays a small central concentration.

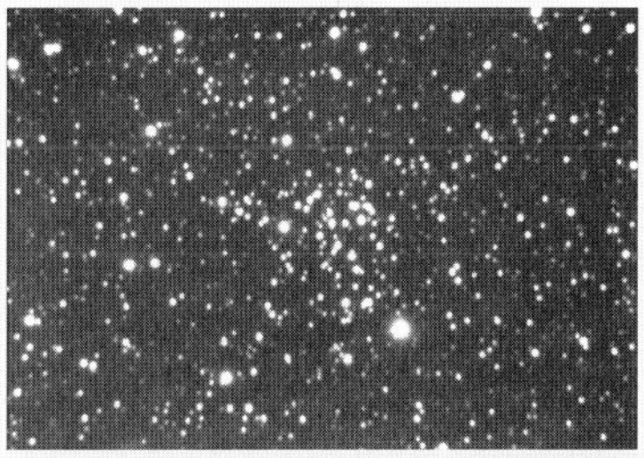

Telescope Aperture:	4" f/5	4" f/9	6" f/7	6" f/9	8" f/6.3	8" f/10	10" f/6.3	10" f/10	12" f/6.3	12" f/10
FOV(35mm film):	2.7° x 4.1°	1.50° x 2.26°	1.29° x 1.93°	1.0° x 1.50°	1.07° x 1.61°	0.68° x 1.02°	0.86° x 1.29°	0.54° x 0.81°	0.72° x 1.07°	0.45° x 0.68°

Boo

Crosses Prime Meridian: May thru June

SPRING

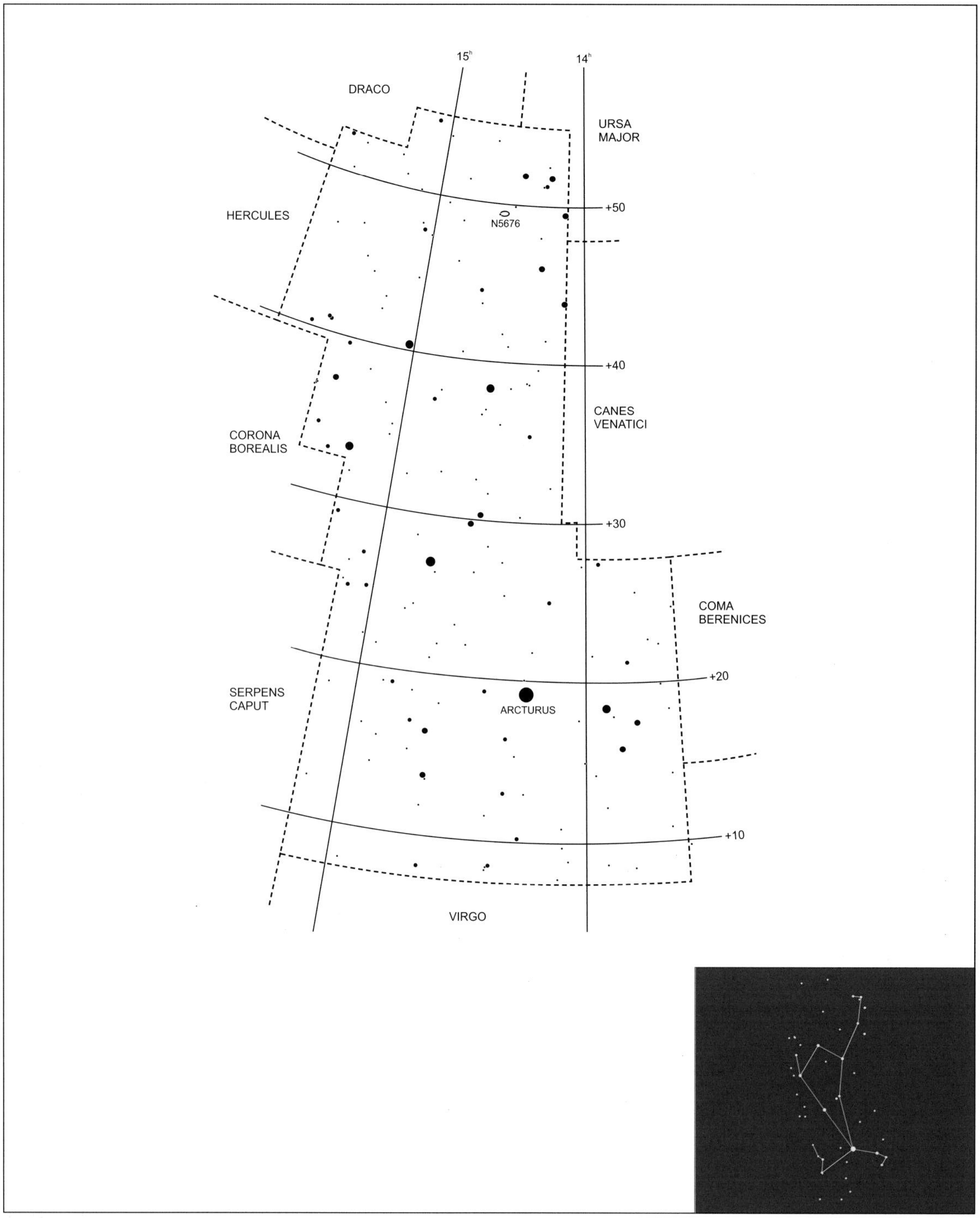

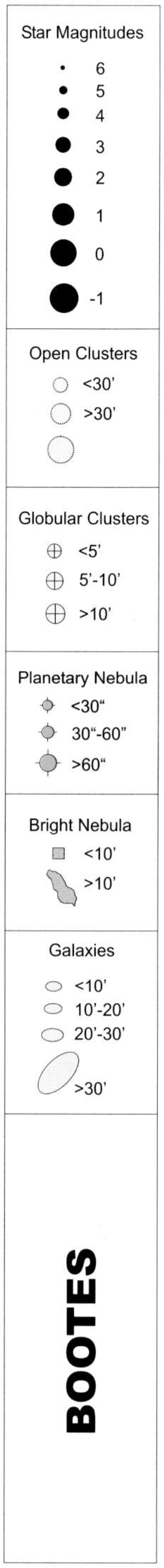

Constellation Facts:

Bootes; (bow-OH-tease)

Bootes, the Herdsman.
The constellation rises in the northeast, passes thru the meridian overhead, and sets in the northwest.
On clear nights, the fainter stars north of Arcturus appear in the general shape of an ice cream cone or, a kite.
The constellation is 907 square degrees.

Constellation is visible from 90° N to 35° S. Partially visible from 35° S to 90° S.

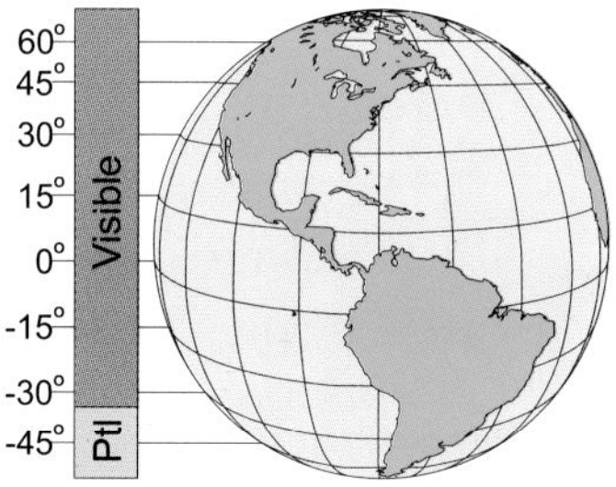

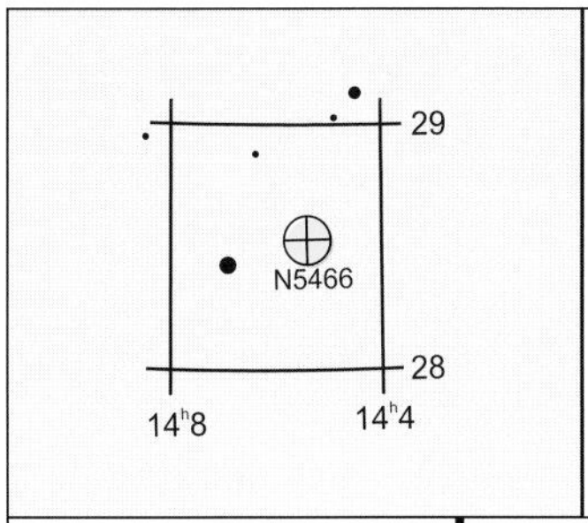

NGC 5466

RA:	$14^h\ 05^m\ 33.5^s$	Con:	Bootes
Dec:	28° 31' 52"	Type:	Globular Cluster
Size:	11.0'	Mag:	9.1

NGC 5466 is a large loose globular cluster, composed of faint stars. Object is located 2° northeast of 11 Bootis.

Telescope Aperture:	4" f/5	4" f/9	6" f/7	6" f/9	8" f/6.3	8" f/10	10" f/6.3	10" f/10	12" f/6.3	12" f/10
FOV(35mm film):	2.7° x 4.1°	1.50° x 2.26°	1.29° x 1.93°	1.0° x 1.50°	1.07° x 1.61°	0.68° x 1.02°	0.86° x 1.29°	0.54° x 0.81°	0.72° x 1.07°	0.45° x 0.68°

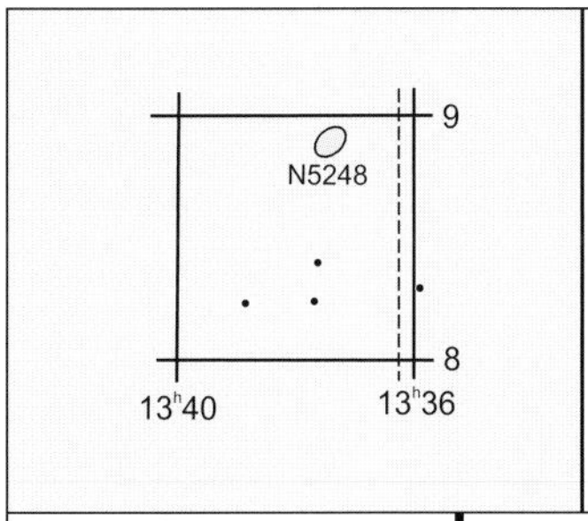

NGC 5248

RA:	$13^h\ 37^m\ 33.7^s$	Con:	Bootes
Dec:	08° 52' 44"	Type:	Spiral Galaxy
Size:	6.0' x 5.0'	Mag:	10.2

NGC 5248 in located in the farthest southwestern corner of the constellation. It is the constellation's brightest galaxy, and is an Sc-type spiral galaxy.

Telescope Aperture:	4" f/5	4" f/9	6" f/7	6" f/9	8" f/6.3	8" f/10	10" f/6.3	10" f/10	12" f/6.3	12" f/10
FOV(35mm film):	2.7° x 4.1°	1.50° x 2.26°	1.29° x 1.93°	1.0° x 1.50°	1.07° x 1.61°	0.68° x 1.02°	0.86° x 1.29°	0.54° x 0.81°	0.72° x 1.07°	0.45° x 0.68°

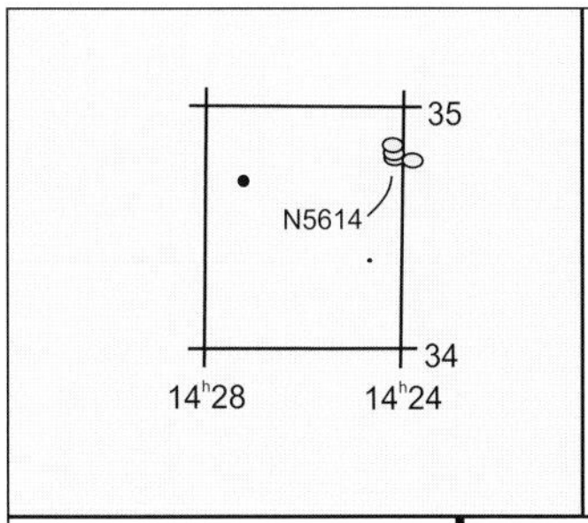

NGC 5614

RA:	$14^h 24^m\ 9.5^s$	Con:	Bootes
Dec:	34° 51' 54"	Type:	Spiral Galaxy
Size:	2.0' x 1.8'	Mag:	11.7

NGC 5614 is located in the central region of the constellation. The spiral galaxy is located 2° east of A Bootis.

Telescope Aperture:	4" f/5	4" f/9	6" f/7	6" f/9	8" f/6.3	8" f/10	10" f/6.3	10" f/10	12" f/6.3	12" f/10
FOV(35mm film):	2.7° x 4.1°	1.50° x 2.26°	1.29° x 1.93°	1.0° x 1.50°	1.07° x 1.61°	0.68° x 1.02°	0.86° x 1.29°	0.54° x 0.81°	0.72° x 1.07°	0.45° x 0.68°

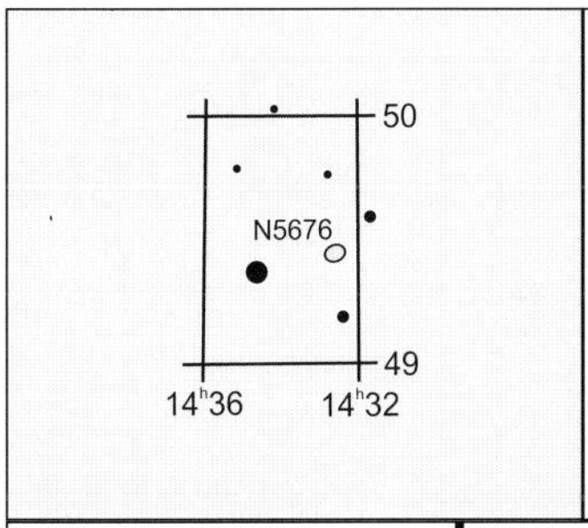

NGC 5676

RA:	$14^h 32^m\ 51.2^s$	Con:	Bootes
Dec:	49° 27' 58"	Type:	Spiral Galaxy
Size:	3.4' x 1.4'	Mag:	10.9

NGC 5676 is an Sc-type spiral galaxy located 1° southeast of NGC 5660.

Telescope Aperture:	4" f/5	4" f/9	6" f/7	6" f/9	8" f/6.3	8" f/10	10" f/6.3	10" f/10	12" f/6.3	12" f/10
FOV(35mm film):	2.7° x 4.1°	1.50° x 2.26°	1.29° x 1.93°	1.0° x 1.50°	1.07° x 1.61°	0.68° x 1.02°	0.86° x 1.29°	0.54° x 0.81°	0.72° x 1.07°	0.45° x 0.68°

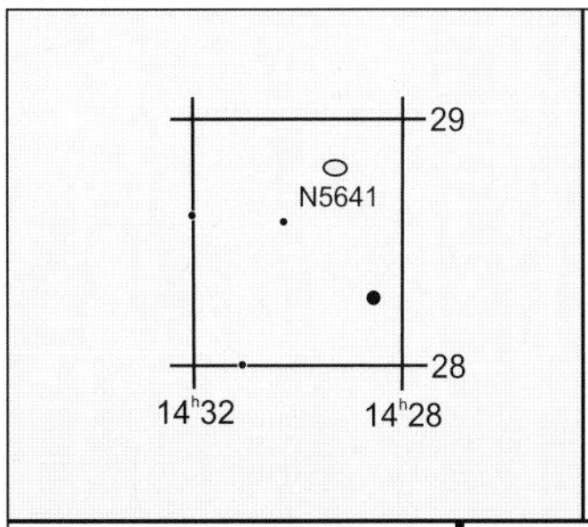

NGC 5641

RA:	$14^h\ 29^m\ 21.6^s$	Con:	Bootes
Dec:	25° 18' 51"	Type:	Barred Spiral
Size:	4.5' x 1.0'	Mag:	13.0

NGC 5641 is located 5° east of NGC 5466. It is a type Sb-barred spiral galaxy.

Telescope Aperture:	4" f/5	4" f/9	6" f/7	6" f/9	8" f/6.3	8" f/10	10" f/6.3	10" f/10	12" f/6.3	12" f/10
FOV(35mm film):	2.7° x 4.1°	1.50° x 2.26°	1.29° x 1.93°	1.0° x 1.50°	1.07° x 1.61°	0.68° x 1.02°	0.86° x 1.29°	0.54° x 0.81°	0.72° x 1.07°	0.45° x 0.68°

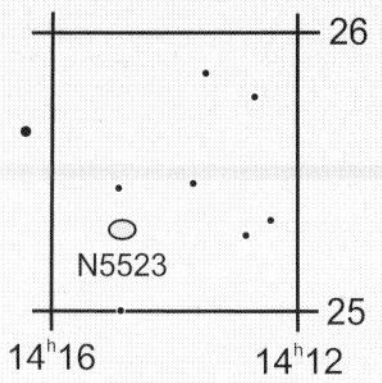

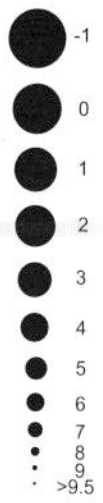

NGC 5523

RA:	$14^h\ 14^m\ 51.6^s$	Con:	Bootes
Dec:	25° 18' 51"	Type:	Spiral Galaxy
Size:	4.5' x 1.0'	Mag:	13.0

NGC 5523 is an edge-on spiral galaxy that forms an equilateral triangle with NGC 5641 and NGC 5466. Object is located 1° east-northeast of the star 12 Bootis

Telescope Aperture:	4" f/5	4" f/9	6" f/7	6" f/9	8" f/6.3	8" f/10	10" f/6.3	10" f/10	12" f/6.3	12" f/10
FOV(35mm film):	2.7° x 4.1°	1.50° x 2.26°	1.29° x 1.93°	1.0° x 1.50°	1.07° x 1.61°	0.68° x 1.02°	0.86° x 1.29°	0.54° x 0.81°	0.72° x 1.07°	0.45° x 0.68°

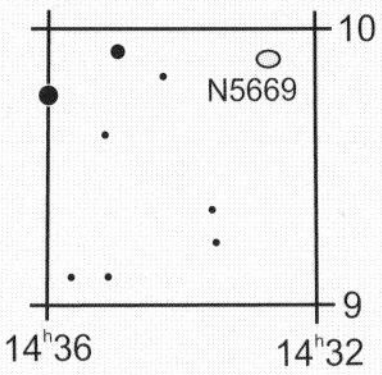

NGC 5669

RA:	$14^h\ 32^m\ 46.0^s$	Con:	Bootes
Dec:	09° 52' 48"	Type:	Amorphous Galaxy
Size:	4.0' x 3.6'	Mag:	12.0

NGC 5669 resides approximately 11° east and slightly south of a prominent curved row of stars. At the end of the row move another 3° northeast. Object is a type Sc-galaxy.

Telescope Aperture:	4" f/5	4" f/9	6" f/7	6" f/9	8" f/6.3	8" f/10	10" f/6.3	10" f/10	12" f/6.3	12" f/10
FOV(35mm film):	2.7° x 4.1°	1.50° x 2.26°	1.29° x 1.93°	1.0° x 1.50°	1.07° x 1.61°	0.68° x 1.02°	0.86° x 1.29°	0.54° x 0.81°	0.72° x 1.07°	0.45° x 0.68°

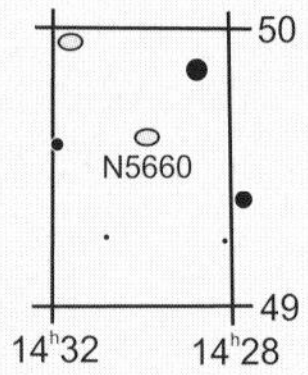

NGC 5660

RA:	$14^h 29^m\ 51.1^s$	Con:	Bootes
Dec:	49° 36' 58"	Type:	Spiral Galaxy
Size:	2.4'	Mag:	11.8

NGC 5660 is an Sc-type spiral galaxy oriented face-on to our line of site. The central condensation is 30" across.

Telescope Aperture:	4" f/5	4" f/9	6" f/7	6" f/9	8" f/6.3	8" f/10	10" f/6.3	10" f/10	12" f/6.3	12" f/10
FOV(35mm film):	2.7° x 4.1°	1.50° x 2.26°	1.29° x 1.93°	1.0° x 1.50°	1.07° x 1.61°	0.68° x 1.02°	0.86° x 1.29°	0.54° x 0.81°	0.72° x 1.07°	0.45° x 0.68°

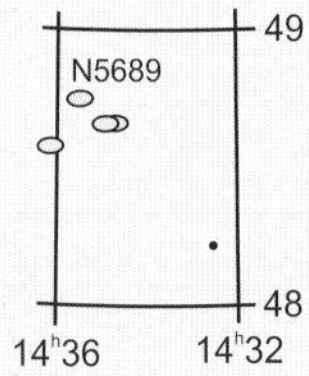

NGC 5689

RA:	$14^h 35^m\ 33.2^s$	Con:	Bootes
Dec:	48° 44' 58"	Type:	Barred Spiral
Size:	2.5' x 0.7'	Mag:	11.9

NGC 5689 is located 1° southeast of NGC 5676. It's a barred spiral galaxy.

Telescope Aperture:	4" f/5	4" f/9	6" f/7	6" f/9	8" f/6.3	8" f/10	10" f/6.3	10" f/10	12" f/6.3	12" f/10
FOV(35mm film):	2.7° x 4.1°	1.50° x 2.26°	1.29° x 1.93°	1.0° x 1.50°	1.07° x 1.61°	0.68° x 1.02°	0.86° x 1.29°	0.54° x 0.81°	0.72° x 1.07°	0.45° x 0.68°

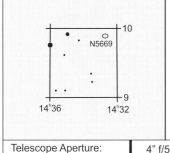

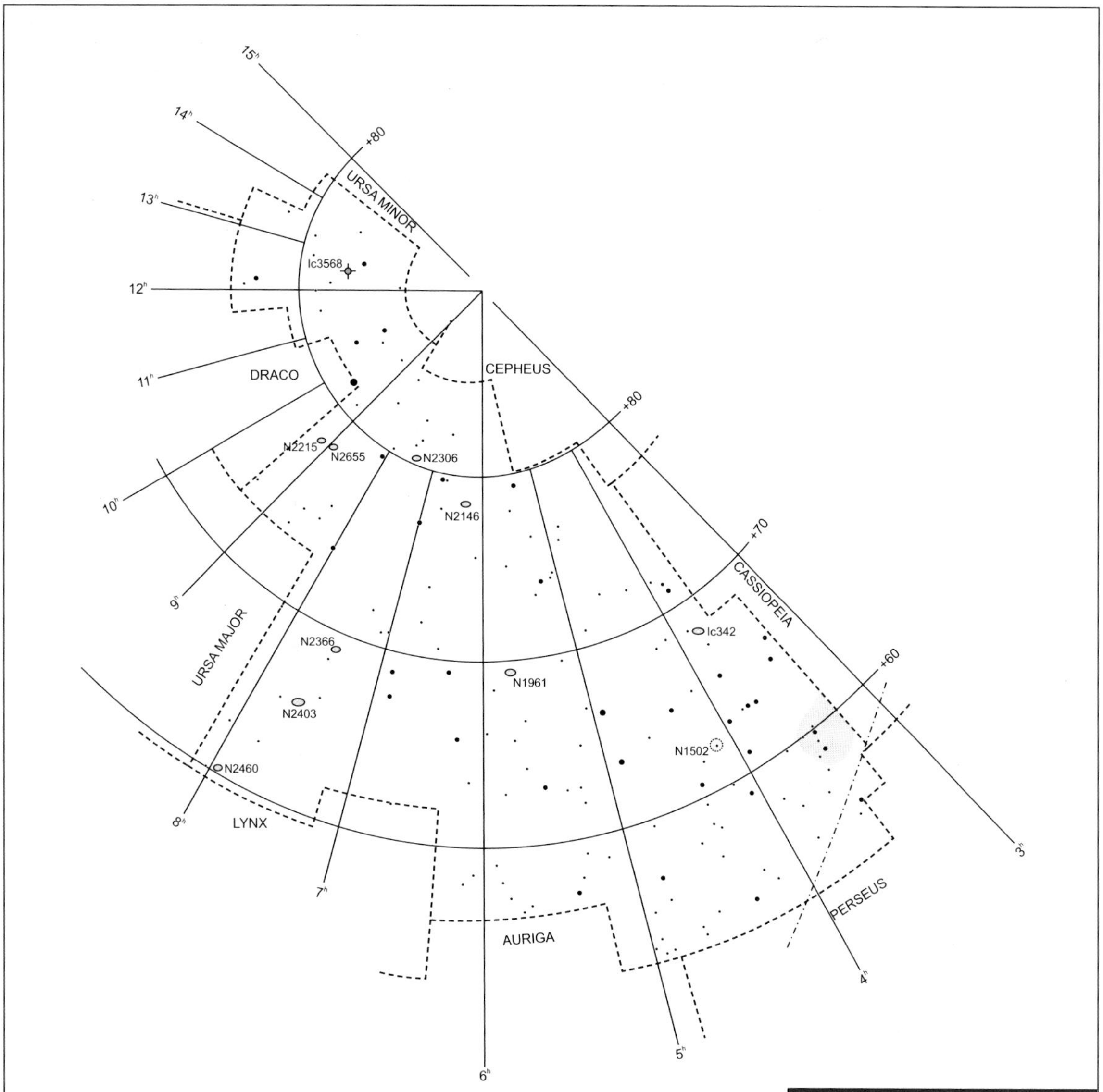

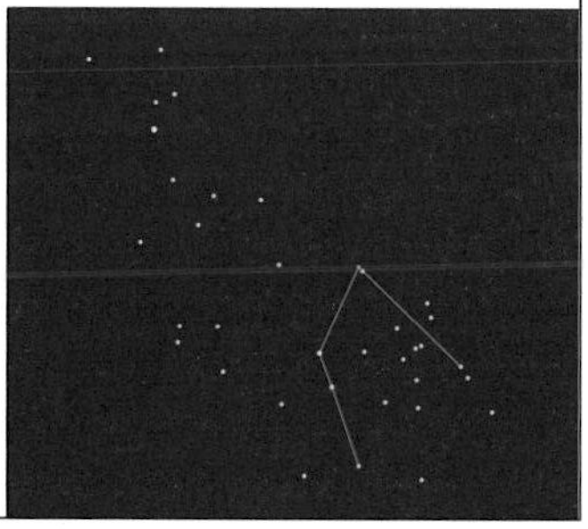

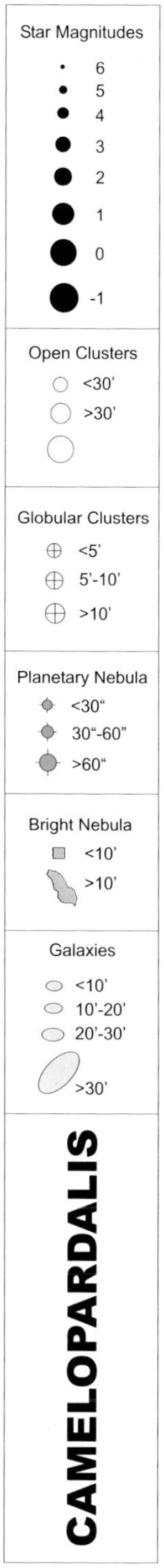

Constellation Facts:

Camelopardalis;

Is a faint northern circumpolar constellation.

Constellation is visible from 90° N to 3° S. Partially visible from 3° S to 50° S.

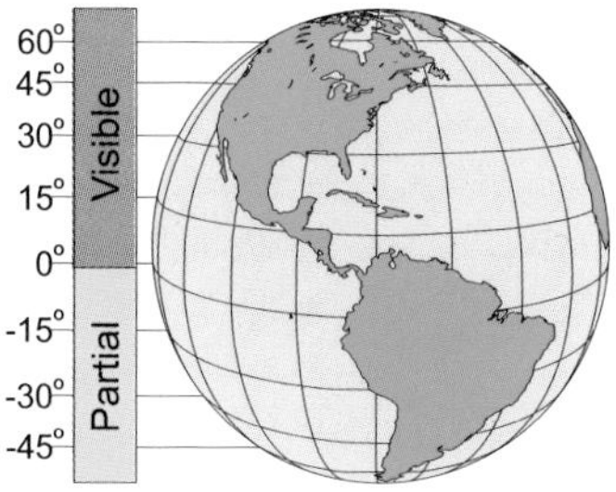

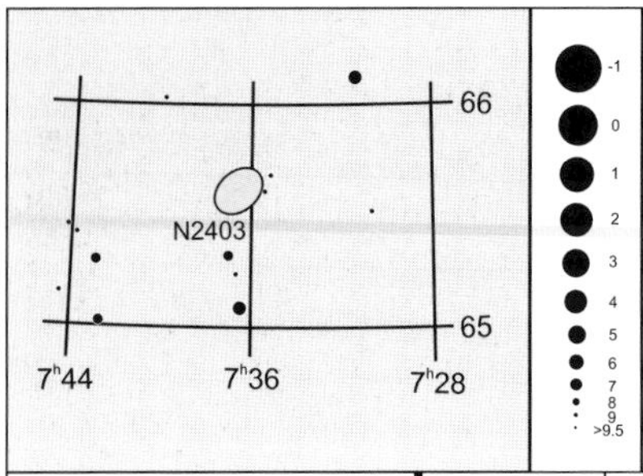

NGC 2403

RA:	$07^h\ 36^m\ 57.6^s$	Con:	Camelopardalis
Dec:	65° 35' 51"	Type:	Spiral Galaxy
Size:	18.0' x 10.1'	Mag:	8.4

NGC 2403 is found in the southeastern corner of the constellation. Object is a magnificent spiral galaxy appearing almost face-on to our line of sight.

Telescope Aperture:	4" f/5	4" f/9	6" f/7	6" f/9	8" f/6.3	8" f/10	10" f/6.3	10" f/10	12" f/6.3	12" f/10
FOV(35mm film):	2.7° x 4.1°	1.50° x 2.26°	1.29° x 1.93°	1.0° x 1.50°	1.07° x 1.61°	0.68° x 1.02°	0.86° x 1.29°	0.54° x 0.81°	0.72° x 1.07°	0.45° x 0.68°

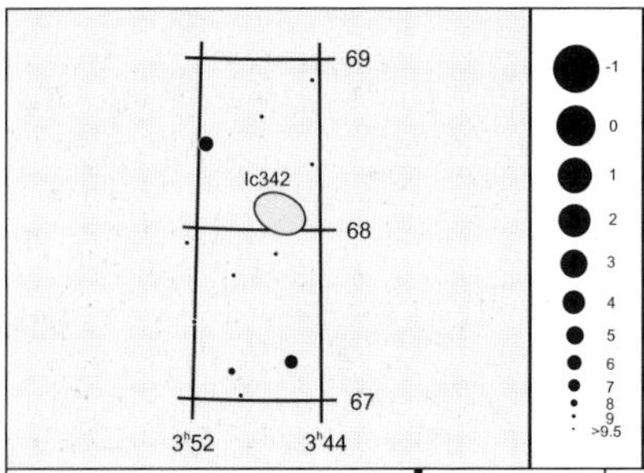

IC 342

RA:	$03^h\ 46^m\ 49.7^s$	Con:	Camelopardalis
Dec:	68° 05' 45"	Type:	Barred Spiral
Size:	21.2' x 20.7'	Mag:	9.1

IC 342 is a large, round barred spiral galaxy, obscured by the thick Milky Way in front of it. It is located 4 hours of right ascension west and a little north of NGC 2403.

Telescope Aperture:	4" f/5	4" f/9	6" f/7	6" f/9	8" f/6.3	8" f/10	10" f/6.3	10" f/10	12" f/6.3	12" f/10
FOV(35mm film):	2.7° x 4.1°	1.50° x 2.26°	1.29° x 1.93°	1.0° x 1.50°	1.07° x 1.61°	0.68° x 1.02°	0.86° x 1.29°	0.54° x 0.81°	0.72° x 1.07°	0.45° x 0.68°

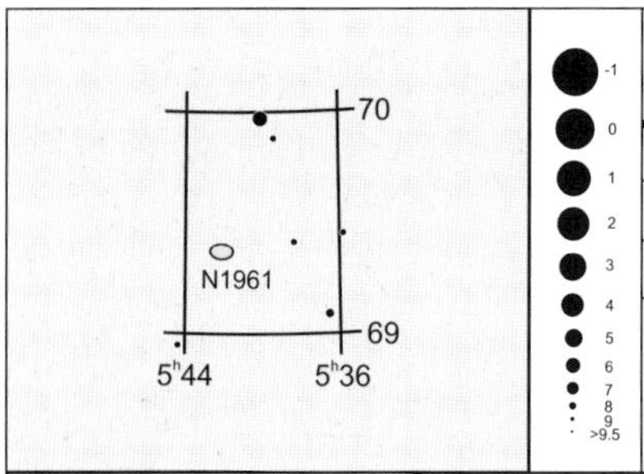

NGC 1961

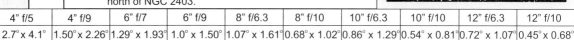

RA:	$05^h\ 42^m\ 10.4^s$	Con:	Camelopardalis
Dec:	69° 22' 54"	Type:	Peculiar Galaxy
Size:	3.9' x 1.9'	Mag:	11.1

NGC 1961 is located midway between NGC 2403 and IC 342. The galaxy is a type-Sb peculiar system.

Telescope Aperture:	4" f/5	4" f/9	6" f/7	6" f/9	8" f/6.3	8" f/10	10" f/6.3	10" f/10	12" f/6.3	12" f/10
FOV(35mm film):	2.7° x 4.1°	1.50° x 2.26°	1.29° x 1.93°	1.0° x 1.50°	1.07° x 1.61°	0.68° x 1.02°	0.86° x 1.29°	0.54° x 0.81°	0.72° x 1.07°	0.45° x 0.68°

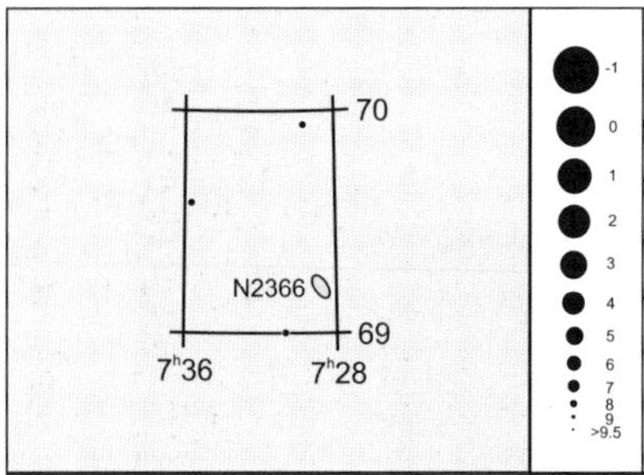

NGC 2366

RA:	$07^h\ 28^m\ 57.8^s$	Con:	Camelopardalis
Dec:	69° 12' 52"	Type:	Irregular Galaxy
Size:	7.8' x 1.9'	Mag:	10.9

NGC 2366 is a strange irregular galaxy. The galaxy is high enough in declination that it is circumpolar from mid-northern latitudes.

Telescope Aperture:	4" f/5	4" f/9	6" f/7	6" f/9	8" f/6.3	8" f/10	10" f/6.3	10" f/10	12" f/6.3	12" f/10
FOV(35mm film):	2.7° x 4.1°	1.50° x 2.26°	1.29° x 1.93°	1.0° x 1.50°	1.07° x 1.61°	0.68° x 1.02°	0.86° x 1.29°	0.54° x 0.81°	0.72° x 1.07°	0.45° x 0.68°

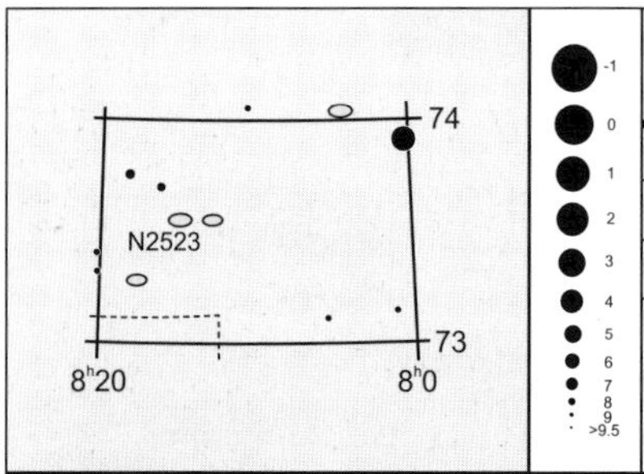

NGC 2523

RA:	$08^h\ 15^m\ 3.5^s$	Con:	Camelopardalis
Dec:	73° 34' 52"	Type:	Barred Spiral
Size:	0.4' x 0.3'	Mag:	12.0

NGC 2523 is an elegant barred spiral galaxy.

Telescope Aperture:	4" f/5	4" f/9	6" f/7	6" f/9	8" f/6.3	8" f/10	10" f/6.3	10" f/10	12" f/6.3	12" f/10
FOV(35mm film):	2.7° x 4.1°	1.50° x 2.26°	1.29° x 1.93°	1.0° x 1.50°	1.07° x 1.61°	0.68° x 1.02°	0.86° x 1.29°	0.54° x 0.81°	0.72° x 1.07°	0.45° x 0.68°

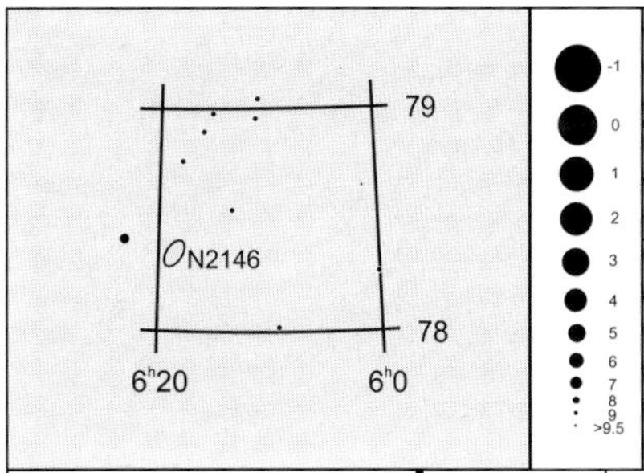

NGC 2146

RA:	$06^h\ 18^m\ 47.0^s$	Con:	Camelopardalis
Dec:	78° 20' 53"	Type:	Barred Spiral
Size:	2.6' x 0.9'	Mag:	10.5

NGC 2146 is a barred spiral galaxy wedged between two scattered groups of stars.

Telescope Aperture:	4" f/5	4" f/9	6" f/7	6" f/9	8" f/6.3	8" f/10	10" f/6.3	10" f/10	12" f/6.3	12" f/10
FOV(35mm film):	2.7° x 4.1°	1.50° x 2.26°	1.29° x 1.93°	1.0° x 1.50°	1.07° x 1.61°	0.68° x 1.02°	0.86° x 1.29°	0.54° x 0.81°	0.72° x 1.07°	0.45° x 0.68°

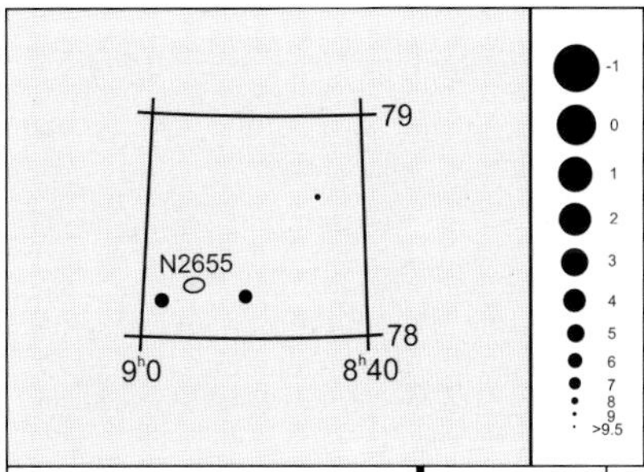

NGC 2655

RA:	$08^h\ 55^m\ 39.0^s$	Con:	Camelopardalis
Dec:	78° 12' 51"	Type:	Barred Spiral
Size:	3.9' x 3.4'	Mag:	10.1

NGC 2655 is located in the northeastern region of the constellation.

Telescope Aperture:	4" f/5	4" f/9	6" f/7	6" f/9	8" f/6.3	8" f/10	10" f/6.3	10" f/10	12" f/6.3	12" f/10
FOV(35mm film):	2.7° x 4.1°	1.50° x 2.26°	1.29° x 1.93°	1.0° x 1.50°	1.07° x 1.61°	0.68° x 1.02°	0.86° x 1.29°	0.54° x 0.81°	0.72° x 1.07°	0.45° x 0.68°

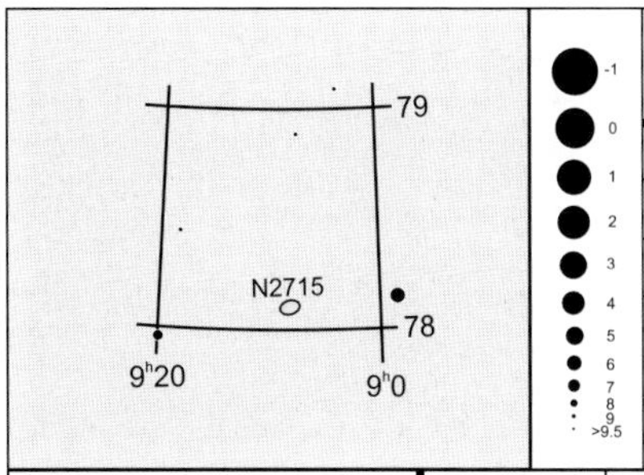

NGC 2715

RA:	$09^h\ 08^m\ 9.0^s$	Con:	Camelopardalis
Dec:	78° 04' 52"	Type:	Spiral Galaxy
Size:	5.0' x 1.4'	Mag:	11.4

NGC 2715 is located 2° from NGC 2655.

Telescope Aperture:	4" f/5	4" f/9	6" f/7	6" f/9	8" f/6.3	8" f/10	10" f/6.3	10" f/10	12" f/6.3	12" f/10
FOV(35mm film):	2.7° x 4.1°	1.50° x 2.26°	1.29° x 1.93°	1.0° x 1.50°	1.07° x 1.61°	0.68° x 1.02°	0.86° x 1.29°	0.54° x 0.81°	0.72° x 1.07°	0.45° x 0.68°

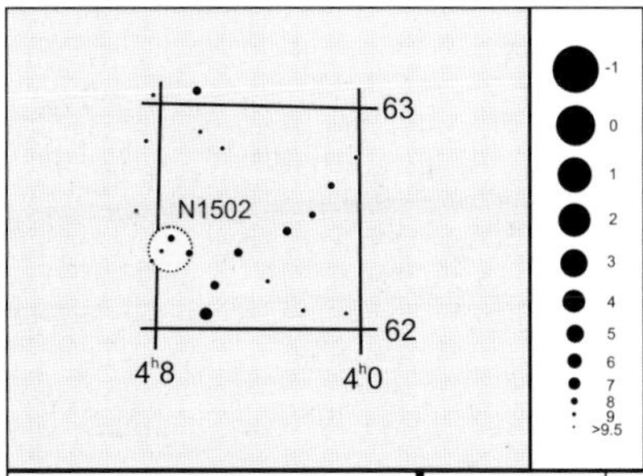

NGC 1502

RA:	$04^h\ 07^m\ 46.5^s$	Con:	Camelopardalis
Dec:	62° 19' 59"	Type:	Open Cluster
Size:	8.0'	Mag:	5.7

NGC 1502 is a rich open cluster containing 45 stars.

Telescope Aperture:	4" f/5	4" f/9	6" f/7	6" f/9	8" f/6.3	8" f/10	10" f/6.3	10" f/10	12" f/6.3	12" f/10
FOV(35mm film):	2.7° x 4.1°	1.50° x 2.26°	1.29° x 1.93°	1.0° x 1.50°	1.07° x 1.61°	0.68° x 1.02°	0.86° x 1.29°	0.54° x 0.81°	0.72° x 1.07°	0.45° x 0.68°

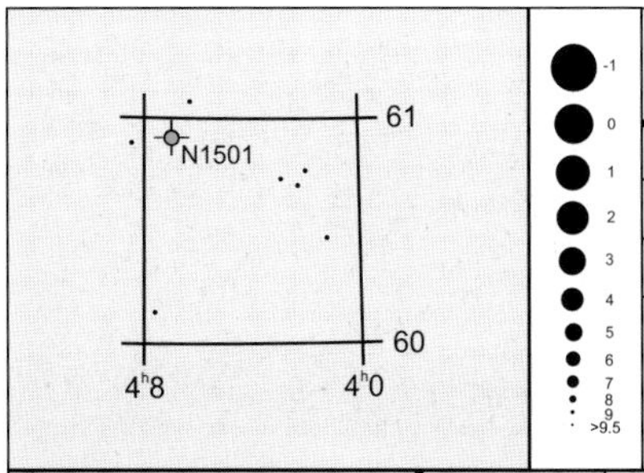

NGC 1501

RA:	$04^h\ 07^m\ 4.4^s$	Con:	Camelopardalis
Dec:	60° 54' 59"	Type:	Planetary Nebula
Size:	0.9'	Mag:	13.0

NGC 1501 is a small typical planetary nebula.

Telescope Aperture:	4" f/5	4" f/9	6" f/7	6" f/9	8" f/6.3	8" f/10	10" f/6.3	10" f/10	12" f/6.3	12" f/10
FOV(35mm film):	2.7° x 4.1°	1.50° x 2.26°	1.29° x 1.93°	1.0° x 1.50°	1.07° x 1.61°	0.68° x 1.02°	0.86° x 1.29°	0.54° x 0.81°	0.72° x 1.07°	0.45° x 0.68°

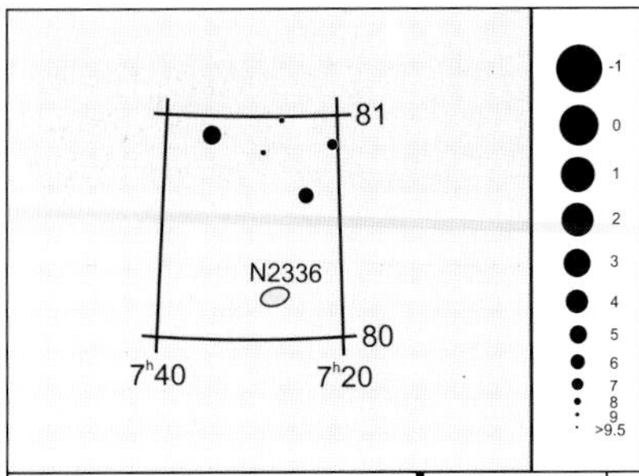

NGC 2336

RA:	07^h 27^m 10.4^s	Con:	Camelopardalis
Dec:	80° 10' 52"	Type:	Spiral Galaxy
Size:	6.4' x 3.2'	Mag:	10.5

NGC2336 is found in the northernmost region of the constellation.

Telescope Aperture:	4" f/5	4" f/9	6" f/7	6" f/9	8" f/6.3	8" f/10	10" f/6.3	10" f/10	12" f/6.3	12" f/10
FOV(35mm film):	2.7°x 4.1°	1.50°x 2.26°	1.29° x 1.93°	1.0° x 1.50°	1.07° x 1.61°	0.68°x 1.02°	0.86° x 1.29°	0.54° x 0.81°	0.72° x 1.07°	0.45°x 0.68°

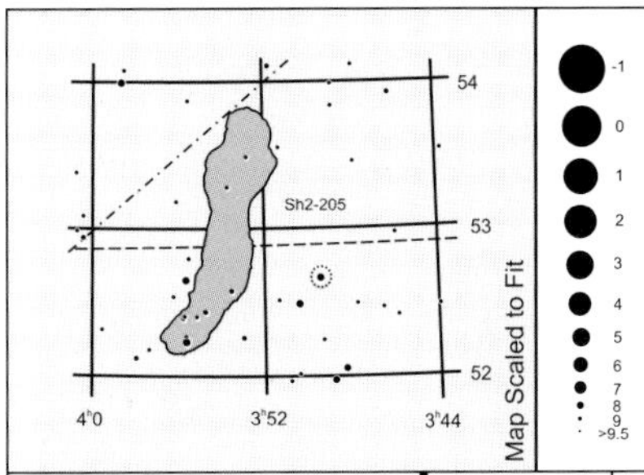

Sh2-205 (Peanut Nebula)

RA:	03^h 56^m 3^s	Con:	Camelopardalis
Dec:	53° 12' 00"	Type:	Emission Nebula
Size:	100' x 30'	Mag:	

Sh2-205 is a faint and diffuse emission nebula with fairly uniform surface brightness.

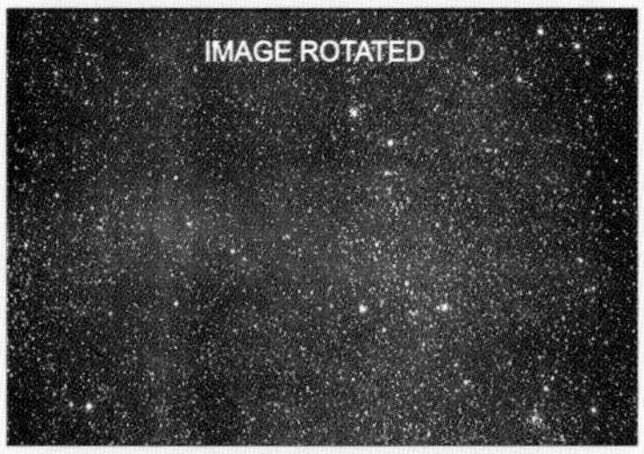

Telescope Aperture:	4" f/5	4" f/9	6" f/7	6" f/9	8" f/6.3	8" f/10	10" f/6.3	10" f/10	12" f/6.3	12" f/10
FOV(35mm film):	2.7°x 4.1°	1.50°x 2.26°	1.29° x 1.93°	1.0° x 1.50°	1.07° x 1.61°	0.68°x 1.02°	0.86° x 1.29°	0.54° x 0.81°	0.72° x 1.07°	0.45°x 0.68°

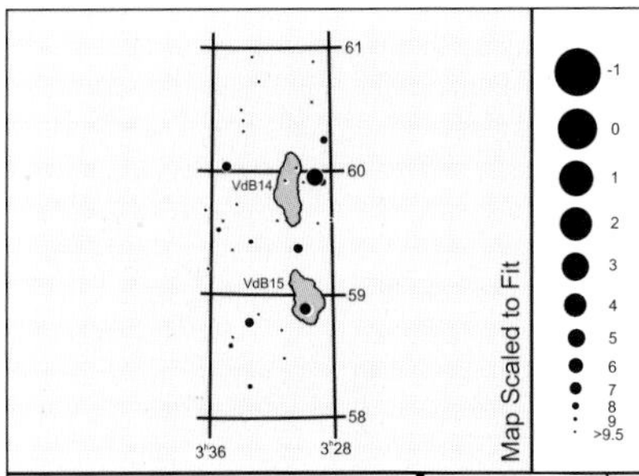

VdB14/15

RA:	03^h 29^m 2^s	Con:	Camelopardalis
Dec:	59° 57' 00"	Type:	Reflection Nebula
Size:	20.0' x 8.0'	Mag:	

VdB-14/15 are reflection nebulae.

Telescope Aperture:	4" f/5	4" f/9	6" f/7	6" f/9	8" f/6.3	8" f/10	10" f/6.3	10" f/10	12" f/6.3	12" f/10
FOV(35mm film):	2.7°x 4.1°	1.50°x 2.26°	1.29° x 1.93°	1.0° x 1.50°	1.07° x 1.61°	0.68°x 1.02°	0.86° x 1.29°	0.54° x 0.81°	0.72° x 1.07°	0.45°x 0.68°

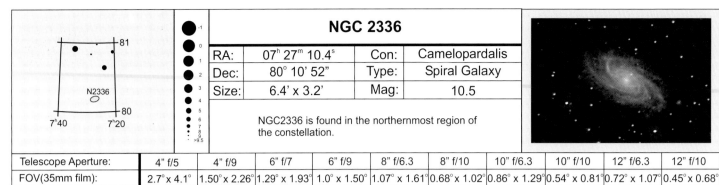

Cnc

Crosses Prime Meridian: February thru March

WINTER

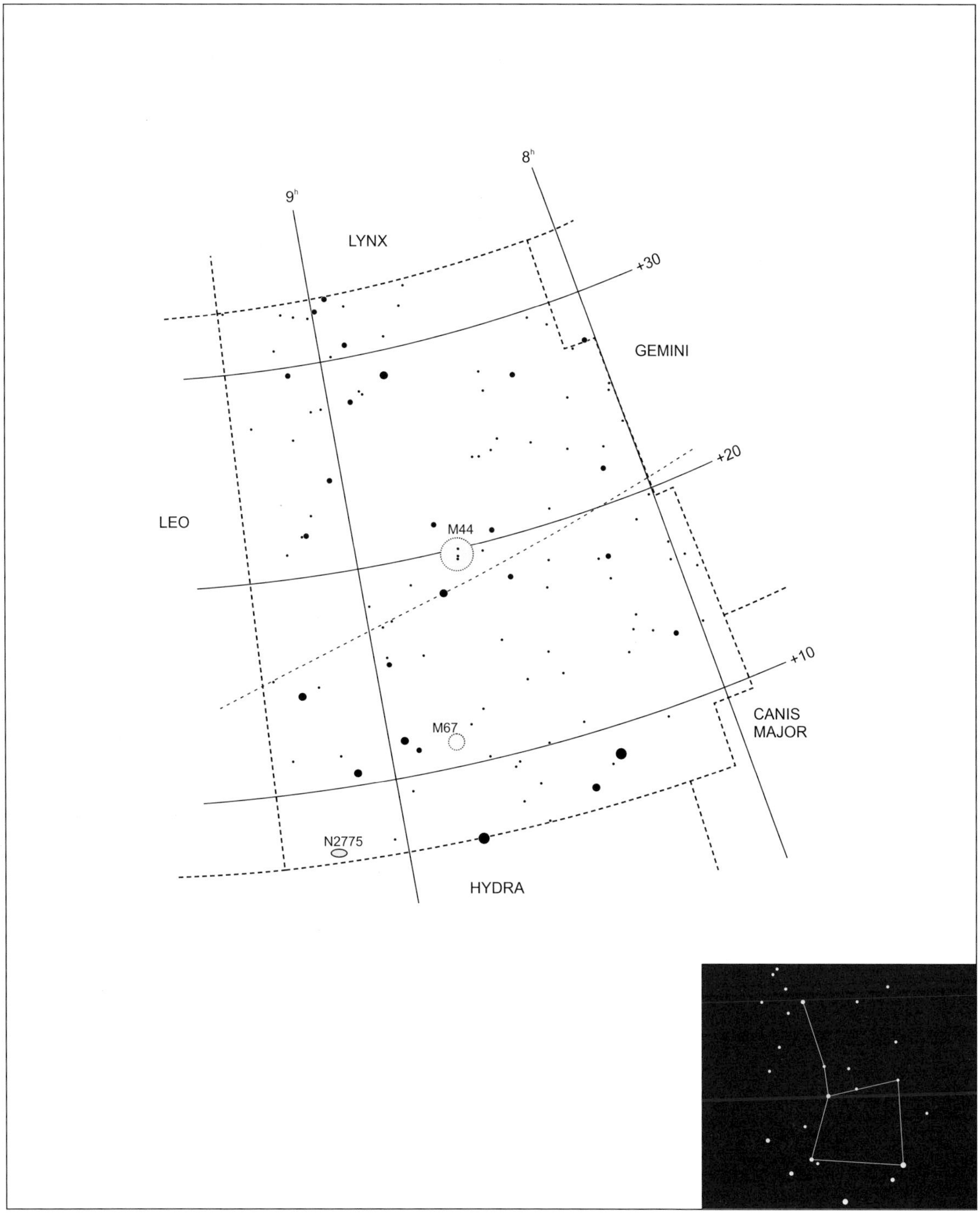

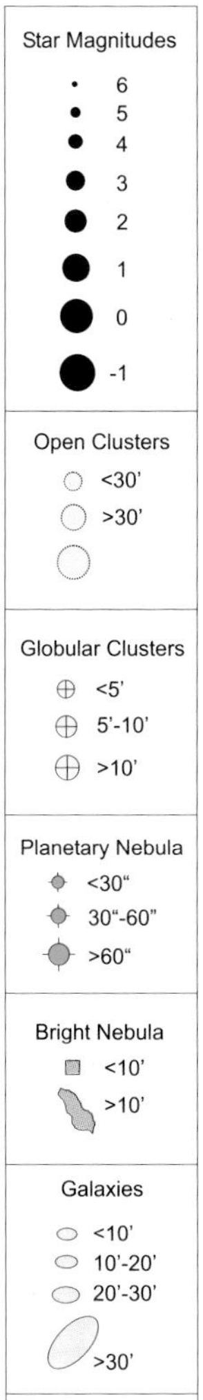

CANCER

Constellation Facts:

Cancer; (KAN-surr)

Cancer, the Crab.
Cancer tracks across the sky from east to west and passes the meridian approximately halfway between the horizon and the zenith.
Cancer is the faintest zodiacal constellation, but contains the Praesepe (Beehive cluster).
Cancer is 506 square degrees in size.

Constellation is visible from 90° N to 57° S. Partially visible from 57° S to 90° S.

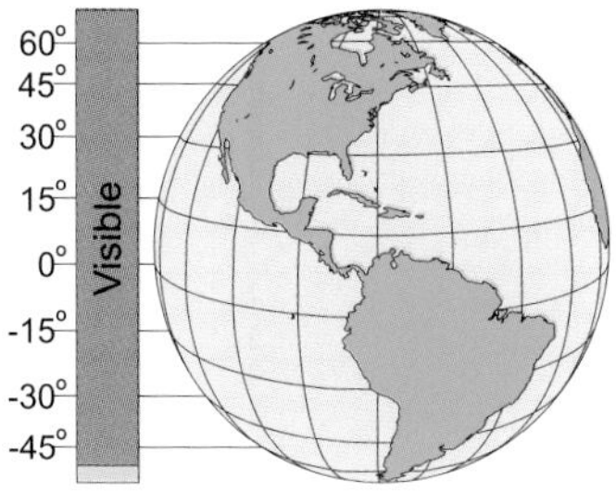

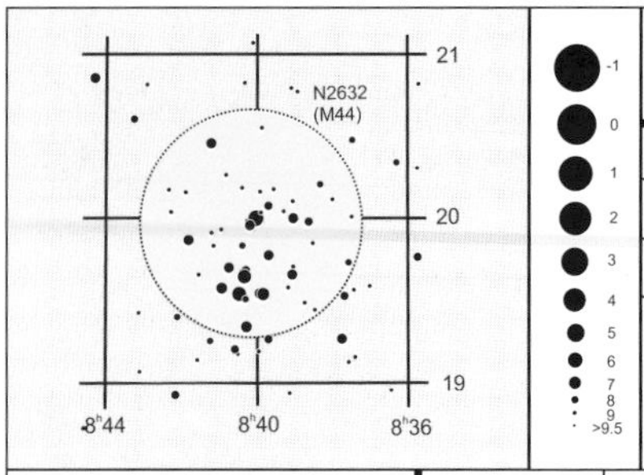

M44 (NGC 2632) "Beehive Cluster"

RA:	$08^h\ 40^m\ 8.8^s$	Con:	Cancer
Dec:	19° 58' 49"	Type:	Open Cluster
Size:	95.0'	Mag:	3.1

M44 (NGC 2632) is commonly called the Beehive Cluster, a bright, scattered open cluster.

Telescope Aperture:	4" f/5	4" f/9	6" f/7	6" f/9	8" f/6.3	8" f/10	10" f/6.3	10" f/10	12" f/6.3	12" f/10
FOV(35mm film):	2.7° x 4.1°	1.50° x 2.26°	1.29° x 1.93°	1.0° x 1.50°	1.07° x 1.61°	0.68° x 1.02°	0.86° x 1.29°	0.54° x 0.81°	0.72° x 1.07°	0.45° x 0.68°

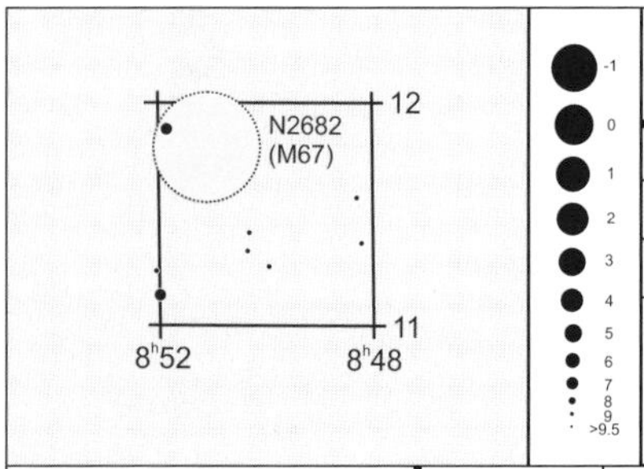

M67 (NGC 2682)

RA:	$08^h\ 50^m\ 26.7^s$	Con:	Cancer
Dec:	11° 48' 47"	Type:	Open Cluster
Size:	30.0'	Mag:	6.9

M67 (NGC 2682) is located 8° south and 2° east of M44. The open cluster contains 200 stars compacted into a 30.0' diameter.

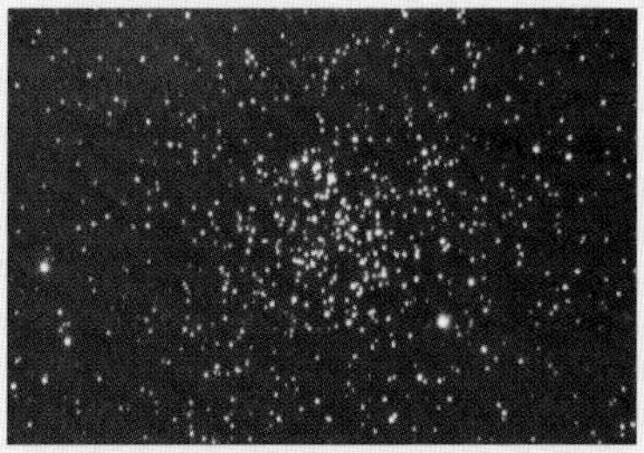

Telescope Aperture:	4" f/5	4" f/9	6" f/7	6" f/9	8" f/6.3	8" f/10	10" f/6.3	10" f/10	12" f/6.3	12" f/10
FOV(35mm film):	2.7° x 4.1°	1.50° x 2.26°	1.29° x 1.93°	1.0° x 1.50°	1.07° x 1.61°	0.68° x 1.02°	0.86° x 1.29°	0.54° x 0.81°	0.72° x 1.07°	0.45° x 0.68°

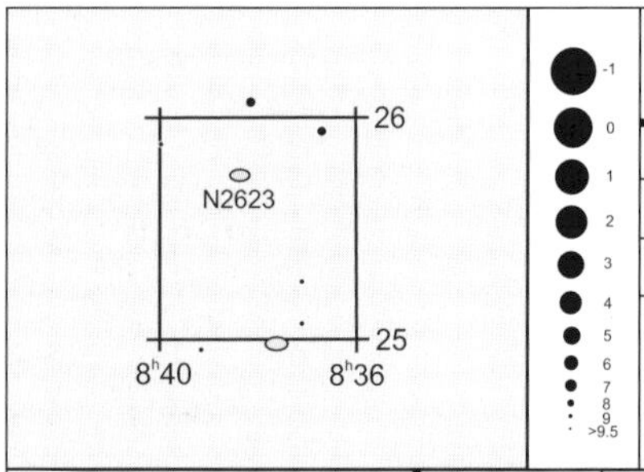

NGC 2623

RA:	$08^h\ 38^m\ 26.9^s$	Con:	Cancer
Dec:	25° 44' 49"	Type:	Peculiar Galaxy
Size:	2.2' x 0.6'	Mag:	13.8

NGC 2623 is a peculiar galaxy found 6° north of the Beehive cluster.

Telescope Aperture:	4" f/5	4" f/9	6" f/7	6" f/9	8" f/6.3	8" f/10	10" f/6.3	10" f/10	12" f/6.3	12" f/10
FOV(35mm film):	2.7° x 4.1°	1.50° x 2.26°	1.29° x 1.93°	1.0° x 1.50°	1.07° x 1.61°	0.68° x 1.02°	0.86° x 1.29°	0.54° x 0.81°	0.72° x 1.07°	0.45° x 0.68°

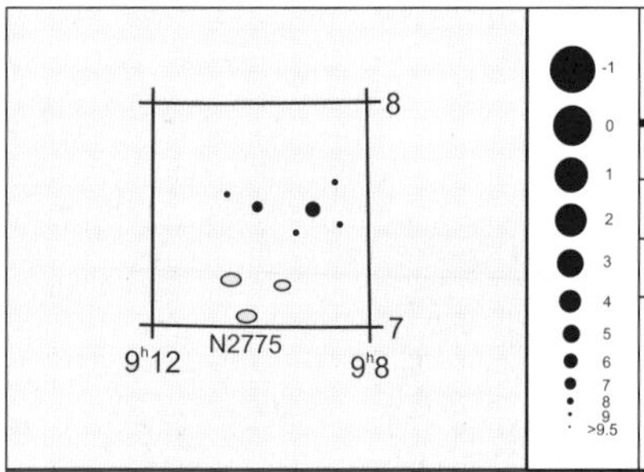

NGC 2775

RA:	$09^h\ 10^m\ 20.6^s$	Con:	Cancer
Dec:	07° 01' 45"	Type:	Spiral Galaxy
Size:	4.5' x 3.5'	Mag:	10.3

NGC 2775 is found near the Cancer/Hydra border. It is specifically located 6° southwest of M67.

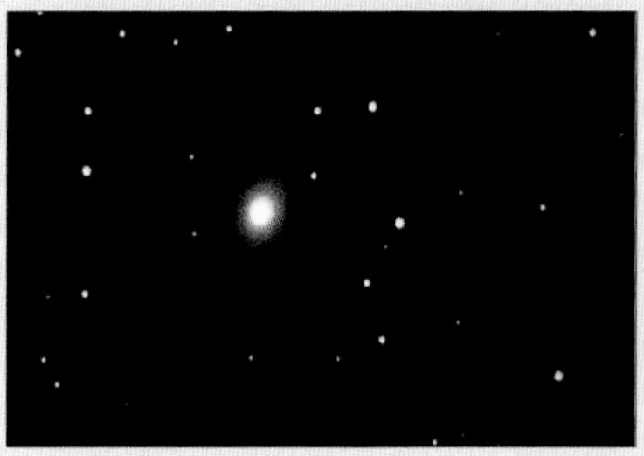

Telescope Aperture:	4" f/5	4" f/9	6" f/7	6" f/9	8" f/6.3	8" f/10	10" f/6.3	10" f/10	12" f/6.3	12" f/10
FOV(35mm film):	2.7° x 4.1°	1.50° x 2.26°	1.29° x 1.93°	1.0° x 1.50°	1.07° x 1.61°	0.68° x 1.02°	0.86° x 1.29°	0.54° x 0.81°	0.72° x 1.07°	0.45° x 0.68°

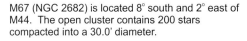

CVn

Crosses Prime Meridian: April thru May

SPRING

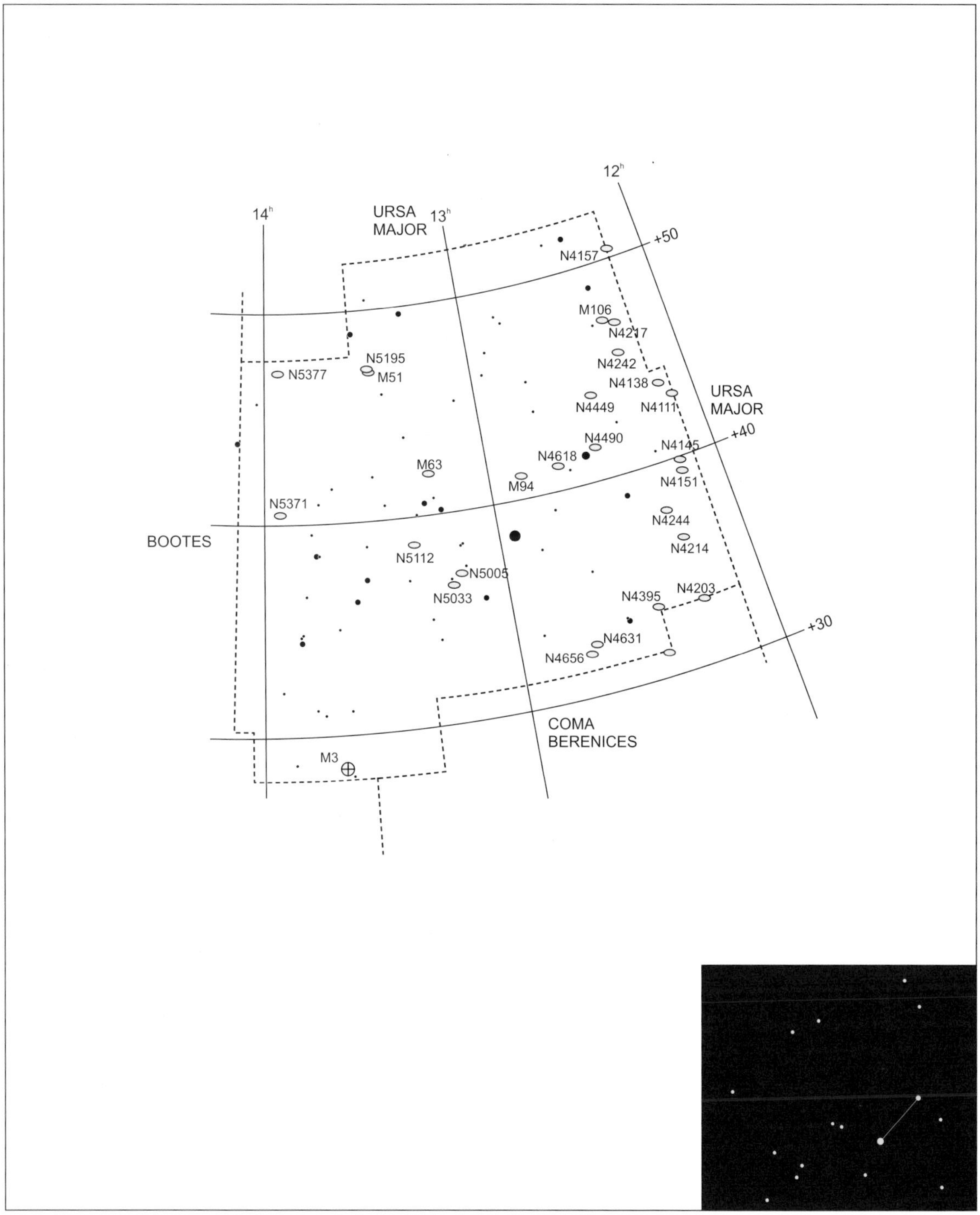

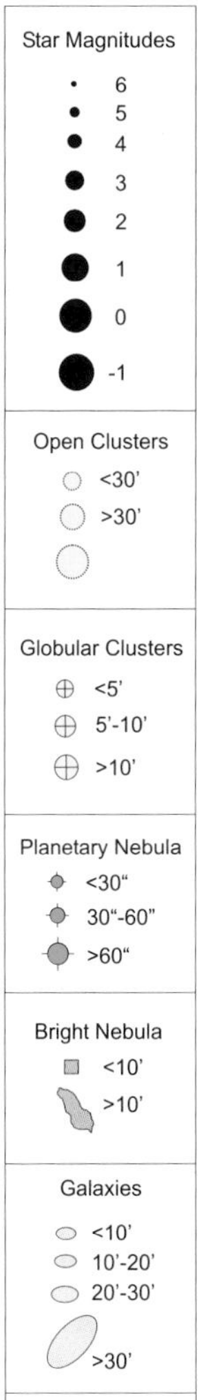

CANES VENATICI

Constellation Facts:

Canes Venatici; (KAY-nees vee-NAT-ih-sigh)

Canes Venatici, the Hunting Dogs.
The constellation rises in the northeast, moves across the meridian overhead, and sets in the northwest.
Constellation lies in one of the most transparent parts of the Milky Way, about 90° from the plane. This provides a clear view of galaxies beyond the Milky Way.
Constellation covers 465 square degrees.

Constellation is visible from 90° N to 37° S. Partially visible from 37° S to 90° S.

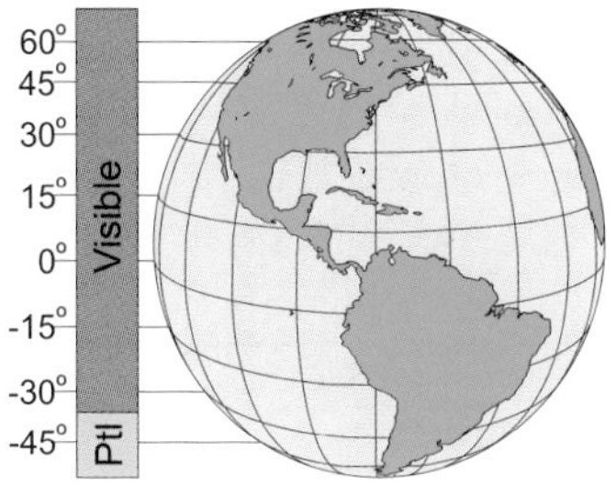

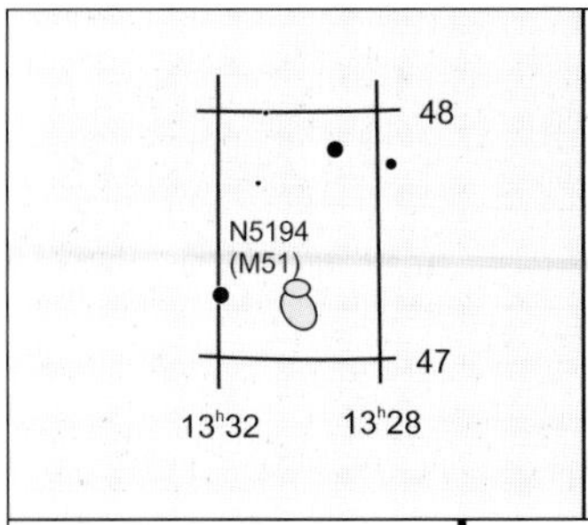

M51 (NGC 5194)

RA:	13h 29m 57.1s	Con:	Canes Venatici
Dec:	47° 11' 55"	Type:	Spiral Galaxy
Size:	9.0' x 8.0'	Mag:	8.4

M51 is one of the best galaxies in the night sky. It is a large face-on spiral galaxy known as the "Whirlpool Galaxy".

Telescope Aperture:	4" f/5	4" f/9	6" f/7	6" f/9	8" f/6.3	8" f/10	10" f/6.3	10" f/10	12" f/6.3	12" f/10
FOV(35mm film):	2.7° x 4.1°	1.50° x 2.26°	1.29° x 1.93°	1.0° x 1.50°	1.07° x 1.61°	0.68° x 1.02°	0.86° x 1.29°	0.54° x 0.81°	0.72° x 1.07°	0.45° x 0.68°

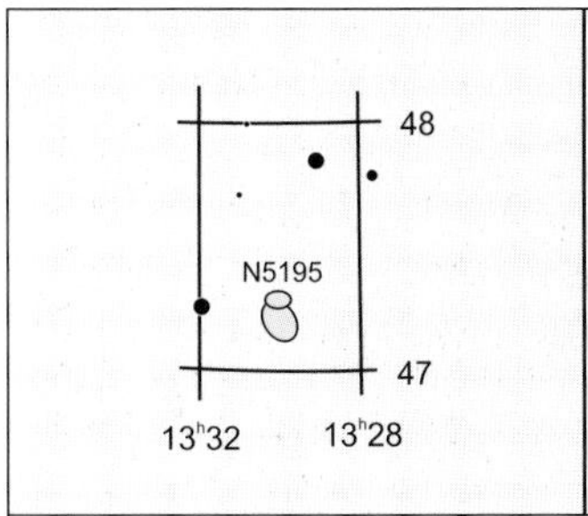

NGC 5195

RA:	13h 30m 3.1s	Con:	Canes Venatici
Dec:	47° 15' 55"	Type:	Irregular Galaxy
Size:	4.5' x 3.5'	Mag:	9.6

NGC 5195 is a small irregular galaxy that is interacting with M51.

Telescope Aperture:	4" f/5	4" f/9	6" f/7	6" f/9	8" f/6.3	8" f/10	10" f/6.3	10" f/10	12" f/6.3	12" f/10
FOV(35mm film):	2.7° x 4.1°	1.50° x 2.26°	1.29° x 1.93°	1.0° x 1.50°	1.07° x 1.61°	0.68° x 1.02°	0.86° x 1.29°	0.54° x 0.81°	0.72° x 1.07°	0.45° x 0.68°

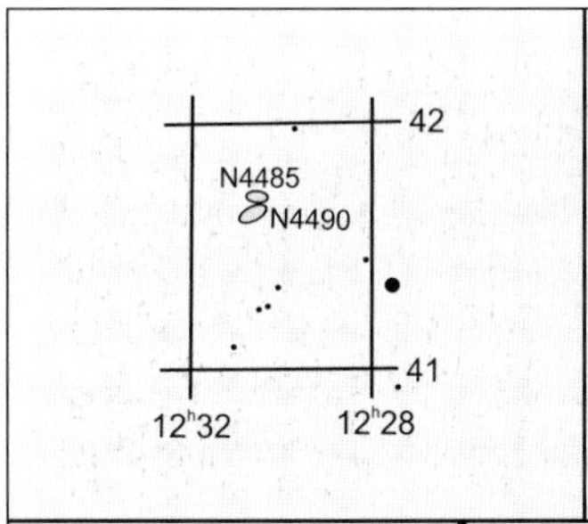

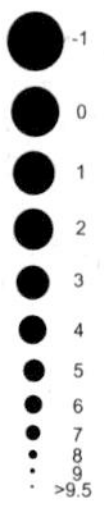

NGC 4485

RA:	12h 30m 33.1s	Con:	Canes Venatici
Dec:	41° 41' 51"	Type:	Irregular Galaxy
Size:	1.6' x 0.8'	Mag:	12.0

NGC 4485 is located 3' north of NGC 4490. It is classified as either a peculiar galaxy or an irregular galaxy.

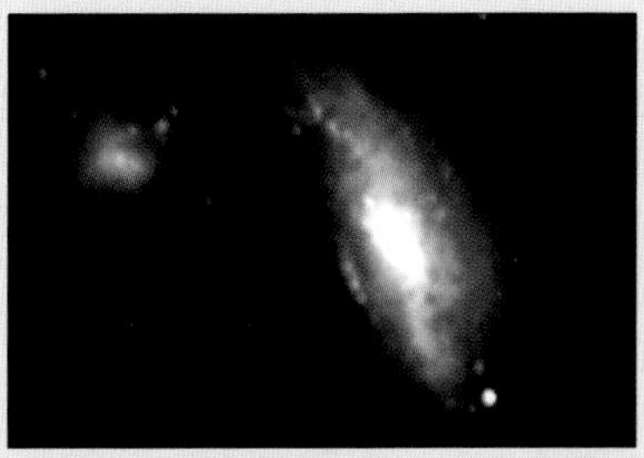

Telescope Aperture:	4" f/5	4" f/9	6" f/7	6" f/9	8" f/6.3	8" f/10	10" f/6.3	10" f/10	12" f/6.3	12" f/10
FOV(35mm film):	2.7° x 4.1°	1.50° x 2.26°	1.29° x 1.93°	1.0° x 1.50°	1.07° x 1.61°	0.68° x 1.02°	0.86° x 1.29°	0.54° x 0.81°	0.72° x 1.07°	0.45° x 0.68°

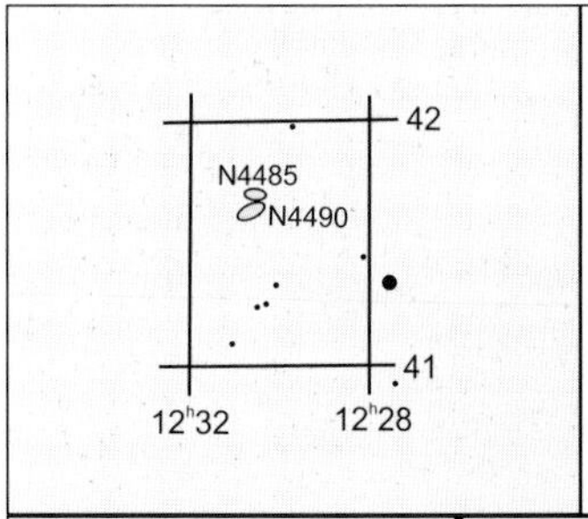

NGC 4490

RA:	12h 30m 39.1s	Con:	Canes Venatici
Dec:	41° 37' 51"	Type:	Spiral Galaxy
Size:	6.0' x 3.0'	Mag:	9.8

NGC 4490 appears as an elongated spiral galaxy with a bright core. Both NGC 4490 and NGC 4485 are an interacting pair of galaxies.

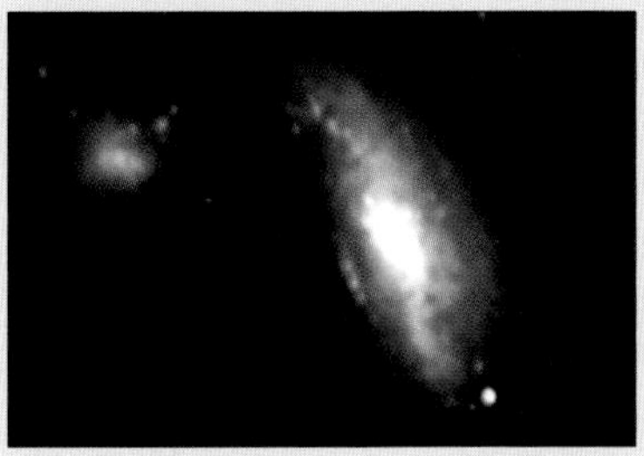

Telescope Aperture:	4" f/5	4" f/9	6" f/7	6" f/9	8" f/6.3	8" f/10	10" f/6.3	10" f/10	12" f/6.3	12" f/10
FOV(35mm film):	2.7° x 4.1°	1.50° x 2.26°	1.29° x 1.93°	1.0° x 1.50°	1.07° x 1.61°	0.68° x 1.02°	0.86° x 1.29°	0.54° x 0.81°	0.72° x 1.07°	0.45° x 0.68°

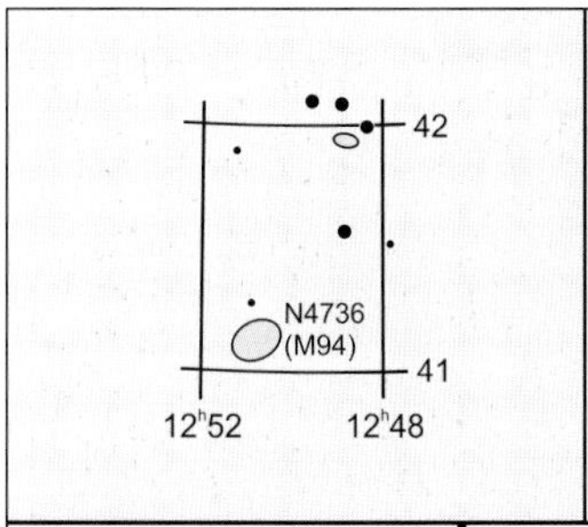

M94 (NGC 4736)

RA:	12h 50m 57.1s	Con:	Canes Venatici
Dec:	41° 06' 52"	Type:	Spiral Galaxy
Size:	14.0' x 13.0'	Mag:	8.2

M92 (NGC 4736) is found almost due east of the NGC 4485/4490 pair of galaxies. M92 is a bright compact, tightly-wound spiral galaxy.

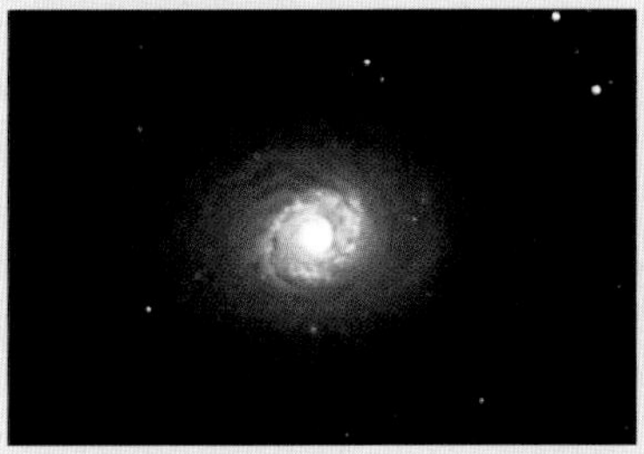

Telescope Aperture:	4" f/5	4" f/9	6" f/7	6" f/9	8" f/6.3	8" f/10	10" f/6.3	10" f/10	12" f/6.3	12" f/10
FOV(35mm film):	2.7° x 4.1°	1.50° x 2.26°	1.29° x 1.93°	1.0° x 1.50°	1.07° x 1.61°	0.68° x 1.02°	0.86° x 1.29°	0.54° x 0.81°	0.72° x 1.07°	0.45° x 0.68°

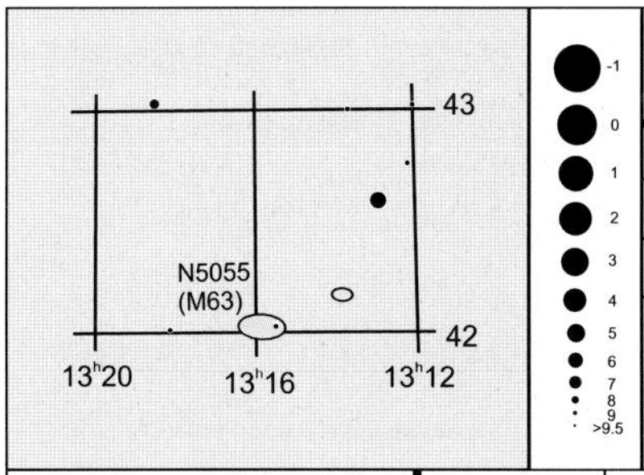

M63 (NGC 5055)

RA:	$13^h\ 15^m\ 51.2^s$	Con:	Canes Venatici
Dec:	42° 01’ 53”	Type:	Spiral Galaxy
Size:	13.0’ x 8.0’	Mag:	8.6

M63 the “Sunflower Galaxy” is found east of M94. It is a multiple-arm spiral galaxy with a bright core.

Telescope Aperture:	4” f/5	4” f/9	6” f/7	6” f/9	8” f/6.3	8” f/10	10” f/6.3	10” f/10	12” f/6.3	12” f/10
FOV(35mm film):	2.7° x 4.1°	1.50° x 2.26°	1.29° x 1.93°	1.0° x 1.50°	1.07° x 1.61°	0.68° x 1.02°	0.86° x 1.29°	0.54° x 0.81°	0.72° x 1.07°	0.45° x 0.68°

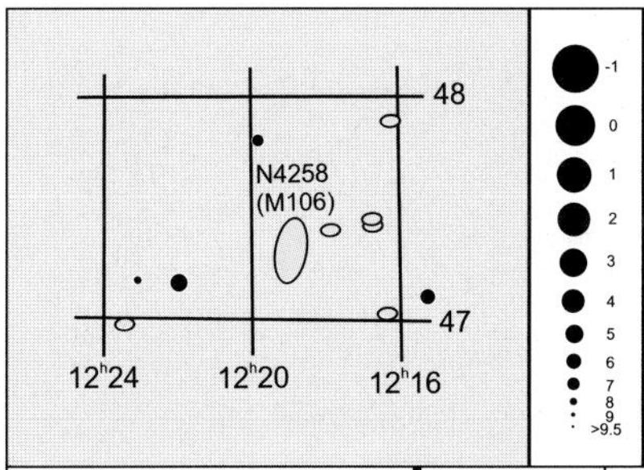

M106 (NGC 4258)

RA:	$12^h\ 19^m\ 3.0^s$	Con:	Canes Venatici
Dec:	47° 17’ 52”	Type:	Spiral Galaxy
Size:	12.0’ x 4.0’	Mag:	8.3

M106 is located in the northwest corner of the constellation. The object displays high surface brightness and is inclined slightly to our line of vision.

Telescope Aperture:	4” f/5	4” f/9	6” f/7	6” f/9	8” f/6.3	8” f/10	10” f/6.3	10” f/10	12” f/6.3	12” f/10
FOV(35mm film):	2.7° x 4.1°	1.50° x 2.26°	1.29° x 1.93°	1.0° x 1.50°	1.07° x 1.61°	0.68° x 1.02°	0.86° x 1.29°	0.54° x 0.81°	0.72° x 1.07°	0.45° x 0.68°

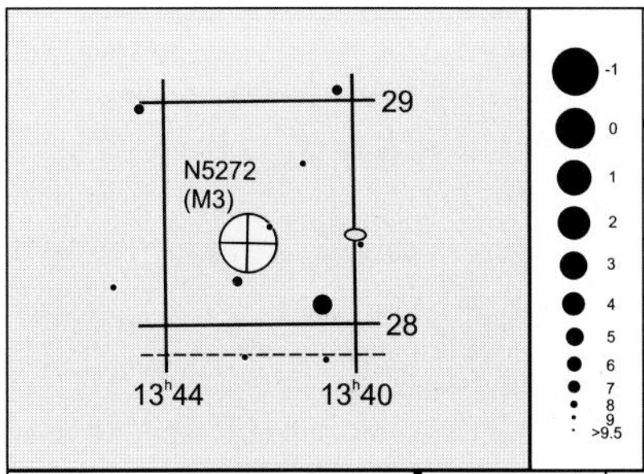

M3 (NGC 5272)

RA:	$13^h\ 42^m\ 15.5^s$	Con:	Canes Venatici
Dec:	28° 22’ 51”	Type:	Globular Cluster
Size:	16.2’	Mag:	6.4

M3 is a highly resolved globular cluster comprised of more than a million stars. The object lies in the southeastern part of the constellation on the border with Bootes.

Telescope Aperture:	4” f/5	4” f/9	6” f/7	6” f/9	8” f/6.3	8” f/10	10” f/6.3	10” f/10	12” f/6.3	12” f/10
FOV(35mm film):	2.7° x 4.1°	1.50° x 2.26°	1.29° x 1.93°	1.0° x 1.50°	1.07° x 1.61°	0.68° x 1.02°	0.86° x 1.29°	0.54° x 0.81°	0.72° x 1.07°	0.45° x 0.68°

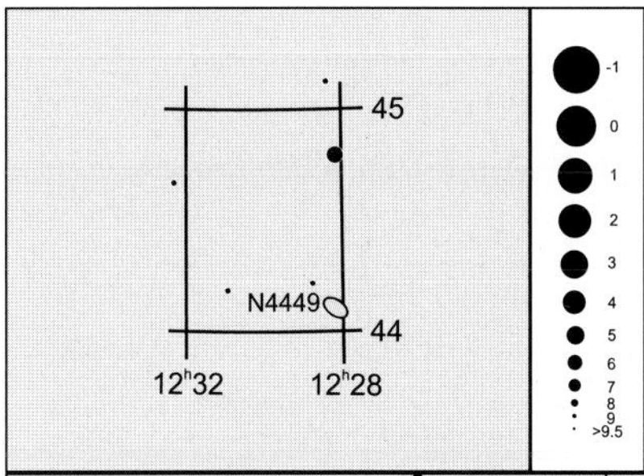

NGC 4449

RA:	$12^h\ 28^m\ 15.1^s$	Con:	Canes Venatici
Dec:	44° 05’ 52”	Type:	Irregular Galaxy
Size:	5.5’ x 4.5’	Mag:	9.4

NGC 4449 is positioned halfway between M106 and NGC 4485/4490. Object has a peculiar box-like structure.

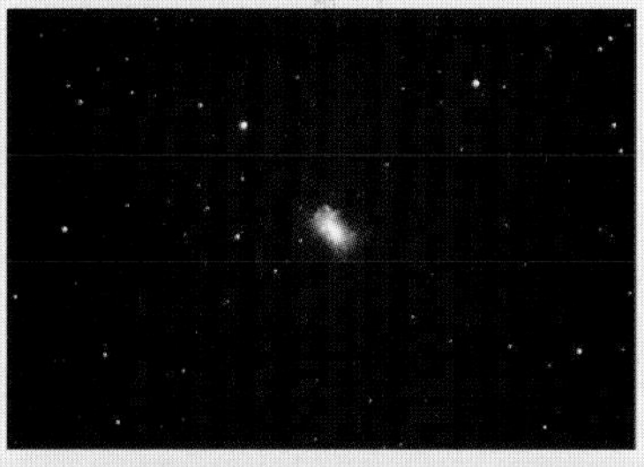

Telescope Aperture:	4” f/5	4” f/9	6” f/7	6” f/9	8” f/6.3	8” f/10	10” f/6.3	10” f/10	12” f/6.3	12” f/10
FOV(35mm film):	2.7° x 4.1°	1.50° x 2.26°	1.29° x 1.93°	1.0° x 1.50°	1.07° x 1.61°	0.68° x 1.02°	0.86° x 1.29°	0.54° x 0.81°	0.72° x 1.07°	0.45° x 0.68°

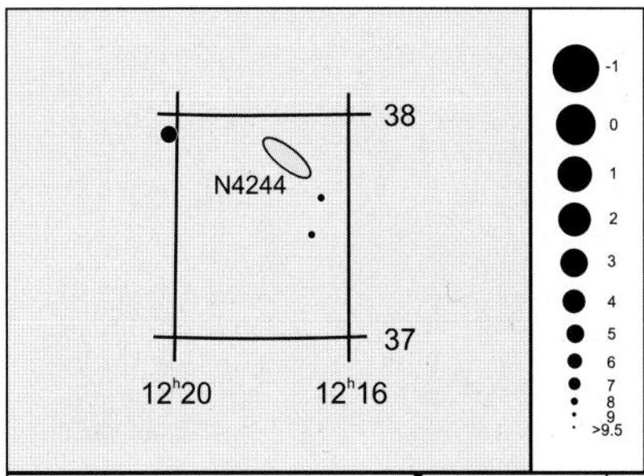

NGC 4244

RA:	$12^h\ 17^m\ 33.1^s$	Con:	Canes Venatici
Dec:	37° 48’ 50”	Type:	Spiral Galaxy
Size:	16.0’ x 1.8’	Mag:	10.2

NGC 4244 is an edge-on spiral galaxy

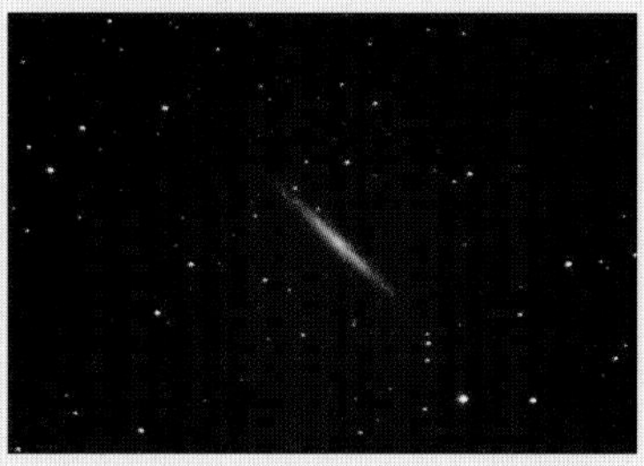

Telescope Aperture:	4” f/5	4” f/9	6” f/7	6” f/9	8” f/6.3	8” f/10	10” f/6.3	10” f/10	12” f/6.3	12” f/10
FOV(35mm film):	2.7° x 4.1°	1.50° x 2.26°	1.29° x 1.93°	1.0° x 1.50°	1.07° x 1.61°	0.68° x 1.02°	0.86° x 1.29°	0.54° x 0.81°	0.72° x 1.07°	0.45° x 0.68°

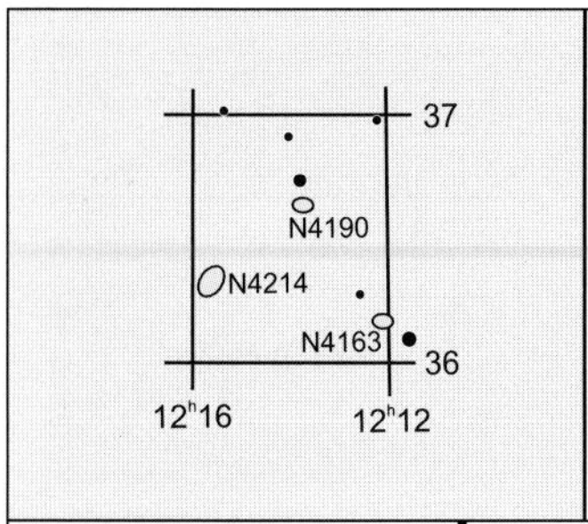

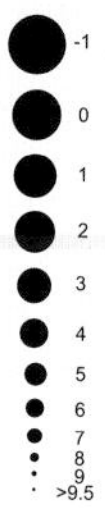

NGC 4214

RA:	$12^h\ 15^m\ 39.1^s$	Con:	Canes Venatici
Dec:	36° 19' 50"	Type:	Spiral Galaxy
Size:	7.5' x 6.0'	Mag:	9.7

NGC 4214 is located south and west of NGC 4244.

Telescope Aperture:	4" f/5	4" f/9	6" f/7	6" f/9	8" f/6.3	8" f/10	10" f/6.3	10" f/10	12" f/6.3	12" f/10
FOV(35mm film):	2.7° x 4.1°	1.50° x 2.26°	1.29° x 1.93°	1.0° x 1.50°	1.07° x 1.61°	0.68° x 1.02°	0.86° x 1.29°	0.54° x 0.81°	0.72° x 1.07°	0.45° x 0.68°

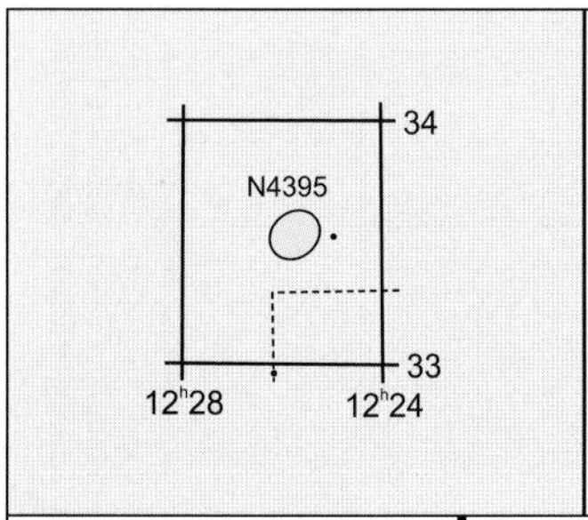

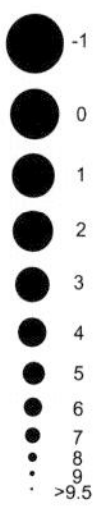

NGC 4395

RA:	$12^h\ 25^m\ 51.2^s$	Con:	Canes Venatici
Dec:	33° 32' 48"	Type:	Spiral Galaxy
Size:	12.0' x 11.0'	Mag:	10.2

NGC 4395 lies at the southwestern edge of the constellation. Majestic spiral galaxy with 3 arms visible. Galaxy has a near circular shape and has low surface brightness.

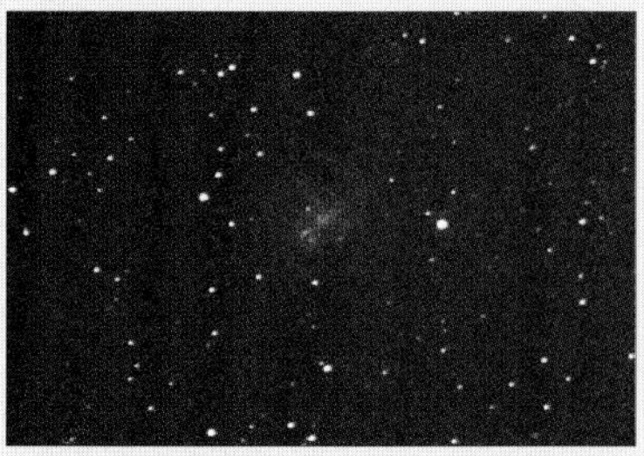

Telescope Aperture:	4" f/5	4" f/9	6" f/7	6" f/9	8" f/6.3	8" f/10	10" f/6.3	10" f/10	12" f/6.3	12" f/10
FOV(35mm film):	2.7° x 4.1°	1.50° x 2.26°	1.29° x 1.93°	1.0° x 1.50°	1.07° x 1.61°	0.68° x 1.02°	0.86° x 1.29°	0.54° x 0.81°	0.72° x 1.07°	0.45° x 0.68°

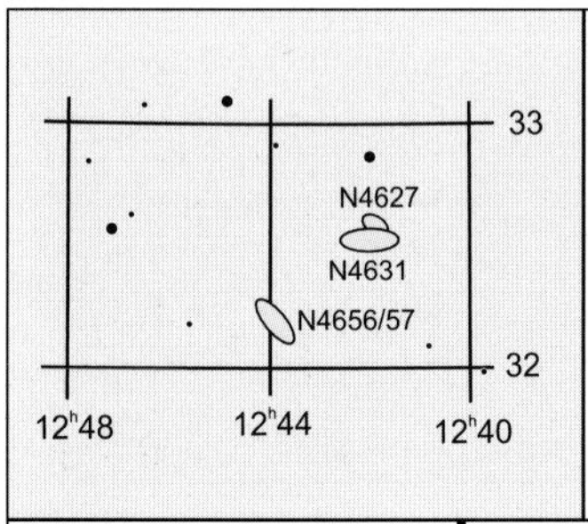

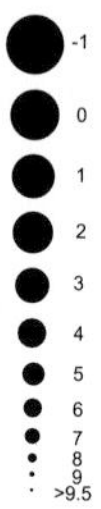

NGC 4631

RA:	$12^h 42^m\ 9.2^s$	Con:	Canes Venatici
Dec:	32° 31' 49"	Type:	Spiral Galaxy
Size:	14.0' x 2.0'	Mag:	9.3

NGC 4631 is located southeast of NGC 4395. It is an edge-on spiral galaxy that is possibly an Sc-type galaxy. Edge-on perspective reveals dust lanes and structure on the periphery.

Telescope Aperture:	4" f/5	4" f/9	6" f/7	6" f/9	8" f/6.3	8" f/10	10" f/6.3	10" f/10	12" f/6.3	12" f/10
FOV(35mm film):	2.7° x 4.1°	1.50° x 2.26°	1.29° x 1.93°	1.0° x 1.50°	1.07° x 1.61°	0.68° x 1.02°	0.86° x 1.29°	0.54° x 0.81°	0.72° x 1.07°	0.45° x 0.68°

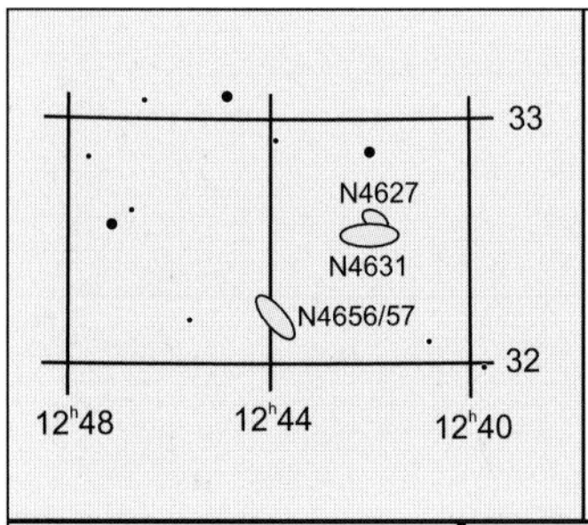

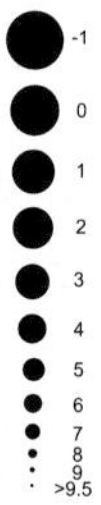

NGC 4656

RA:	$12^h 44^m\ 3.2^s$	Con:	Canes Venatici
Dec:	32° 09' 48"	Type:	Irregular Galaxy
Size:	20.0' x 2.5'	Mag:	10.4

NGC 4656 is a large and irregular galaxy. The galaxy is very elongated and is thought to be interacting with its close companion.

Telescope Aperture:	4" f/5	4" f/9	6" f/7	6" f/9	8" f/6.3	8" f/10	10" f/6.3	10" f/10	12" f/6.3	12" f/10
FOV(35mm film):	2.7° x 4.1°	1.50° x 2.26°	1.29° x 1.93°	1.0° x 1.50°	1.07° x 1.61°	0.68° x 1.02°	0.86° x 1.29°	0.54° x 0.81°	0.72° x 1.07°	0.45° x 0.68°

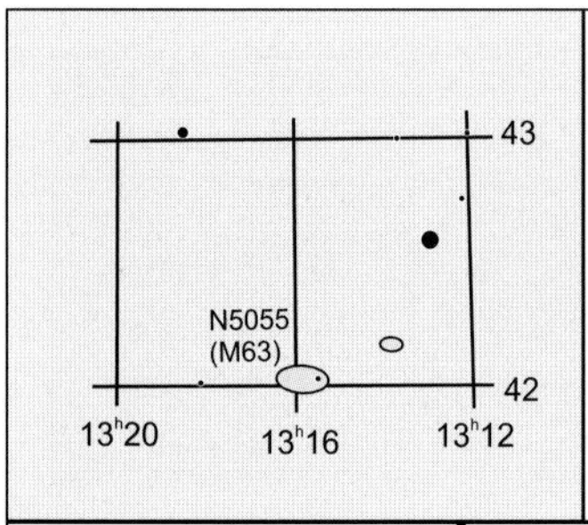

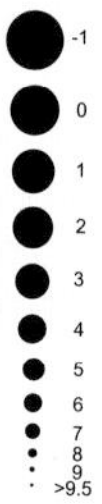

NGC 5005

RA:	$13^h\ 10^m\ 57.2^s$	Con:	Canes Venatici
Dec:	37° 02' 52"	Type:	Spiral Galaxy
Size:	5.0' x 2.2'	Mag:	9.8

NGC 5005 is a small spiral galaxy located in the central region of the constellation.

Telescope Aperture:	4" f/5	4" f/9	6" f/7	6" f/9	8" f/6.3	8" f/10	10" f/6.3	10" f/10	12" f/6.3	12" f/10
FOV(35mm film):	2.7° x 4.1°	1.50° x 2.26°	1.29° x 1.93°	1.0° x 1.50°	1.07° x 1.61°	0.68° x 1.02°	0.86° x 1.29°	0.54° x 0.81°	0.72° x 1.07°	0.45° x 0.68°

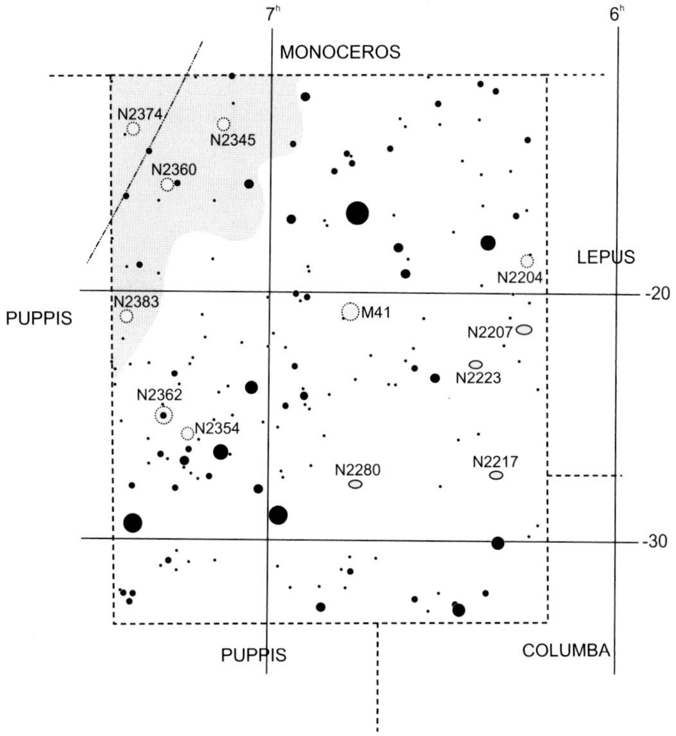

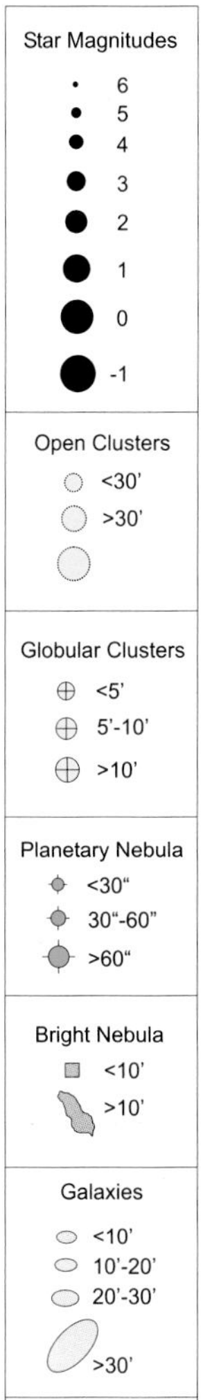

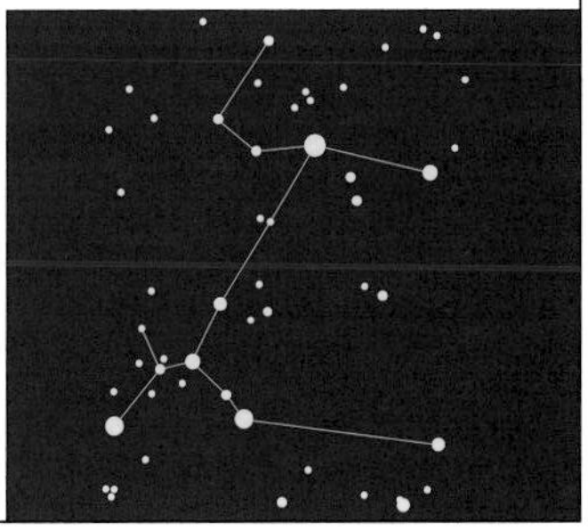

CANIS MAJOR

Constellation Facts:

Canis Major; (KAY-niss MAY-jer)

Canis Major, the Large Dog.
Constellation rises in the southeast and sets in the southwest.
Canis Major is located along the western edge of the Milky Way, and is visible low in the southern sky.
Canis Major covers 380 square degrees.

Constellation is visible from 56° N to 90° S. Partially visible from 56° N to 80° N.

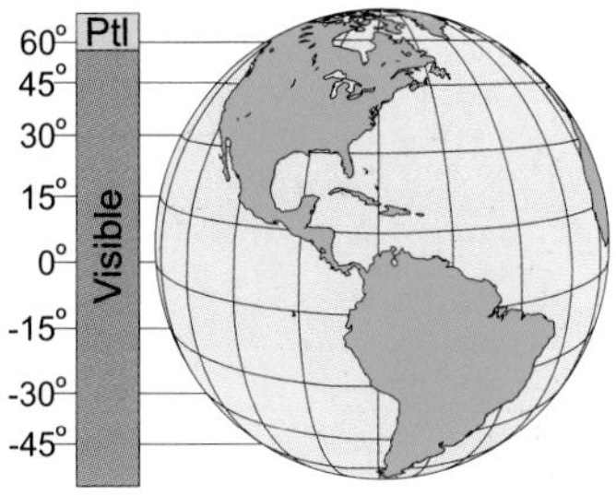

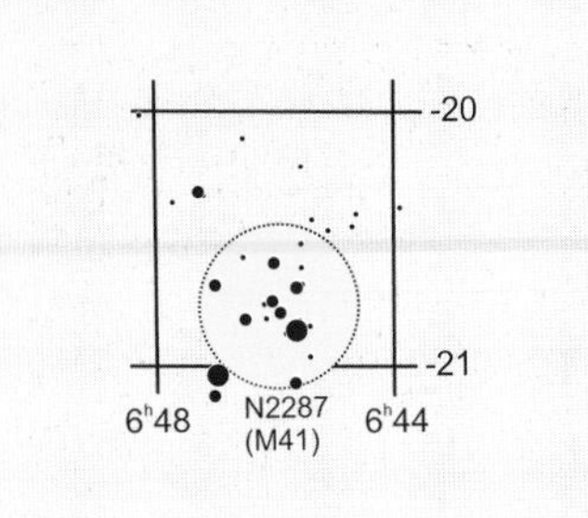

M41 (NGC 2287)

RA:	$06^h\ 46^m\ 1.7^s$	Con:	Canis Major
Dec:	-20° 44' 03"	Type:	Open Cluster
Size:	38.0'	Mag:	4.5

M41 (NGC 2287) is located approximately 4° south of Sirius. Object contains about 100 stars.

Telescope Aperture:	4" f/5	4" f/9	6" f/7	6" f/9	8" f/6.3	8" f/10	10" f/6.3	10" f/10	12" f/6.3	12" f/10
FOV(35mm film):	2.7° x 4.1°	1.50° x 2.26°	1.29° x 1.93°	1.0° x 1.50°	1.07° x 1.61°	0.68° x 1.02°	0.86° x 1.29°	0.54° x 0.81°	0.72° x 1.07°	0.45° x 0.68°

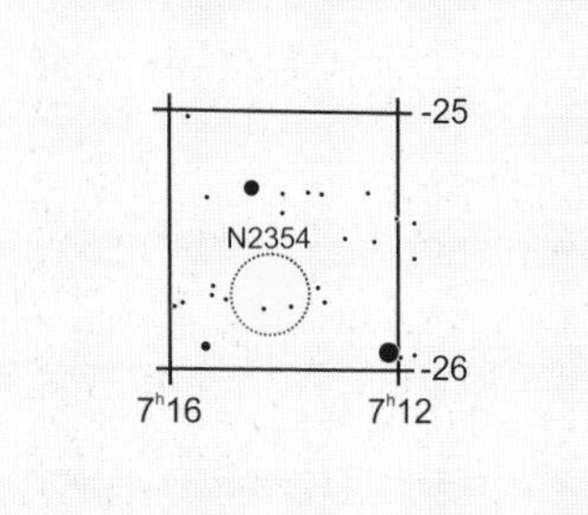

NGC 2354

RA:	$07^h\ 14^m\ 19.5^s$	Con:	Canis Major
Dec:	-25° 44' 06"	Type:	Open Cluster
Size:	20.0'	Mag:	6.5

NGC 2354 is an appealing open cluster located in the hind legs of the "Dog". Object is composed of approximately 60 stars.

Telescope Aperture:	4" f/5	4" f/9	6" f/7	6" f/9	8" f/6.3	8" f/10	10" f/6.3	10" f/10	12" f/6.3	12" f/10
FOV(35mm film):	2.7° x 4.1°	1.50° x 2.26°	1.29° x 1.93°	1.0° x 1.50°	1.07° x 1.61°	0.68° x 1.02°	0.86° x 1.29°	0.54° x 0.81°	0.72° x 1.07°	0.45° x 0.68°

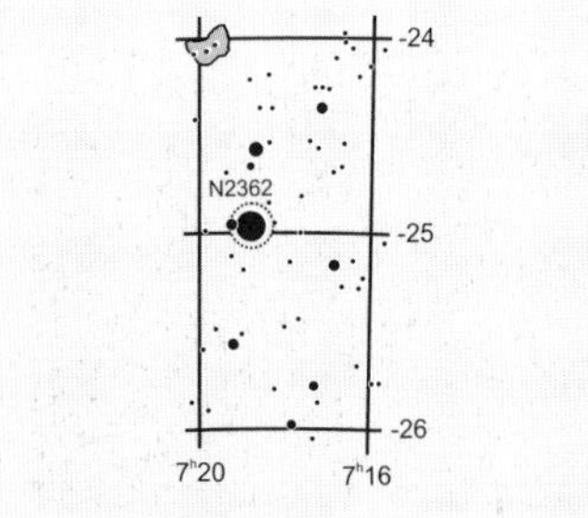

NGC 2362

RA:	$07^h\ 18^m\ 49.5^s$	Con:	Canis Major
Dec:	-24° 57' 07"	Type:	Open Cluster
Size:	8.0'	Mag:	4.1

NGC 2362 is yet another open cluster found in the hind quarters of the "Dog". Object is comprised of approximately 40 stars.

Telescope Aperture:	4" f/5	4" f/9	6" f/7	6" f/9	8" f/6.3	8" f/10	10" f/6.3	10" f/10	12" f/6.3	12" f/10
FOV(35mm film):	2.7° x 4.1°	1.50° x 2.26°	1.29° x 1.93°	1.0° x 1.50°	1.07° x 1.61°	0.68° x 1.02°	0.86° x 1.29°	0.54° x 0.81°	0.72° x 1.07°	0.45° x 0.68°

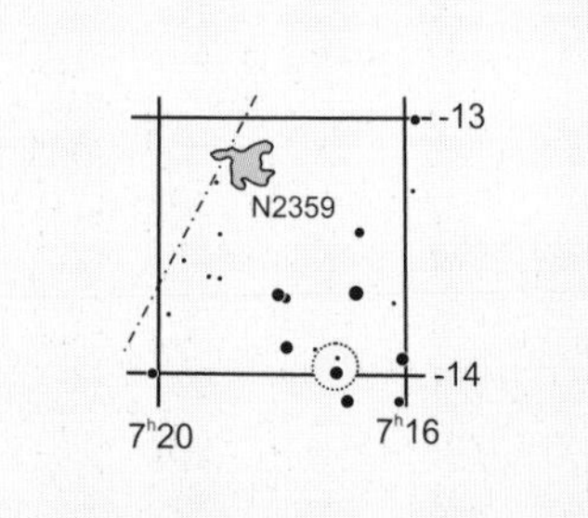

NGC 2359

RA:	$07^h\ 18^m\ 37.9^s$	Con:	Canis Major
Dec:	-13° 12' 07"	Type:	Emission Nebula
Size:	8.0'	Mag:	10.4

NGC 2359 is a dim emission nebula shaped similar to a twisted comet.

Telescope Aperture:	4" f/5	4" f/9	6" f/7	6" f/9	8" f/6.3	8" f/10	10" f/6.3	10" f/10	12" f/6.3	12" f/10
FOV(35mm film):	2.7° x 4.1°	1.50° x 2.26°	1.29° x 1.93°	1.0° x 1.50°	1.07° x 1.61°	0.68° x 1.02°	0.86° x 1.29°	0.54° x 0.81°	0.72° x 1.07°	0.45° x 0.68°

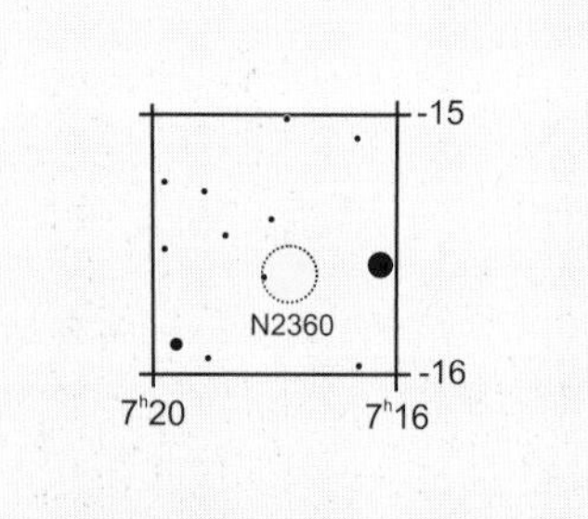

NGC 2360

RA:	$07^h\ 17^m\ 49.9^s$	Con:	Canis Major
Dec:	-15° 37' 07"	Type:	Open Cluster
Size:	13.0'	Mag:	7.2

NGC 2360 is a rich open grouping of about 50 stars.

Telescope Aperture:	4" f/5	4" f/9	6" f/7	6" f/9	8" f/6.3	8" f/10	10" f/6.3	10" f/10	12" f/6.3	12" f/10
FOV(35mm film):	2.7° x 4.1°	1.50° x 2.26°	1.29° x 1.93°	1.0° x 1.50°	1.07° x 1.61°	0.68° x 1.02°	0.86° x 1.29°	0.54° x 0.81°	0.72° x 1.07°	0.45° x 0.68°

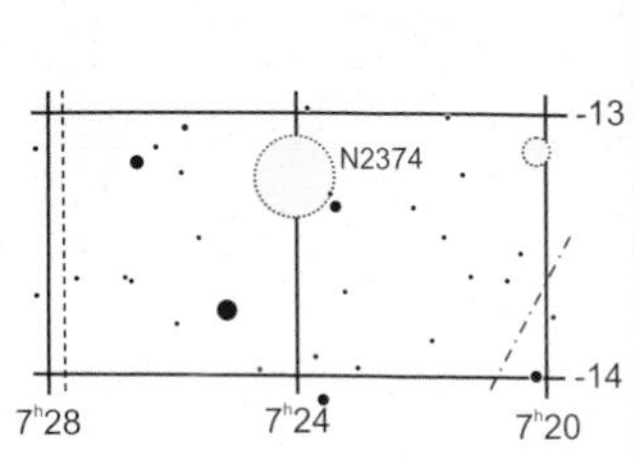

NGC 2374

RA:	07h 24m 2.0s	Con:	Canis Major
Dec:	-13° 16' 07"	Type:	Open Cluster
Size:	19.0'	Mag:	8.0

NGC 2374 is a rich, scattered open cluster located in the northeastern region of the constellation.

Telescope Aperture:	4" f/5	4" f/9	6" f/7	6" f/9	8" f/6.3	8" f/10	10" f/6.3	10" f/10	12" f/6.3	12" f/10
FOV(35mm film):	2.7° x 4.1°	1.50° x 2.26°	1.29° x 1.93°	1.0° x 1.50°	1.07° x 1.61°	0.68° x 1.02°	0.86° x 1.29°	0.54° x 0.81°	0.72° x 1.07°	0.45° x 0.68°

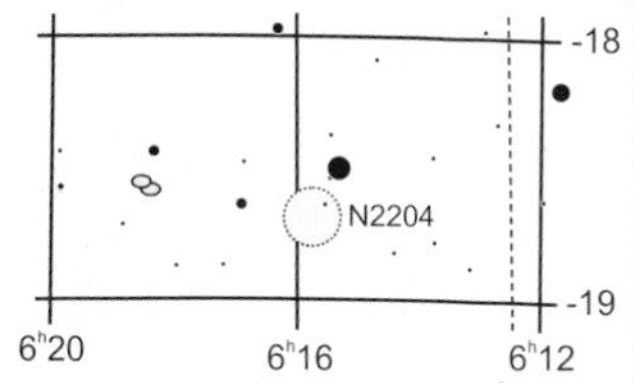

NGC 2204

RA:	06h 15m 43.8s	Con:	Canis Major
Dec:	-18° 38' 59"	Type:	Open Cluster
Size:	13.0'	Mag:	8.6

NGC 2204 is a rich collection of approximately 20 stars.

Telescope Aperture:	4" f/5	4" f/9	6" f/7	6" f/9	8" f/6.3	8" f/10	10" f/6.3	10" f/10	12" f/6.3	12" f/10
FOV(35mm film):	2.7° x 4.1°	1.50° x 2.26°	1.29° x 1.93°	1.0° x 1.50°	1.07° x 1.61°	0.68° x 1.02°	0.86° x 1.29°	0.54° x 0.81°	0.72° x 1.07°	0.45° x 0.68°

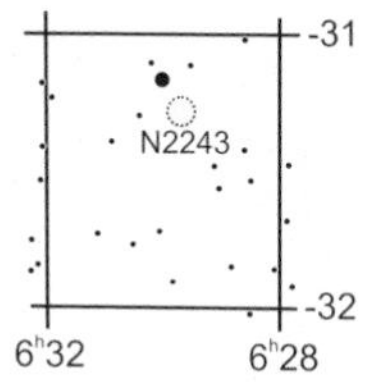

NGC 2243

RA:	06h 29m 49.2s	Con:	Canis Major
Dec:	-31° 17' 00"	Type:	Open Cluster
Size:	5.0'	Mag:	9.4

NGC 2243 is a faint, small rich collection of 50 stars.

Telescope Aperture:	4" f/5	4" f/9	6" f/7	6" f/9	8" f/6.3	8" f/10	10" f/6.3	10" f/10	12" f/6.3	12" f/10
FOV(35mm film):	2.7° x 4.1°	1.50° x 2.26°	1.29° x 1.93°	1.0° x 1.50°	1.07° x 1.61°	0.68° x 1.02°	0.86° x 1.29°	0.54° x 0.81°	0.72° x 1.07°	0.45° x 0.68°

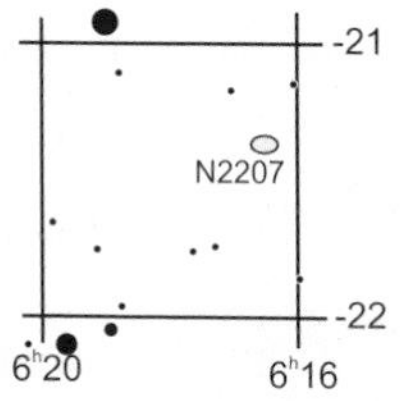

NGC 2207

RA:	06h 16m 25.7s	Con:	Canis Major
Dec:	-21° 21' 59"	Type:	Galaxy
Size:	3.5' x 2.0'	Mag:	10.7

NGC 2207 is possibly a double or interacting pair of galaxies.

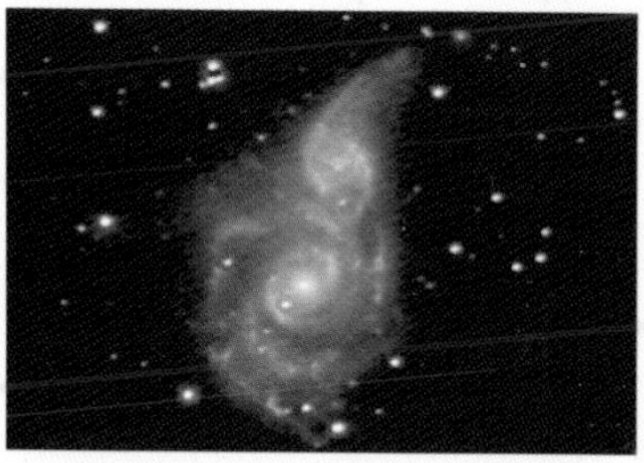

Telescope Aperture:	4" f/5	4" f/9	6" f/7	6" f/9	8" f/6.3	8" f/10	10" f/6.3	10" f/10	12" f/6.3	12" f/10
FOV(35mm film):	2.7° x 4.1°	1.50° x 2.26°	1.29° x 1.93°	1.0° x 1.50°	1.07° x 1.61°	0.68° x 1.02°	0.86° x 1.29°	0.54° x 0.81°	0.72° x 1.07°	0.45° x 0.68°

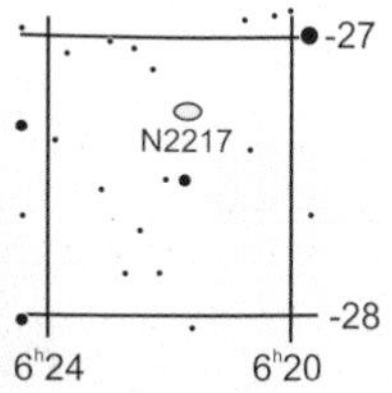

NGC 2217

RA:	06h 21m 43.4s	Con:	Canis Major
Dec:	-27° 13' 59"	Type:	Barred Spiral
Size:	3.3' x 3.0'	Mag:	10.4

NGC 2217 is a barred spiral galaxy that appears in the eyepiece as an elongated disk with a bright core.

Telescope Aperture:	4" f/5	4" f/9	6" f/7	6" f/9	8" f/6.3	8" f/10	10" f/6.3	10" f/10	12" f/6.3	12" f/10
FOV(35mm film):	2.7° x 4.1°	1.50° x 2.26°	1.29° x 1.93°	1.0° x 1.50°	1.07° x 1.61°	0.68° x 1.02°	0.86° x 1.29°	0.54° x 0.81°	0.72° x 1.07°	0.45° x 0.68°

Cap

Crosses Prime Meridian:
August thru September

SUMMER

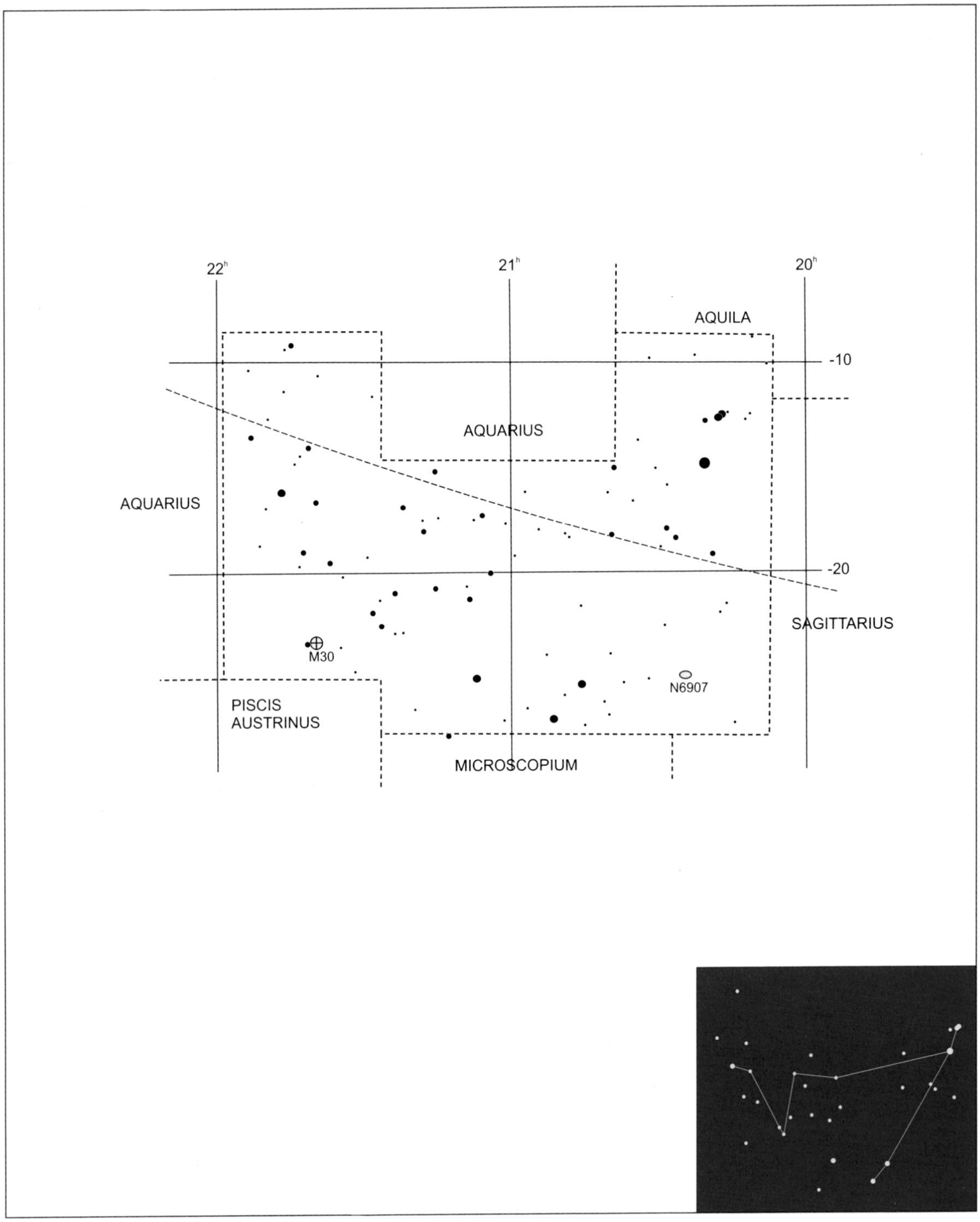

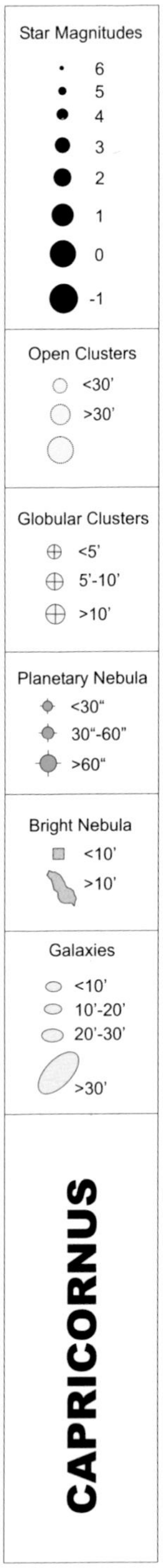

Constellation Facts:

Capricornus; (CAP-rih-CORE-nus)

Capricornus, the Sea Goat.
The constellation rises in the southeast, remains near the southern horizon before setting in the southwest.
On clear nights the constellation is an elegant symmetrical figure of faint fourth and fifth magnitude stars.
Constellation covers 414 square degrees.

Constellation is visible from 62° N to 90° S. Partially visible from 62° N to 80° N.

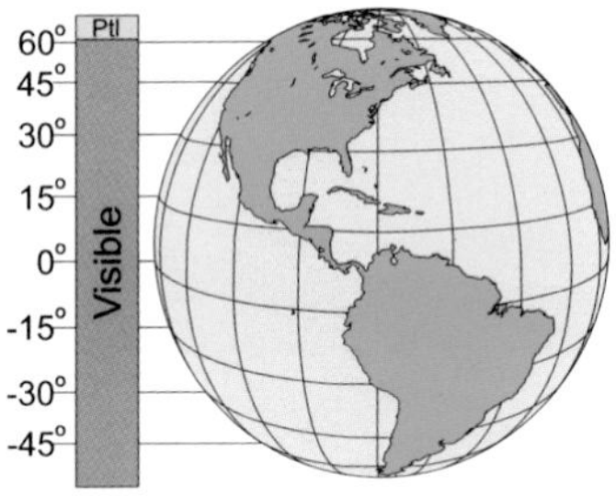

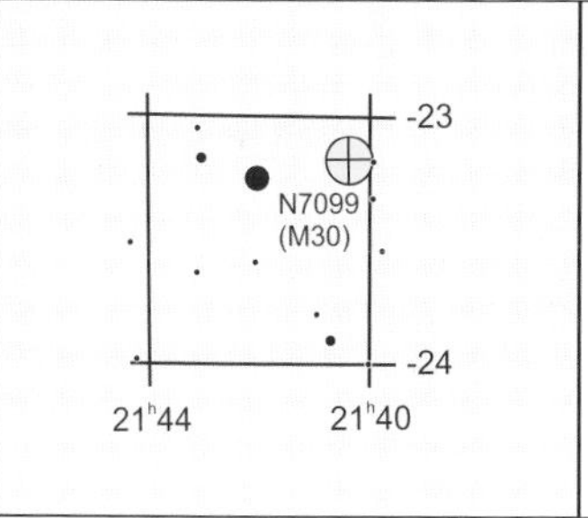

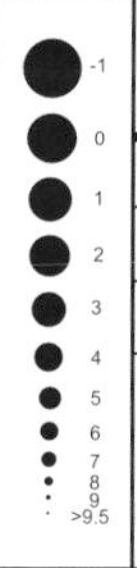

M30 (NGC 7099)

RA:	$21^h\ 40^m\ 29.2^s$	Con:	Capricornus
Dec:	-23° 10' 32"	Type:	Globular Cluster
Size:	11.0'	Mag:	7.5

M30 (NGC 7099) is a highly resolved globular cluster.

Telescope Aperture:	4" f/5	4" f/9	6" f/7	6" f/9	8" f/6.3	8" f/10	10" f/6.3	10" f/10	12" f/6.3	12" f/10
FOV(35mm film):	2.7° x 4.1°	1.50° x 2.26°	1.29° x 1.93°	1.0° x 1.50°	1.07° x 1.61°	0.68° x 1.02°	0.86° x 1.29°	0.54° x 0.81°	0.72° x 1.07°	0.45° x 0.68°

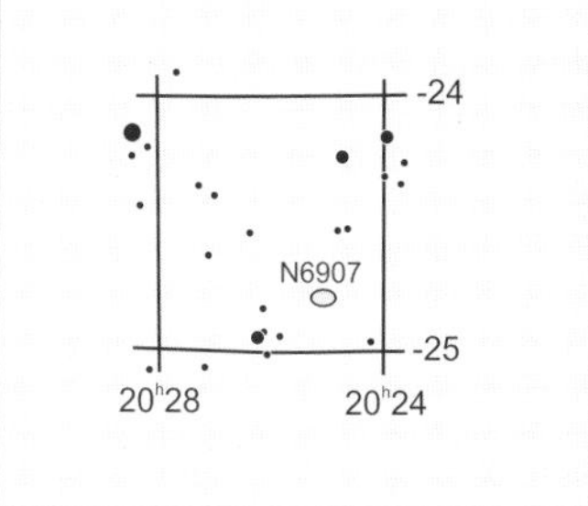

NGC 6907

RA:	$20^h\ 25^m\ 11.7^s$	Con:	Capricornus
Dec:	-24° 48' 40"	Type:	Barred Spiral
Size:	3.0' x 2.5'	Mag:	11.3

NGC 6907 is a barred spiral galaxy.

Telescope Aperture:	4" f/5	4" f/9	6" f/7	6" f/9	8" f/6.3	8" f/10	10" f/6.3	10" f/10	12" f/6.3	12" f/10
FOV(35mm film):	2.7° x 4.1°	1.50° x 2.26°	1.29° x 1.93°	1.0° x 1.50°	1.07° x 1.61°	0.68° x 1.02°	0.86° x 1.29°	0.54° x 0.81°	0.72° x 1.07°	0.45° x 0.68°

Cas

Crosses Prime Meridian:
October thru December

AUTUMN

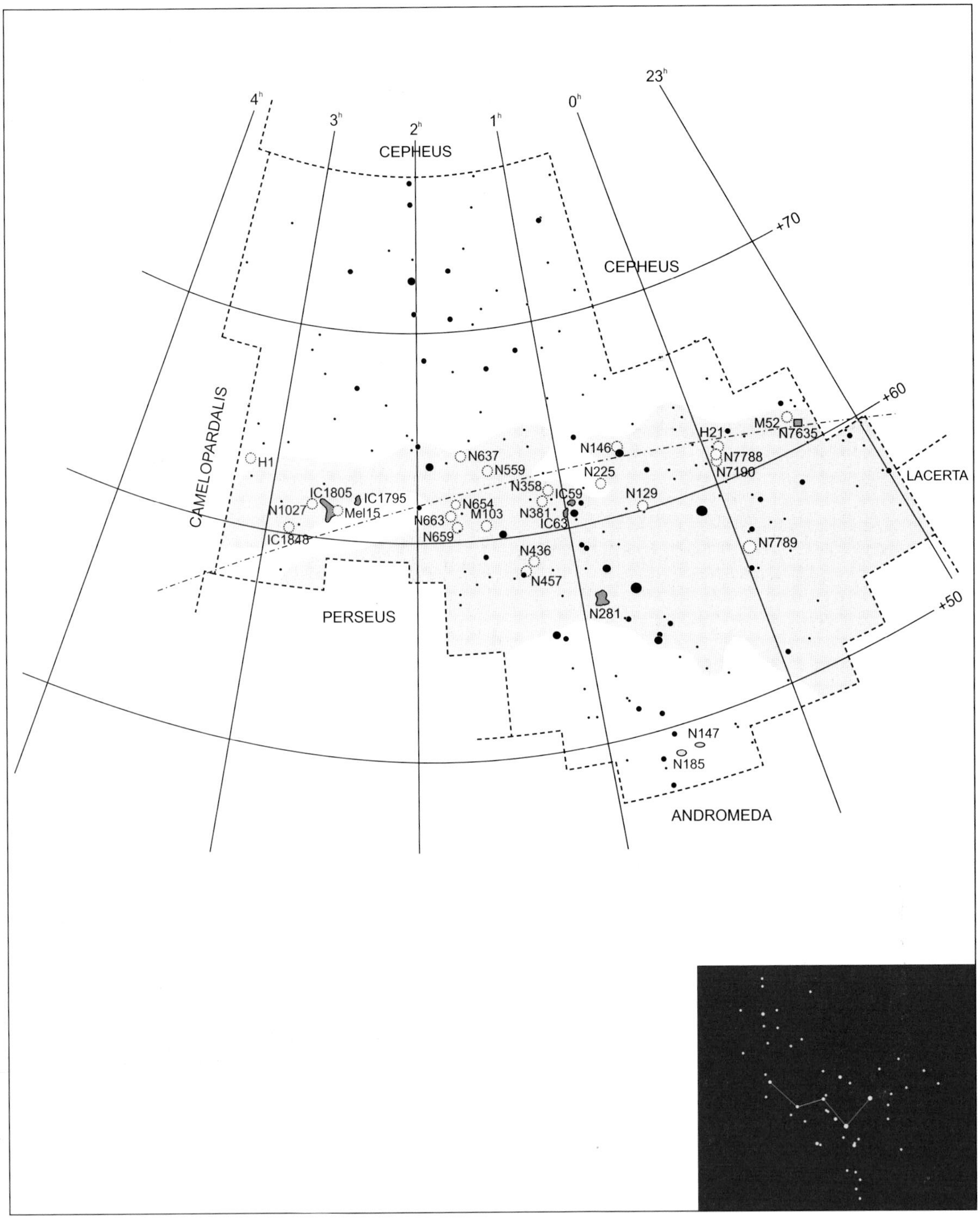

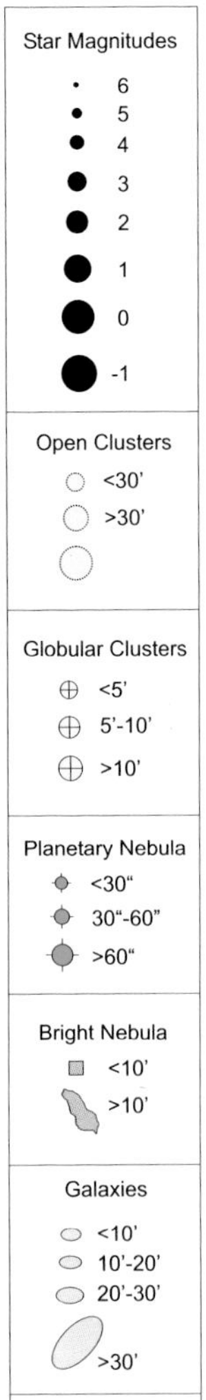

CASSIOPEIA

Constellation Facts:
Cassiopeia; (KASS-ee-oh-PEE-ah)

Cassiopeia, the Queen.
North circumpolar constellation, thus it is above the horizon throughout the year.
The "W" of Cassiopeia is one of the most familiar asterisms in the night sky. The constellation circles the northern sky in a counterclockwise direction.
Cassiopeia is opposite the Galactic center in Sagittarius.
The constellation covers 598 square degrees.

Constellation is visible from 90° N to 12° S. Partially visible from 12° S to 70° S.

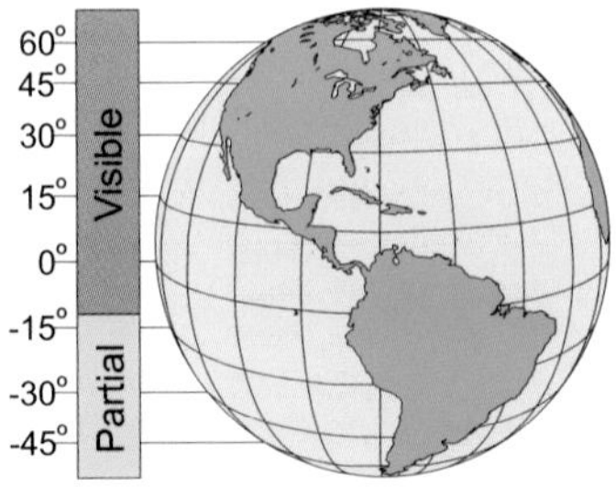

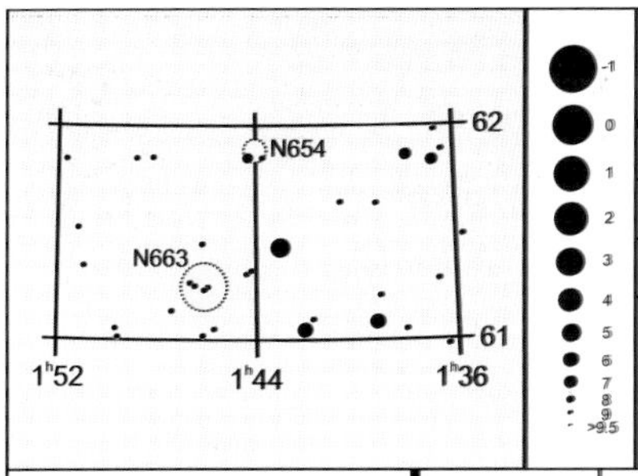

NGC 654

RA:	01^h 44^m 10.8^s	Con:	Cassiopeia
Dec:	61° 53' 05"	Type:	Open Cluster
Size:	5.0'	Mag:	6.5

NGC 654 is a dense open cluster.

Telescope Aperture:	4" f/5	4" f/9	6" f/7	6" f/9	8" f/6.3	8" f/10	10" f/6.3	10" f/10	12" f/6.3	12" f/10
FOV(35mm film):	2.7° x 4.1°	1.50° x 2.26°	1.29° x 1.93°	1.0° x 1.50°	1.07° x 1.61°	0.68° x 1.02°	0.86° x 1.29°	0.54° x 0.81°	0.72° x 1.07°	0.45° x 0.68°

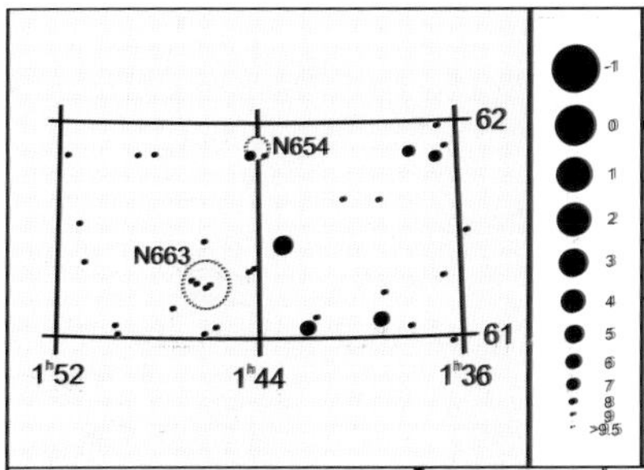

NGC 663

RA:	01^h 46^m 4.8^s	Con:	Cassiopeia
Dec:	61° 15' 05"	Type:	Open Cluster
Size:	16.0'	Mag:	7.1

NGC 663 is a dense open cluster

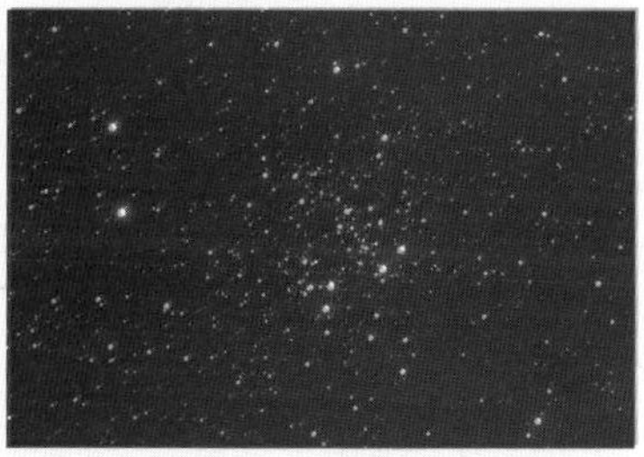

Telescope Aperture:	4" f/5	4" f/9	6" f/7	6" f/9	8" f/6.3	8" f/10	10" f/6.3	10" f/10	12" f/6.3	12" f/10
FOV(35mm film):	2.7° x 4.1°	1.50° x 2.26°	1.29° x 1.93°	1.0° x 1.50°	1.07° x 1.61°	0.68° x 1.02°	0.86° x 1.29°	0.54° x 0.81°	0.72° x 1.07°	0.45° x 0.68°

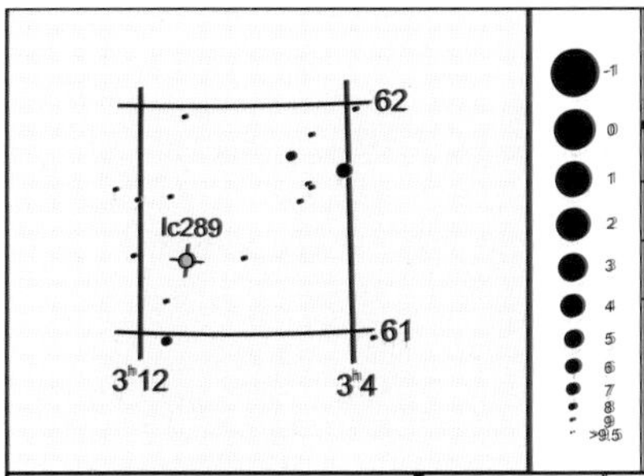

IC 289

RA:	03^h 10^m 22.6^s	Con:	Cassiopeia
Dec:	61° 19' 01"	Type:	Planetary Nebula
Size:	0.6'	Mag:	12.0

IC 289 is a small and indistinct planetary nebula.

Telescope Aperture:	4" f/5	4" f/9	6" f/7	6" f/9	8" f/6.3	8" f/10	10" f/6.3	10" f/10	12" f/6.3	12" f/10
FOV(35mm film):	2.7° x 4.1°	1.50° x 2.26°	1.29° x 1.93°	1.0° x 1.50°	1.07° x 1.61°	0.68° x 1.02°	0.86° x 1.29°	0.54° x 0.81°	0.72° x 1.07°	0.45° x 0.68°

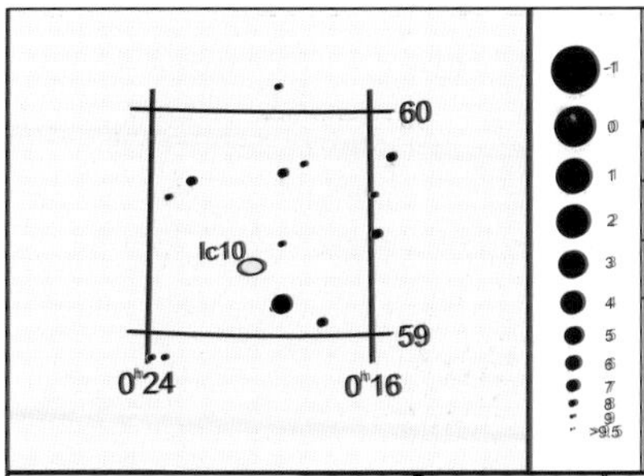

IC 10

RA:	00^h 20^m 28.7^s	Con:	Cassiopeia
Dec:	59° 18' 09"	Type:	Galaxy
Size:	5.0' x 4.0'	Mag:	10.0

Telescope Aperture:	4" f/5	4" f/9	6" f/7	6" f/9	8" f/6.3	8" f/10	10" f/6.3	10" f/10	12" f/6.3	12" f/10
FOV(35mm film):	2.7° x 4.1°	1.50° x 2.26°	1.29° x 1.93°	1.0° x 1.50°	1.07° x 1.61°	0.68° x 1.02°	0.86° x 1.29°	0.54° x 0.81°	0.72° x 1.07°	0.45° x 0.68°

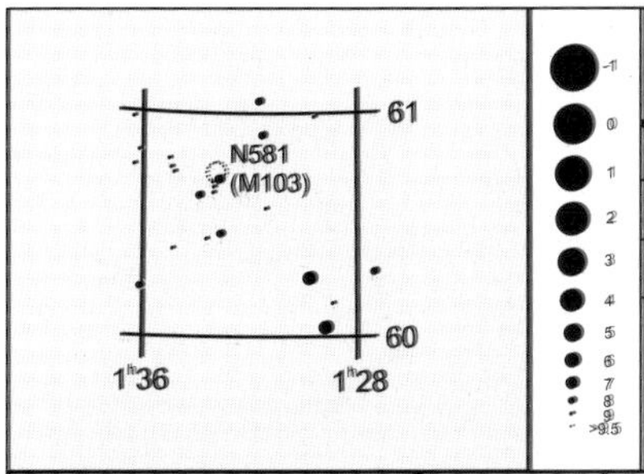

M103 (NGC 581)

RA:	01^h 33^m 16.7^s	Con:	Cassiopeia
Dec:	60° 42' 06"	Type:	Open Cluster
Size:	6.0'	Mag:	7.4

M103 is a rich open cluster.

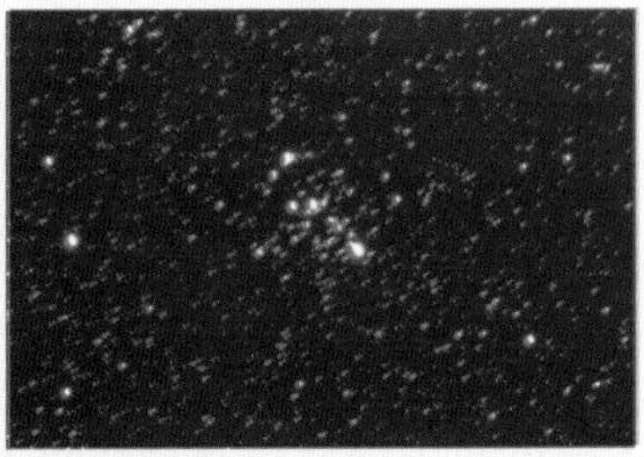

Telescope Aperture:	4" f/5	4" f/9	6" f/7	6" f/9	8" f/6.3	8" f/10	10" f/6.3	10" f/10	12" f/6.3	12" f/10
FOV(35mm film):	2.7° x 4.1°	1.50° x 2.26°	1.29° x 1.93°	1.0° x 1.50°	1.07° x 1.61°	0.68° x 1.02°	0.86° x 1.29°	0.54° x 0.81°	0.72° x 1.07°	0.45° x 0.68°

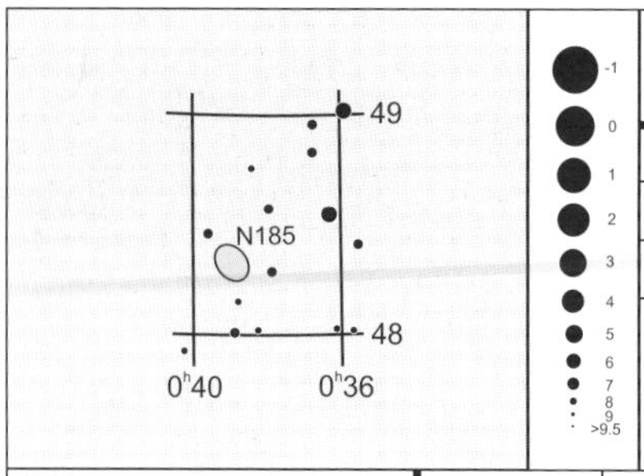

NGC 185

RA:	$00^h\ 39^m\ 4.4^s$	Con:	Cassiopeia
Dec:	48° 20' 11"	Type:	Elliptical Galaxy
Size:	3.0' x 2.5'	Mag:	9.2

NGC 185 is located in the extreme southern end of the constellation. Object is a dwarf elliptical galaxy.

Telescope Aperture:	4" f/5	4" f/9	6" f/7	6" f/9	8" f/6.3	8" f/10	10" f/6.3	10" f/10	12" f/6.3	12" f/10
FOV(35mm film):	2.7° x 4.1°	1.50° x 2.26°	1.29° x 1.93°	1.0° x 1.50°	1.07° x 1.61°	0.68° x 1.02°	0.86° x 1.29°	0.54° x 0.81°	0.72° x 1.07°	0.45° x 0.68°

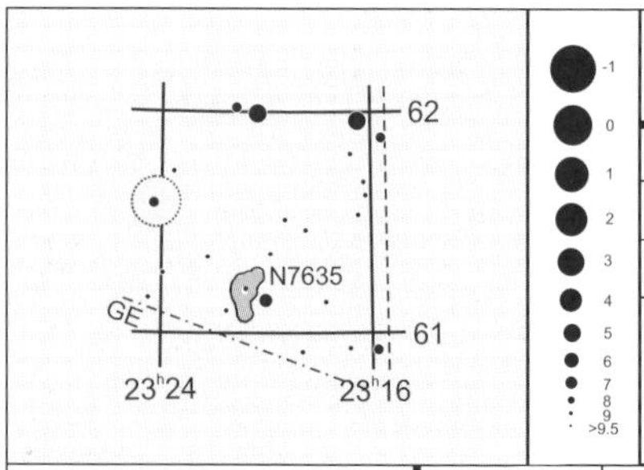

NGC 7635 "Bubble Nebula"

RA:	$23^h\ 20^m\ 46.6^s$	Con:	Cassiopeia
Dec:	61° 12' 10"	Type:	Emission Nebula
Size:	15.0'	Mag:	

NGC 7635 the "Bubble Nebula" is located 1° southwest of M52. It is a faint emission nebula.

Telescope Aperture:	4" f/5	4" f/9	6" f/7	6" f/9	8" f/6.3	8" f/10	10" f/6.3	10" f/10	12" f/6.3	12" f/10
FOV(35mm film):	2.7° x 4.1°	1.50° x 2.26°	1.29° x 1.93°	1.0° x 1.50°	1.07° x 1.61°	0.68° x 1.02°	0.86° x 1.29°	0.54° x 0.81°	0.72° x 1.07°	0.45° x 0.68°

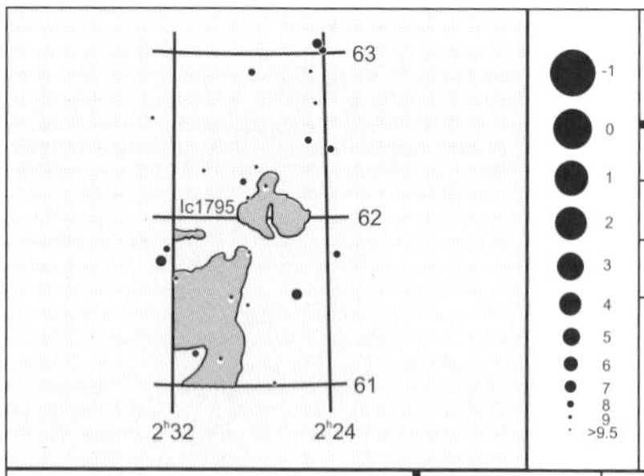

IC 1795

RA:	$02^h\ 26^m\ 34.8^s$	Con:	Cassiopeia
Dec:	62° 04' 03"	Type:	Emission Nebula
Size:	20.0'	Mag:	

IC 1795 is yet another emission nebula located in Cassiopeia.

Telescope Aperture:	4" f/5	4" f/9	6" f/7	6" f/9	8" f/6.3	8" f/10	10" f/6.3	10" f/10	12" f/6.3	12" f/10
FOV(35mm film):	2.7° x 4.1°	1.50° x 2.26°	1.29° x 1.93°	1.0° x 1.50°	1.07° x 1.61°	0.68° x 1.02°	0.86° x 1.29°	0.54° x 0.81°	0.72° x 1.07°	0.45° x 0.68°

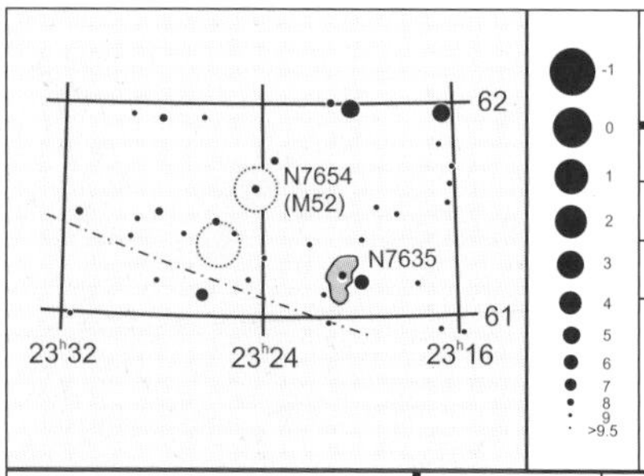

M52 (NGC 7654)

RA:	$23^h\ 24^m\ 16.7^s$	Con:	Cassiopeia
Dec:	61° 35' 10"	Type:	Open Cluster
Size:	13.0'	Mag:	6.9

M52 (NGC 7654) is a dense open cluster containing approximately 100 stars.

Telescope Aperture:	4" f/5	4" f/9	6" f/7	6" f/9	8" f/6.3	8" f/10	10" f/6.3	10" f/10	12" f/6.3	12" f/10
FOV(35mm film):	2.7° x 4.1°	1.50° x 2.26°	1.29° x 1.93°	1.0° x 1.50°	1.07° x 1.61°	0.68° x 1.02°	0.86° x 1.29°	0.54° x 0.81°	0.72° x 1.07°	0.45° x 0.68°

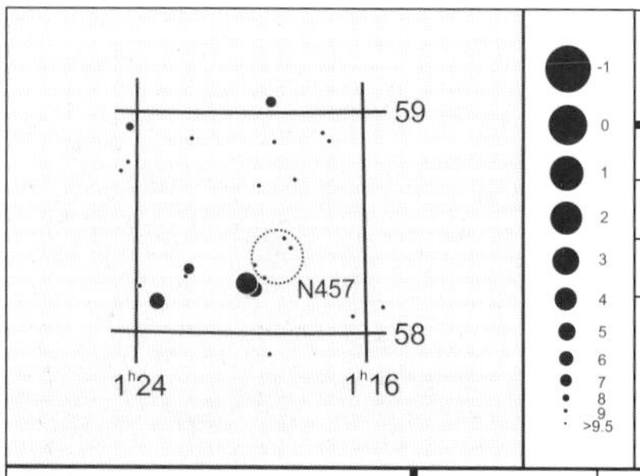

NGC 457 "Owl Cluster"

RA:	$01^h\ 19^m\ 10.7^s$	Con:	Cassiopeia
Dec:	58° 20' 07"	Type:	Open Cluster
Size:	13.0'	Mag:	6.4

NGC 457 is the brightest open cluster in Cassiopeia. The cluster contains approximately 80 stars, and is commonly called the "Owl Cluster".

Telescope Aperture:	4" f/5	4" f/9	6" f/7	6" f/9	8" f/6.3	8" f/10	10" f/6.3	10" f/10	12" f/6.3	12" f/10
FOV(35mm film):	2.7° x 4.1°	1.50° x 2.26°	1.29° x 1.93°	1.0° x 1.50°	1.07° x 1.61°	0.68° x 1.02°	0.86° x 1.29°	0.54° x 0.81°	0.72° x 1.07°	0.45° x 0.68°

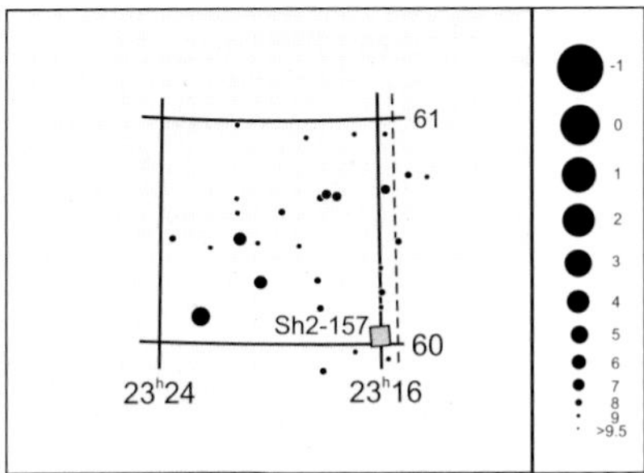

Sh2-157

RA:	00h 39m 4.4s	Con:	Cassiopeia
Dec:	48° 20' 11"	Type:	Emission Nebula
Size:	3.0' x 2.5'	Mag:	9.2

Sh2-157 is located in the extreme southern end of the constellation. Object is a dwarf elliptical galaxy.

Telescope Aperture:	4" f/5	4" f/9	6" f/7	6" f/9	8" f/6.3	8" f/10	10" f/6.3	10" f/10	12" f/6.3	12" f/10
FOV(35mm film):	2.7° x 4.1°	1.50° x 2.26°	1.29° x 1.93°	1.0° x 1.50°	1.07° x 1.61°	0.68° x 1.02°	0.86° x 1.29°	0.54° x 0.81°	0.72° x 1.07°	0.45° x 0.68°

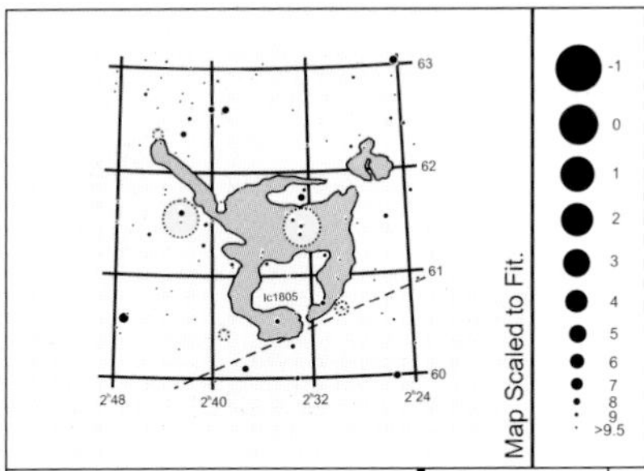

IC 1805

RA:	02h 32m 46.7s	Con:	Cassiopeia
Dec:	61° 27' 03"	Type:	Nebula & Cluster
Size:	60.0'	Mag:	6.0

IC 1805 is a dense open cluster surrounded by nebula located in the southeastern corner of the constellation.

Telescope Aperture:	4" f/5	4" f/9	6" f/7	6" f/9	8" f/6.3	8" f/10	10" f/6.3	10" f/10	12" f/6.3	12" f/10
FOV(35mm film):	2.7° x 4.1°	1.50° x 2.26°	1.29° x 1.93°	1.0° x 1.50°	1.07° x 1.61°	0.68° x 1.02°	0.86° x 1.29°	0.54° x 0.81°	0.72° x 1.07°	0.45° x 0.68°

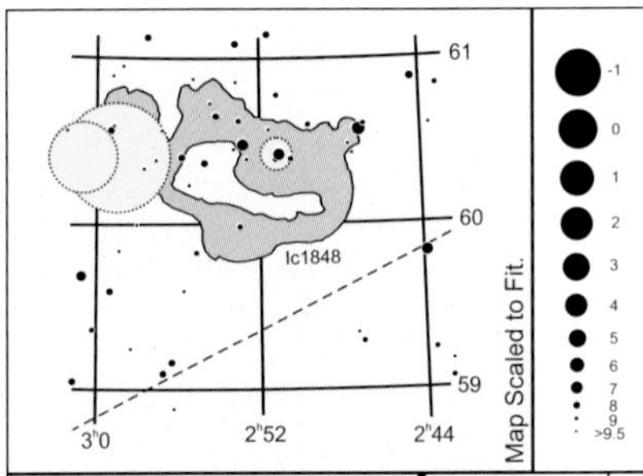

IC 1848

RA:	02h 51m 16.6s	Con:	Cassiopeia
Dec:	60° 26' 02"	Type:	Nebula & Cluster
Size:	60.0'	Mag:	6.0

IC 1848 is located about 2.5° southeast of IC 1805. Cluster of stars surrounded by nebulosity.

Telescope Aperture:	4" f/5	4" f/9	6" f/7	6" f/9	8" f/6.3	8" f/10	10" f/6.3	10" f/10	12" f/6.3	12" f/10
FOV(35mm film):	2.7° x 4.1°	1.50° x 2.26°	1.29° x 1.93°	1.0° x 1.50°	1.07° x 1.61°	0.68° x 1.02°	0.86° x 1.29°	0.54° x 0.81°	0.72° x 1.07°	0.45° x 0.68°

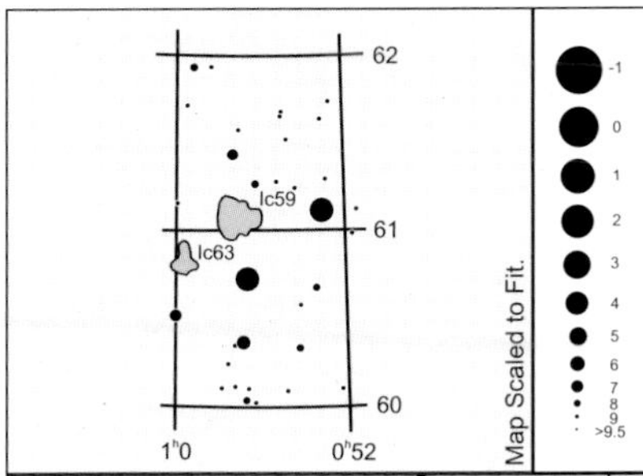

IC 59

RA:	00h 56m 46.8s	Con:	Cassiopeia
Dec:	61° 04' 07"	Type:	Nebula
Size:	10.0'	Mag:	

IC 59 is a faint nebula surrounding the star Gamma Cassiopeiae.

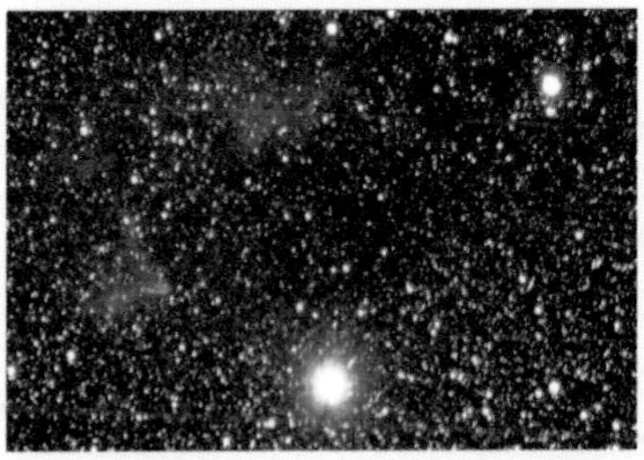

Telescope Aperture:	4" f/5	4" f/9	6" f/7	6" f/9	8" f/6.3	8" f/10	10" f/6.3	10" f/10	12" f/6.3	12" f/10
FOV(35mm film):	2.7° x 4.1°	1.50° x 2.26°	1.29° x 1.93°	1.0° x 1.50°	1.07° x 1.61°	0.68° x 1.02°	0.86° x 1.29°	0.54° x 0.81°	0.72° x 1.07°	0.45° x 0.68°

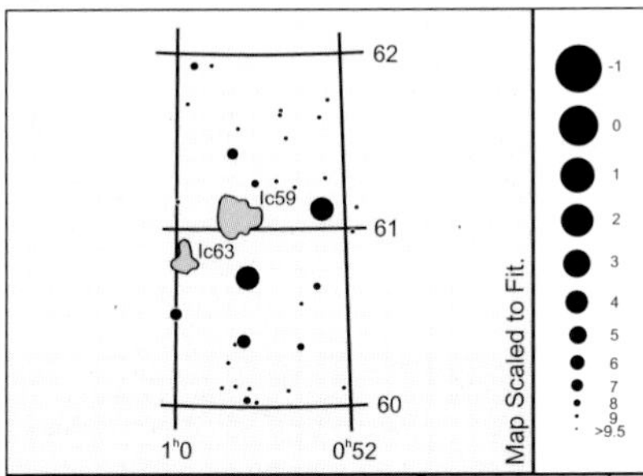

IC 63

RA:	00h 59m 34.8s	Con:	Cassiopeia
Dec:	60° 49' 07"	Type:	Nebula
Size:	10.0'	Mag:	

IC 63 is a companion nebula to IC 59 surrounding the star Gamma Cassiopeiae.

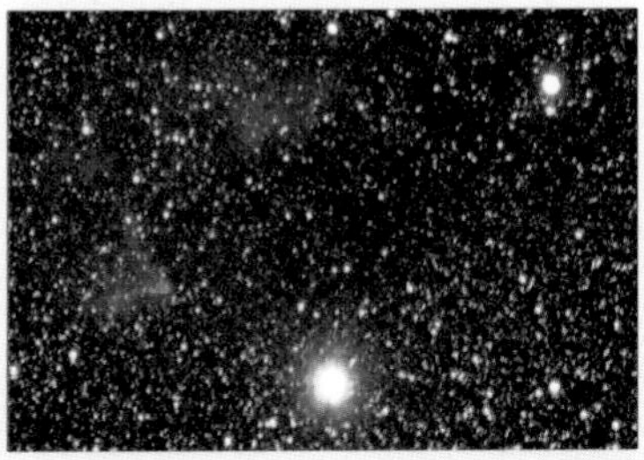

Telescope Aperture:	4" f/5	4" f/9	6" f/7	6" f/9	8" f/6.3	8" f/10	10" f/6.3	10" f/10	12" f/6.3	12" f/10
FOV(35mm film):	2.7° x 4.1°	1.50° x 2.26°	1.29° x 1.93°	1.0° x 1.50°	1.07° x 1.61°	0.68° x 1.02°	0.86° x 1.29°	0.54° x 0.81°	0.72° x 1.07°	0.45° x 0.68°

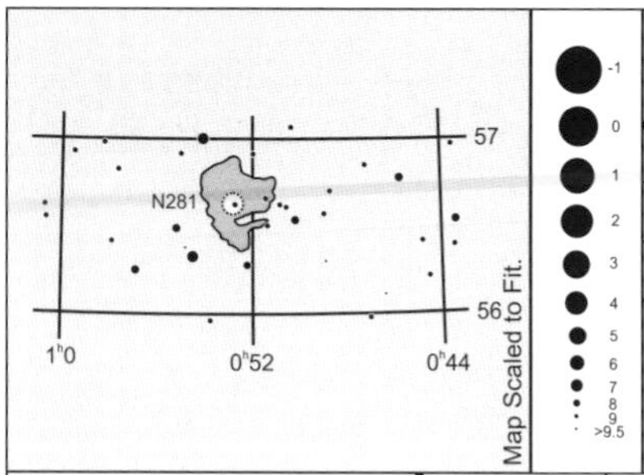

NGC 281

RA:	00h 52m 52.6s	Con:	Cassiopeia
Dec:	56° 37' 08"	Type:	Nebula & Cluster
Size:	35.0'	Mag:	7.0

NGC 281 is an open cluster with an associated faint nebula covering an equilateral triangle of 7th magnitude stars.

Telescope Aperture:	4" f/5	4" f/9	6" f/7	6" f/9	8" f/6.3	8" f/10	10" f/6.3	10" f/10	12" f/6.3	12" f/10
FOV(35mm film):	2.7° x 4.1°	1.50° x 2.26°	1.29° x 1.93°	1.0° x 1.50°	1.07° x 1.61°	0.68° x 1.02°	0.86° x 1.29°	0.54° x 0.81°	0.72° x 1.07°	0.45° x 0.68°

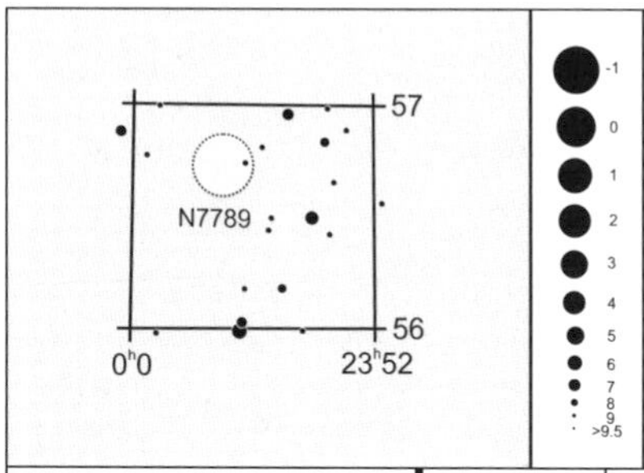

NGC 7789

RA:	23h 57m 4.6s	Con:	Cassiopeia
Dec:	56° 44' 10"	Type:	Open Cluster
Size:	16.0'	Mag:	6.7

NGC 7789 is a dense open cluster found 3° southwest of Beta Cassiopeiae. It is one of the finest open clusters in the night sky, consisting of approximately 300 stars.

Telescope Aperture:	4" f/5	4" f/9	6" f/7	6" f/9	8" f/6.3	8" f/10	10" f/6.3	10" f/10	12" f/6.3	12" f/10
FOV(35mm film):	2.7° x 4.1°	1.50° x 2.26°	1.29° x 1.93°	1.0° x 1.50°	1.07° x 1.61°	0.68° x 1.02°	0.86° x 1.29°	0.54° x 0.81°	0.72° x 1.07°	0.45° x 0.68°

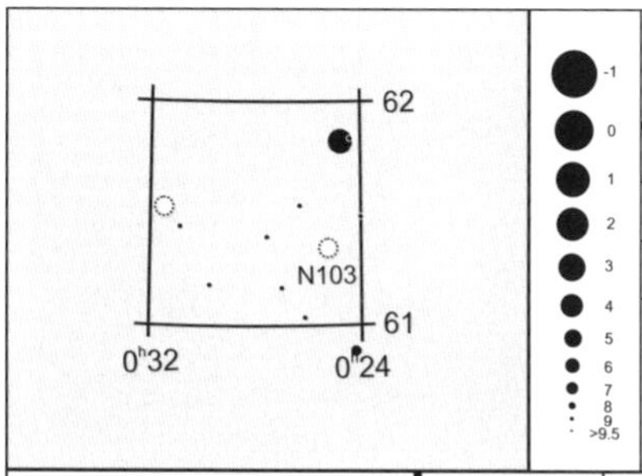

NGC 103

RA:	00h 25m 22.8s	Con:	Cassiopeia
Dec:	61° 21' 08"	Type:	Open Cluster
Size:	5.0'	Mag:	9.8

NGC 103 is a rich open cluster located about 3° northeast of Beta Cassiopeiae. Cluster is comprised of approximately 30 stars.

Telescope Aperture:	4" f/5	4" f/9	6" f/7	6" f/9	8" f/6.3	8" f/10	10" f/6.3	10" f/10	12" f/6.3	12" f/10
FOV(35mm film):	2.7° x 4.1°	1.50° x 2.26°	1.29° x 1.93°	1.0° x 1.50°	1.07° x 1.61°	0.68° x 1.02°	0.86° x 1.29°	0.54° x 0.81°	0.72° x 1.07°	0.45° x 0.68°

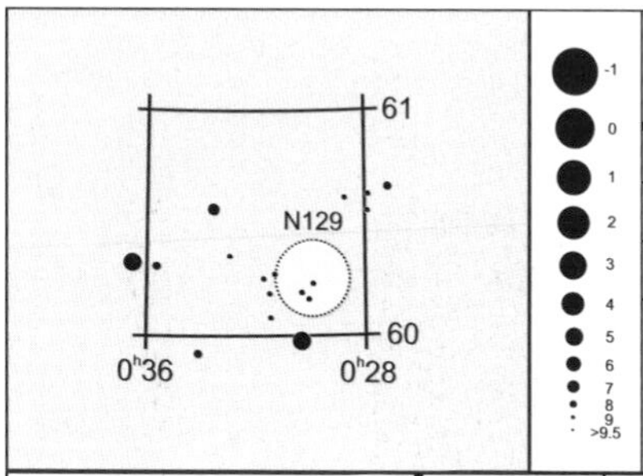

NGC 129

RA:	00h 29m 58.7s	Con:	Cassiopeia
Dec:	60° 14' 08"	Type:	Open Cluster
Size:	21.0'	Mag:	6.5

NGC 129 is a large, bright, scattered open cluster located 1.5° south-southeast of NGC 103. Object is comprised of approximately 35 stars.

Telescope Aperture:	4" f/5	4" f/9	6" f/7	6" f/9	8" f/6.3	8" f/10	10" f/6.3	10" f/10	12" f/6.3	12" f/10
FOV(35mm film):	2.7° x 4.1°	1.50° x 2.26°	1.29° x 1.93°	1.0° x 1.50°	1.07° x 1.61°	0.68° x 1.02°	0.86° x 1.29°	0.54° x 0.81°	0.72° x 1.07°	0.45° x 0.68°

Cep

Crosses Prime Meridian: September thru October

SUMMER

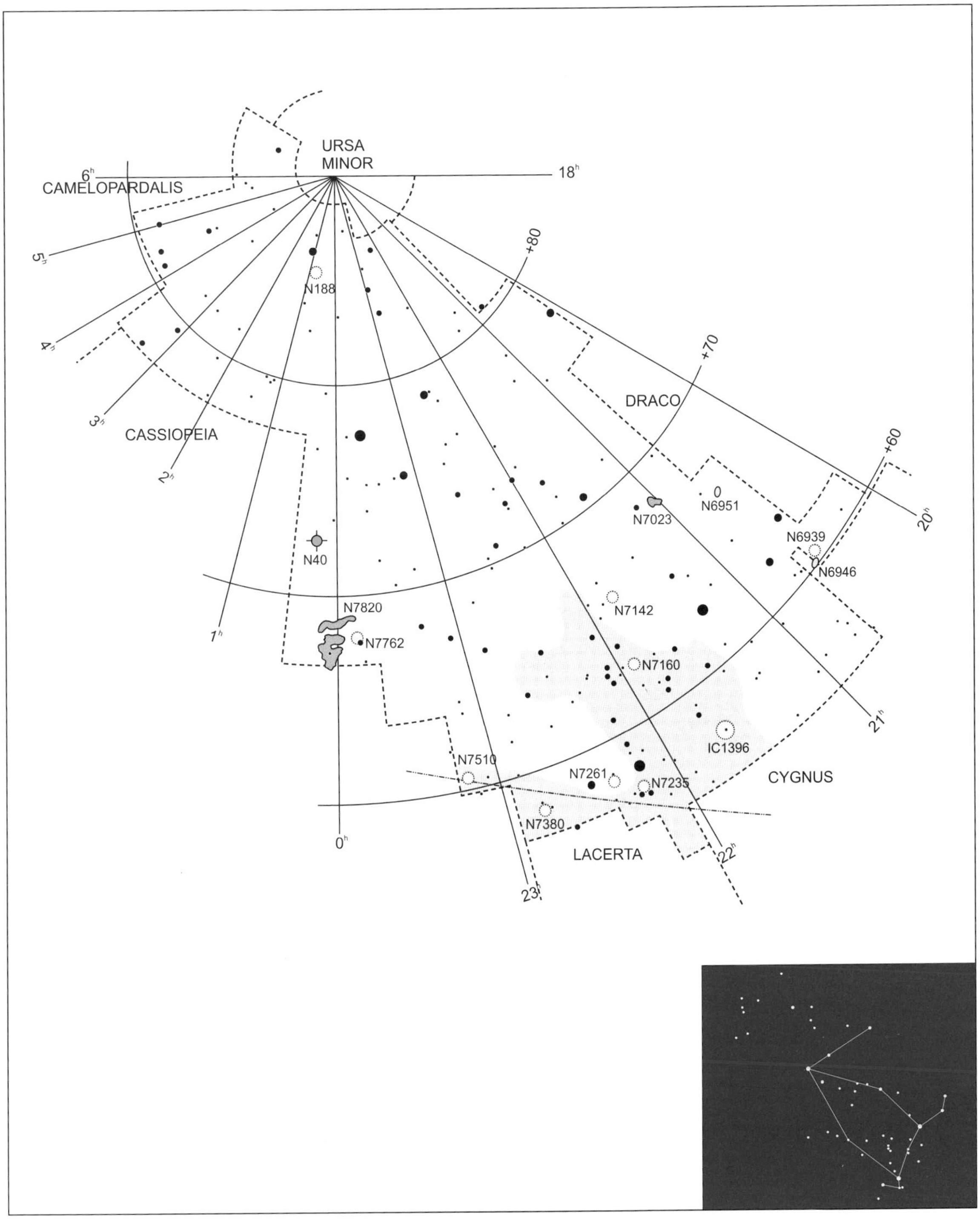

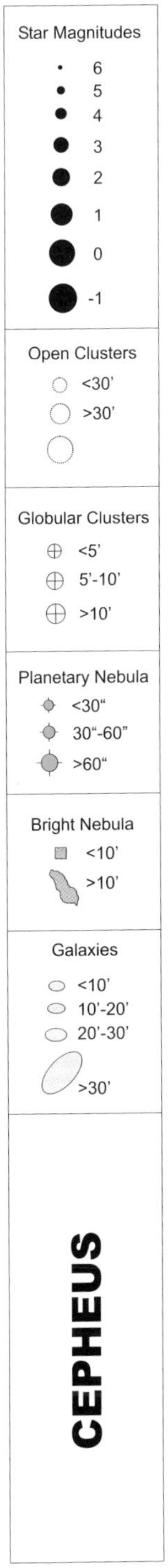

Constellation Facts:

Cepheus; (SEE-fee-us)

Cepheus, the King.
A circumpolar constellation, Cepheus is above the horizon at all times. During the months June thru February, Cepheus is away from the horizon and easily seen during the evening hours.
Faint constellation situated between Draco and Cassiopeia.
Constellation covers 588 square degrees.

Constellation is visible from 90° N to 1° S. Partially visible from 1° S to 60° S.

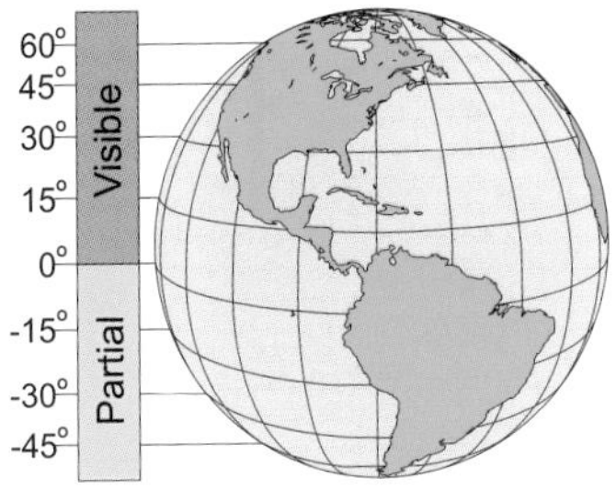

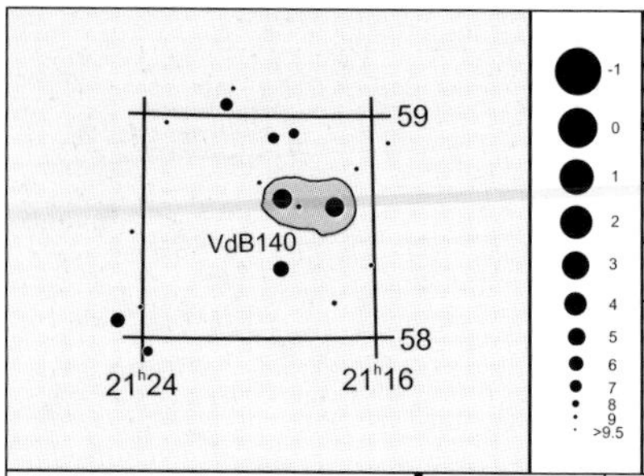

VdB-140

RA:	06h 46m 1.7s	Con:	Cepheus
Dec:	-20° 44' 03"	Type:	Reflection Nebula
Size:	12.0' x 10.0'	Mag:	

VdB-140 displays its brightest regions in the northwest and northeast sectors. Object contains many small absorption patches scattered throughout.

Telescope Aperture:	4" f/5	4" f/9	6" f/7	6" f/9	8" f/6.3	8" f/10	10" f/6.3	10" f/10	12" f/6.3	12" f/10
FOV(35mm film):	2.7° x 4.1°	1.50° x 2.26°	1.29° x 1.93°	1.0° x 1.50°	1.07° x 1.61°	0.68° x 1.02°	0.86° x 1.29°	0.54° x 0.81°	0.72° x 1.07°	0.45° x 0.68°

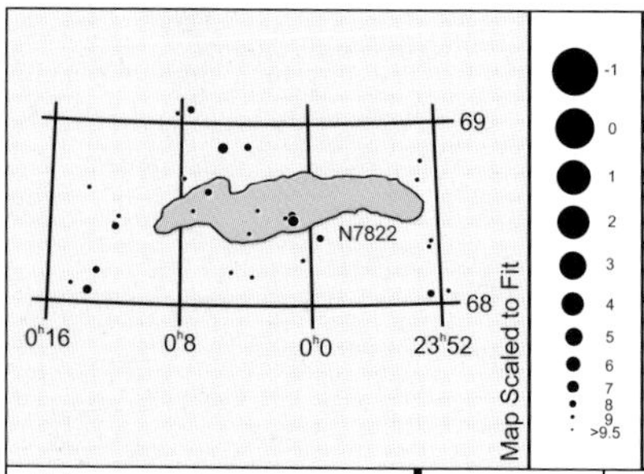

NGC 7822

RA:	00h 03m 41.1s	Con:	Cepheus
Dec:	68° 37' 06"	Type:	Emission Nebula
Size:	60.0'	Mag:	9.0

NGC 7822 is an emission nebula with low surface brightness.

Telescope Aperture:	4" f/5	4" f/9	6" f/7	6" f/9	8" f/6.3	8" f/10	10" f/6.3	10" f/10	12" f/6.3	12" f/10
FOV(35mm film):	2.7° x 4.1°	1.50° x 2.26°	1.29° x 1.93°	1.0° x 1.50°	1.07° x 1.61°	0.68° x 1.02°	0.86° x 1.29°	0.54° x 0.81°	0.72° x 1.07°	0.45° x 0.68°

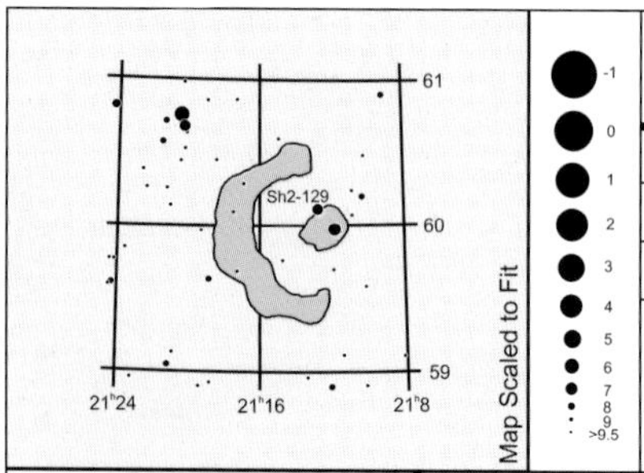

Sh2-129

RA:	20h 53m 35.1s	Con:	Cepheus
Dec:	-12° 31' 39"	Type:	Emission Nebula
Size:	110' x 100'	Mag:	

Sh2-129 is a faint, incomplete ring of nebulosity. Object displays filamentary structure, and is difficult in large scopes.

Telescope Aperture:	4" f/5	4" f/9	6" f/7	6" f/9	8" f/6.3	8" f/10	10" f/6.3	10" f/10	12" f/6.3	12" f/10
FOV(35mm film):	2.7° x 4.1°	1.50° x 2.26°	1.29° x 1.93°	1.0° x 1.50°	1.07° x 1.61°	0.68° x 1.02°	0.86° x 1.29°	0.54° x 0.81°	0.72° x 1.07°	0.45° x 0.68°

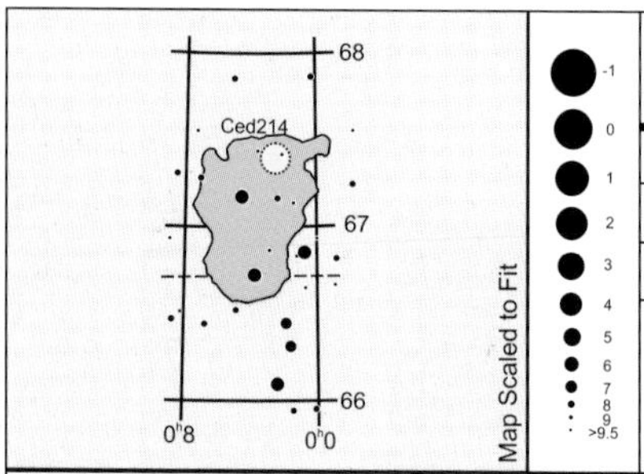

Ced-214

RA:	00h 03m 31.0s	Con:	Cepheus
Dec:	67° 00' 36"	Type:	Emission Nebula
Size:	55' x 40'	Mag:	7.3

Ced-214 is a faint emission nebula.

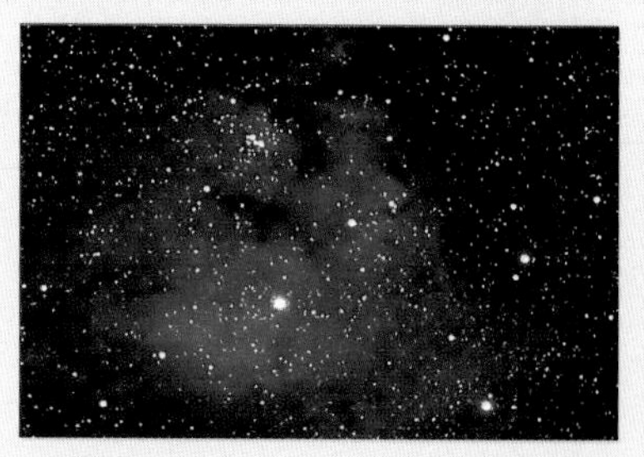

Telescope Aperture:	4" f/5	4" f/9	6" f/7	6" f/9	8" f/6.3	8" f/10	10" f/6.3	10" f/10	12" f/6.3	12" f/10
FOV(35mm film):	2.7° x 4.1°	1.50° x 2.26°	1.29° x 1.93°	1.0° x 1.50°	1.07° x 1.61°	0.68° x 1.02°	0.86° x 1.29°	0.54° x 0.81°	0.72° x 1.07°	0.45° x 0.68°

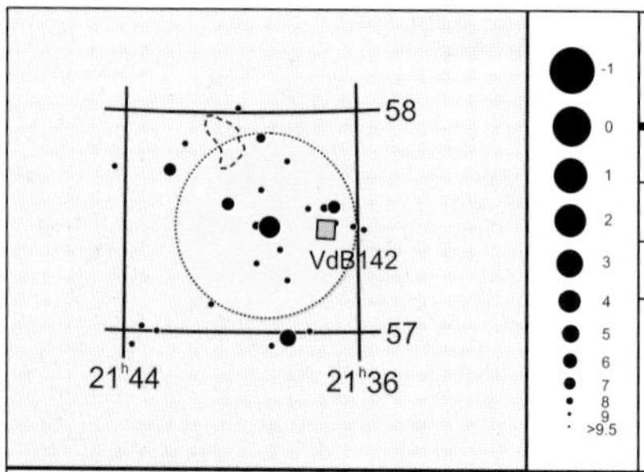

VdB-142

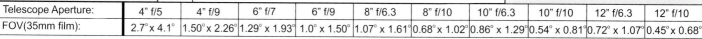

RA:	21h 04m 17.0s	Con:	Cepheus
Dec:	-11° 21' 38"	Type:	Reflection Nebula
Size:	1.0' x 1.0'	Mag:	

VdB-142 is a patch of nebula involving an 8th magnitude star.

Telescope Aperture:	4" f/5	4" f/9	6" f/7	6" f/9	8" f/6.3	8" f/10	10" f/6.3	10" f/10	12" f/6.3	12" f/10
FOV(35mm film):	2.7° x 4.1°	1.50° x 2.26°	1.29° x 1.93°	1.0° x 1.50°	1.07° x 1.61°	0.68° x 1.02°	0.86° x 1.29°	0.54° x 0.81°	0.72° x 1.07°	0.45° x 0.68°

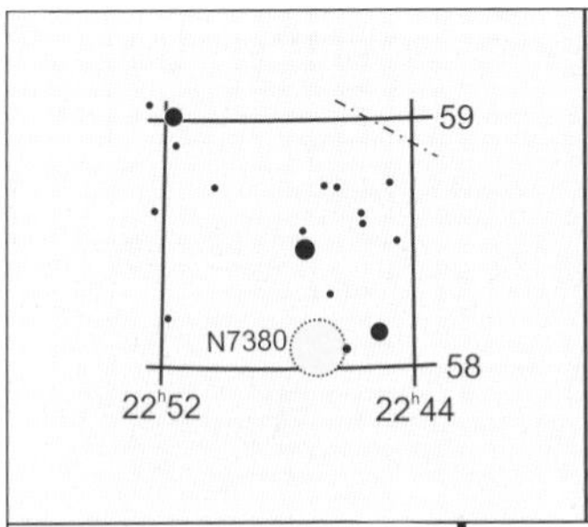

NGC 7380

RA:	$22^h\ 47^m\ 4.5^s$	Con:	Cepheus
Dec:	58° 06' 11"	Type:	Cluster & Nebula
Size:	25.0'	Mag:	7.2

NGC 7380 is an open cluster with associated nebula.

Telescope Aperture:	4" f/5	4" f/9	6" f/7	6" f/9	8" f/6.3	8" f/10	10" f/6.3	10" f/10	12" f/6.3	12" f/10
FOV(35mm film):	2.7° x 4.1°	1.50° x 2.26°	1.29° x 1.93°	1.0° x 1.50°	1.07° x 1.61°	0.68° x 1.02°	0.86° x 1.29°	0.54° x 0.81°	0.72° x 1.07°	0.45° x 0.68°

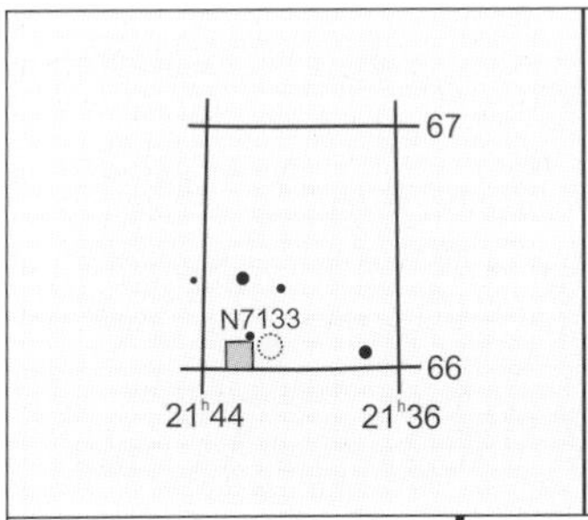

NGC 7133

RA:	$21^h\ 42^m\ 40.4^s$	Con:	Cepheus
Dec:	66° 03' 11"	Type:	Reflection Nebula
Size:	60.0'	Mag:	9.0

NGC 7133 is a faint reflection nebula.

Telescope Aperture:	4" f/5	4" f/9	6" f/7	6" f/9	8" f/6.3	8" f/10	10" f/6.3	10" f/10	12" f/6.3	12" f/10
FOV(35mm film):	2.7° x 4.1°	1.50° x 2.26°	1.29° x 1.93°	1.0° x 1.50°	1.07° x 1.61°	0.68° x 1.02°	0.86° x 1.29°	0.54° x 0.81°	0.72° x 1.07°	0.45° x 0.68°

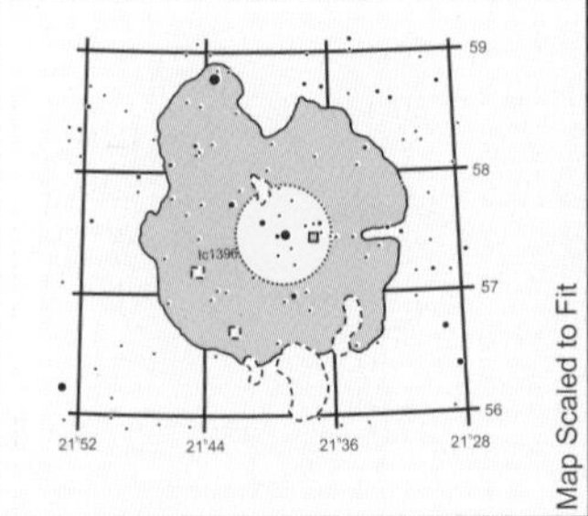

IC 1396

RA:	$21^h\ 39^m\ 10.3^s$	Con:	Cepheus
Dec:	57° 30' 12"	Type:	Emission Nebula
Size:	170' x 140'	Mag:	9.4

IC 1396 is large emission nebula.

Telescope Aperture:	4" f/5	4" f/9	6" f/7	6" f/9	8" f/6.3	8" f/10	10" f/6.3	10" f/10	12" f/6.3	12" f/10
FOV(35mm film):	2.7° x 4.1°	1.50° x 2.26°	1.29° x 1.93°	1.0° x 1.50°	1.07° x 1.61°	0.68° x 1.02°	0.86° x 1.29°	0.54° x 0.81°	0.72° x 1.07°	0.45° x 0.68°

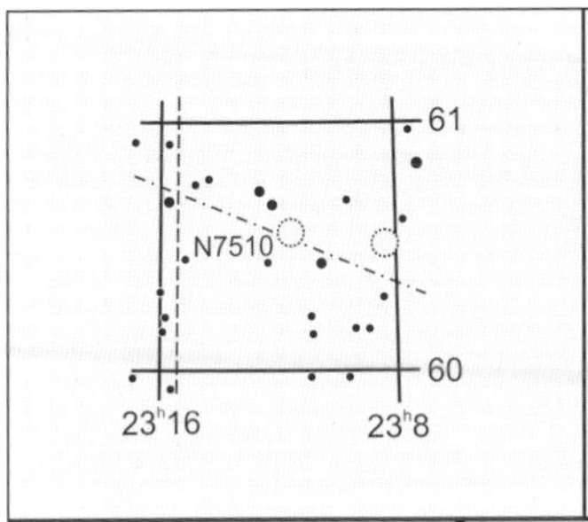

NGC 7510

RA:	$23^h\ 11^m\ 34.6^s$	Con:	Cepheus
Dec:	60° 34' 10"	Type:	Open Cluster
Size:	4.0'	Mag:	7.9

NGC 7510 is a bright compact open cluster comprised of approximately 60 stars.

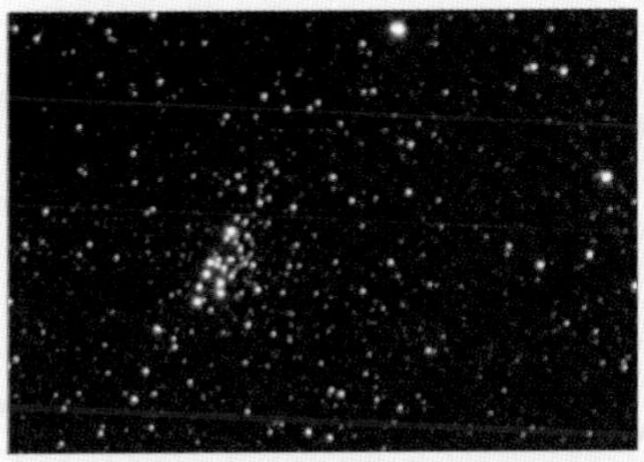

Telescope Aperture:	4" f/5	4" f/9	6" f/7	6" f/9	8" f/6.3	8" f/10	10" f/6.3	10" f/10	12" f/6.3	12" f/10
FOV(35mm film):	2.7° x 4.1°	1.50° x 2.26°	1.29° x 1.93°	1.0° x 1.50°	1.07° x 1.61°	0.68° x 1.02°	0.86° x 1.29°	0.54° x 0.81°	0.72° x 1.07°	0.45° x 0.68°

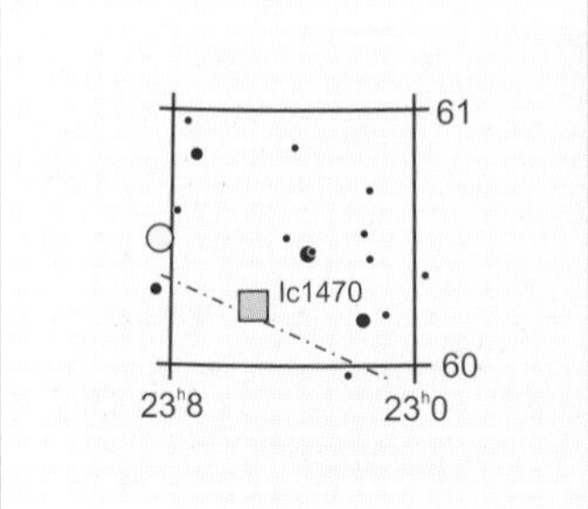

IC 1470

RA:	$23^h\ 05^m\ 16.6^s$	Con:	Cepheus
Dec:	60° 15' 11"	Type:	Emission Nebula
Size:	15.0'	Mag:	8.1

IC 1470 is a small emission nebula.

Telescope Aperture:	4" f/5	4" f/9	6" f/7	6" f/9	8" f/6.3	8" f/10	10" f/6.3	10" f/10	12" f/6.3	12" f/10
FOV(35mm film):	2.7° x 4.1°	1.50° x 2.26°	1.29° x 1.93°	1.0° x 1.50°	1.07° x 1.61°	0.68° x 1.02°	0.86° x 1.29°	0.54° x 0.81°	0.72° x 1.07°	0.45° x 0.68°

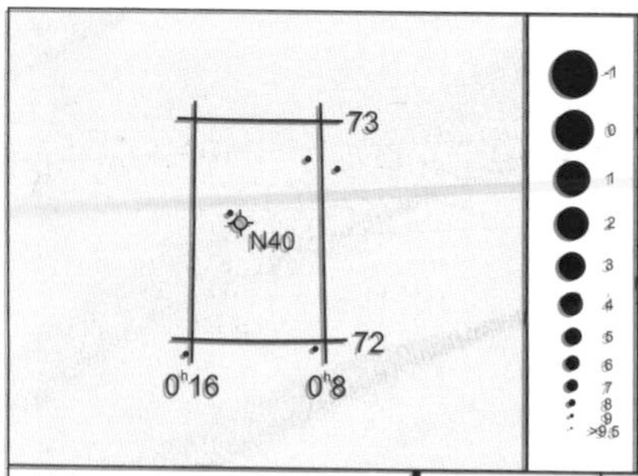

NGC 40

RA:	$00^h\ 13^m\ 5.4^s$	Con:	Cepheus
Dec:	72° 32' 07"	Type:	Planetary Nebula
Size:	0.6'	Mag:	11.0

NGC 40 is one of the finest planetary nebulae found at high declinations. Object displays a planetary nebula disk with a central star.

Telescope Aperture:	4" f/5	4" f/9	6" f/7	6" f/9	8" f/6.3	8" f/10	10" f/6.3	10" f/10	12" f/6.3	12" f/10
FOV(35mm film):	2.7° x 4.1°	1.50° x 2.26°	1.29° x 1.93°	1.0° x 1.50°	1.07° x 1.61°	0.68° x 1.02°	0.86° x 1.29°	0.54° x 0.81°	0.72° x 1.07°	0.45° x 0.68°

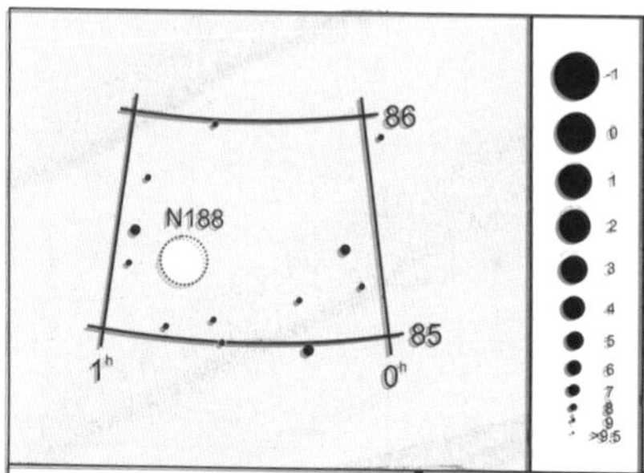

NGC 188

RA:	$00^h\ 44^m\ 10.7^s$	Con:	Cepheus
Dec:	85° 20' 04"	Type:	Open Cluster
Size:	14.0'	Mag:	8.1

NGC 188 is a dense open cluster found near the north celestial pole. Object consists of 120 stars.

Telescope Aperture:	4" f/5	4" f/9	6" f/7	6" f/9	8" f/6.3	8" f/10	10" f/6.3	10" f/10	12" f/6.3	12" f/10
FOV(35mm film):	2.7° x 4.1°	1.50° x 2.26°	1.29° x 1.93°	1.0° x 1.50°	1.07° x 1.61°	0.68° x 1.02°	0.86° x 1.29°	0.54° x 0.81°	0.72° x 1.07°	0.45° x 0.68°

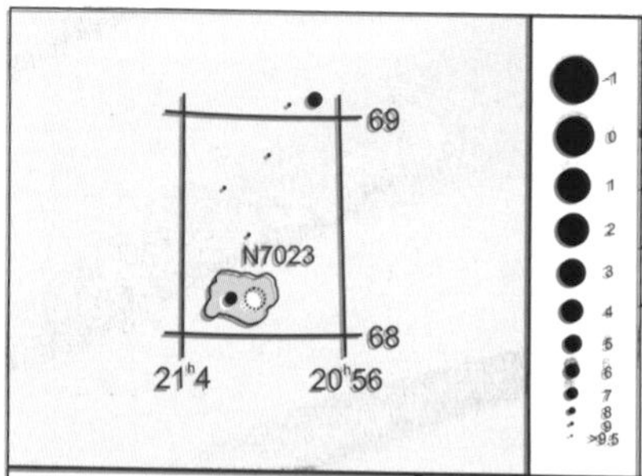

NGC 7023 "Iris Nebula"

RA:	$21^h\ 00^m\ 34.2^s$	Con:	Cepheus
Dec:	68° 10' 11"	Type:	Cluster & Nebula
Size:	18.0'	Mag:	7.0

NGC 7023 is a faint reflection nebula with an associated cluster.

Telescope Aperture:	4" f/5	4" f/9	6" f/7	6" f/9	8" f/6.3	8" f/10	10" f/6.3	10" f/10	12" f/6.3	12" f/10
FOV(35mm film):	2.7° x 4.1°	1.50° x 2.26°	1.29° x 1.93°	1.0° x 1.50°	1.07° x 1.61°	0.68° x 1.02°	0.86° x 1.29°	0.54° x 0.81°	0.72° x 1.07°	0.45° x 0.68°

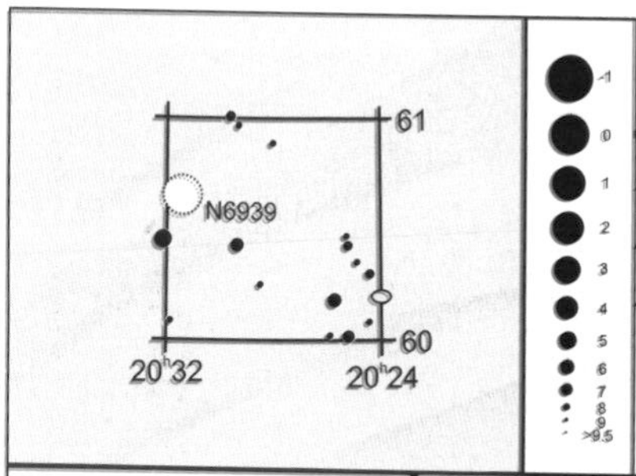

NGC 6939

RA:	$20^h\ 31^m\ 28.1^s$	Con:	Cepheus
Dec:	60° 38' 11"	Type:	Open Cluster
Size:	8.0'	Mag:	7.8

NGC 6939 is located in the far western part of the constellation. Cluster contains approximately 80 stars.

Telescope Aperture:	4" f/5	4" f/9	6" f/7	6" f/9	8" f/6.3	8" f/10	10" f/6.3	10" f/10	12" f/6.3	12" f/10
FOV(35mm film):	2.7° x 4.1°	1.50° x 2.26°	1.29° x 1.93°	1.0° x 1.50°	1.07° x 1.61°	0.68° x 1.02°	0.86° x 1.29°	0.54° x 0.81°	0.72° x 1.07°	0.45° x 0.68°

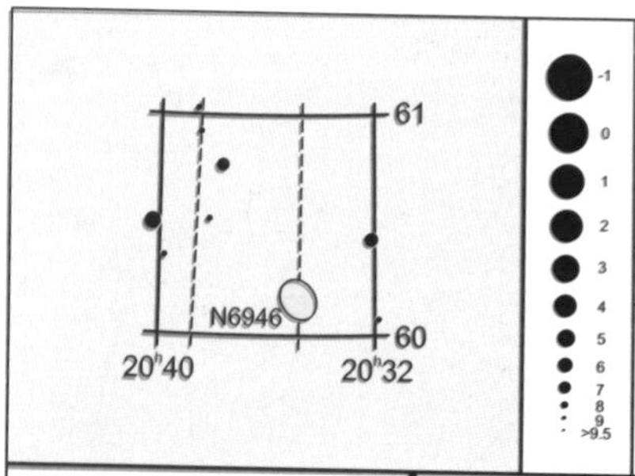

NGC 6946

RA:	$20^h\ 34^m\ 52.1^s$	Con:	Cepheus
Dec:	60° 09' 12"	Type:	Spiral Galaxy
Size:	9.0'	Mag:	8.9

NGC 6946 is a spiral galaxy that is face-on to our line of sight.

Telescope Aperture:	4" f/5	4" f/9	6" f/7	6" f/9	8" f/6.3	8" f/10	10" f/6.3	10" f/10	12" f/6.3	12" f/10
FOV(35mm film):	2.7° x 4.1°	1.50° x 2.26°	1.29° x 1.93°	1.0° x 1.50°	1.07° x 1.61°	0.68° x 1.02°	0.86° x 1.29°	0.54° x 0.81°	0.72° x 1.07°	0.45° x 0.68°

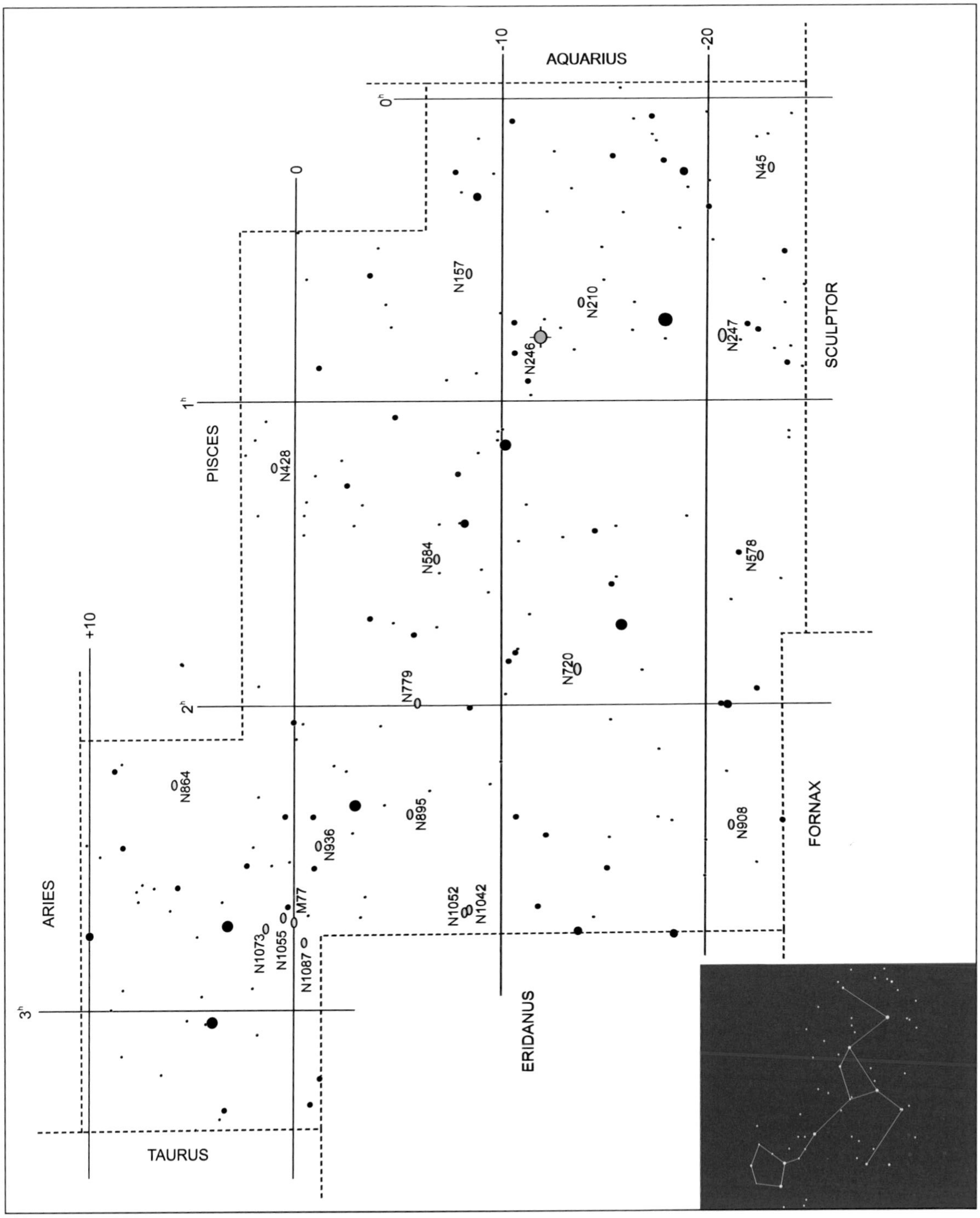

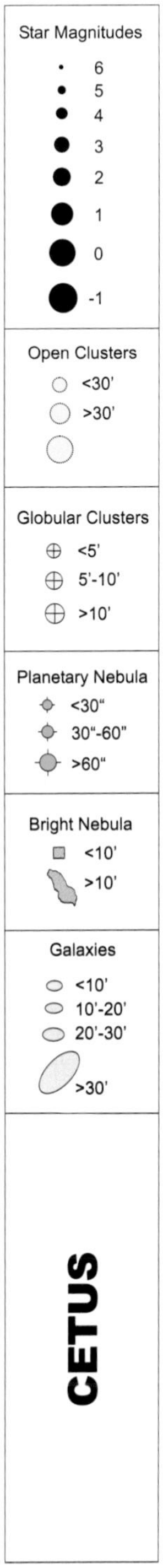

Constellation Facts:

Cetus; (SEE-tus)

Cetus, the Whale.
Cetus is an equatorial constellation, it rises in the east, crosses the meridian halfway between the horizon and the zenith, and sets in the west.
Cetus encompasses a large portion of the sky. •It is comprised of faint stars.
The constellation covers 1231 square degrees.

Constellation is visible from 65° N to 79° S. Partially visible from 65° N to 90° N.

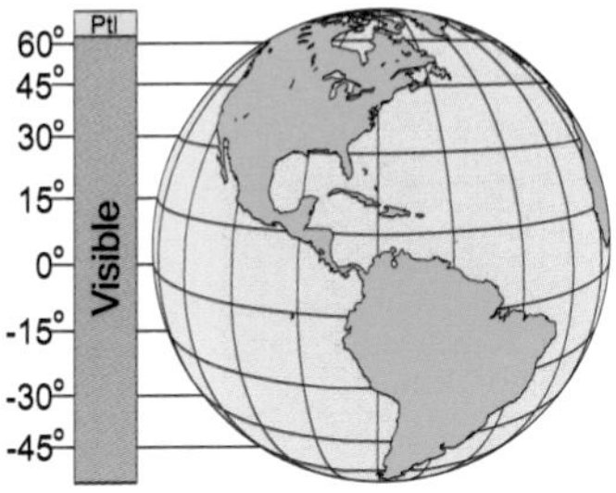

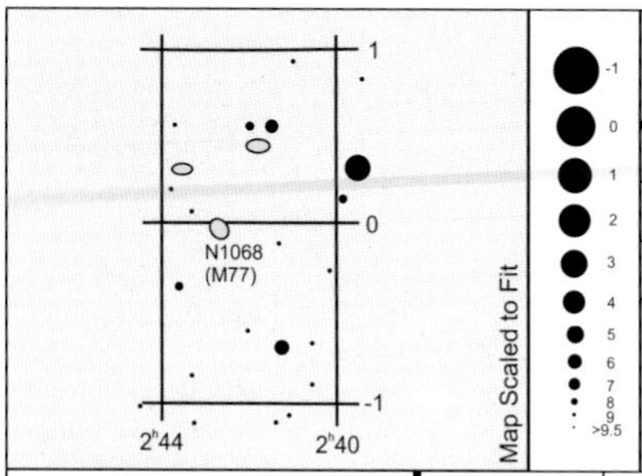

M77 (NGC 1068)

RA:	$02^h\ 42^m\ 45.2^s$	Con:	Cetus
Dec:	-00° 00' 40"	Type:	Seyfert Galaxy
Size:	7.0'	Mag:	8.8

M77 (NGC 1068) is found about 1° southeast of Delta Ceti. Object is a Seyfert Galaxy and is classed as an Sb-type spiral galaxy.

Telescope Aperture:	4" f/5	4" f/9	6" f/7	6" f/9	8" f/6.3	8" f/10	10" f/6.3	10" f/10	12" f/6.3	12" f/10
FOV(35mm film):	2.7° x 4.1°	1.50° x 2.26°	1.29° x 1.93°	1.0° x 1.50°	1.07° x 1.61°	0.68° x 1.02°	0.86° x 1.29°	0.54° x 0.81°	0.72° x 1.07°	0.45° x 0.68°

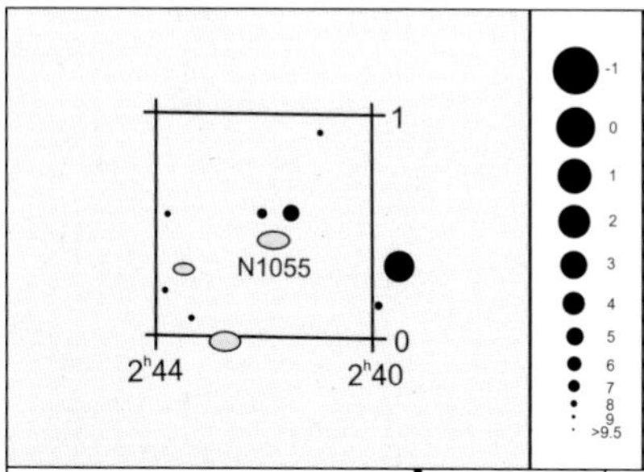

NGC 1055

RA:	$02^h\ 41^m\ 51.2^s$	Con:	Cetus
Dec:	00° 26' 20"	Type:	Spiral Galaxy
Size:	7.0' x 2.5'	Mag:	10.6

NGC 1055 is an edge-on galaxy that is classed as an Sb-type spiral.

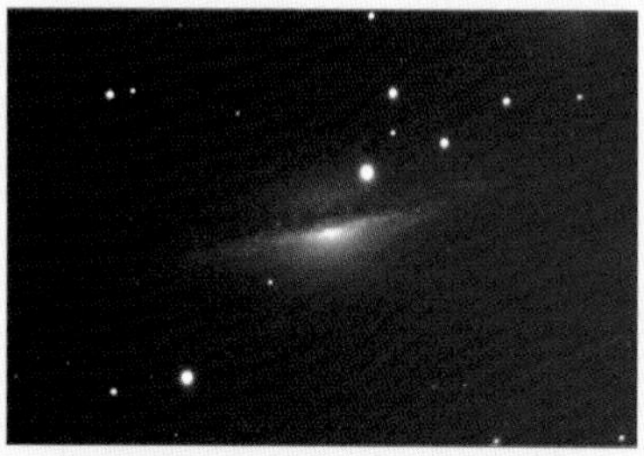

Telescope Aperture:	4" f/5	4" f/9	6" f/7	6" f/9	8" f/6.3	8" f/10	10" f/6.3	10" f/10	12" f/6.3	12" f/10
FOV(35mm film):	2.7° x 4.1°	1.50° x 2.26°	1.29° x 1.93°	1.0° x 1.50°	1.07° x 1.61°	0.68° x 1.02°	0.86° x 1.29°	0.54° x 0.81°	0.72° x 1.07°	0.45° x 0.68°

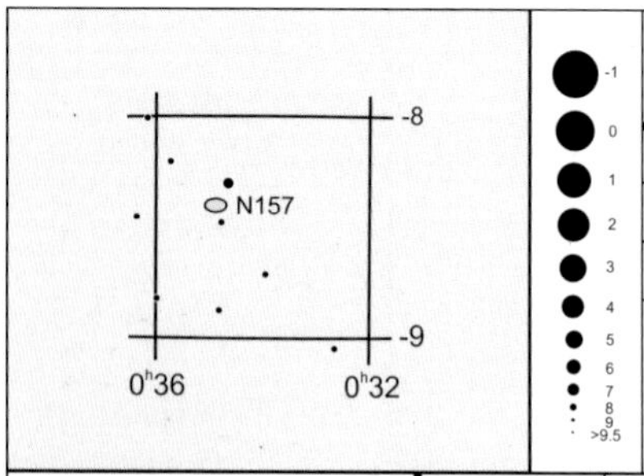

NGC 157

RA:	$00^h\ 34^m\ 51.8^s$	Con:	Cetus
Dec:	-08° 23' 31"	Type:	Spiral Galaxy
Size:	4.5' x 2.7'	Mag:	10.4

NGC 157 is a small spiral galaxy found south of IC 1613. Object displays many arms and is classed as an Sc-type spiral.

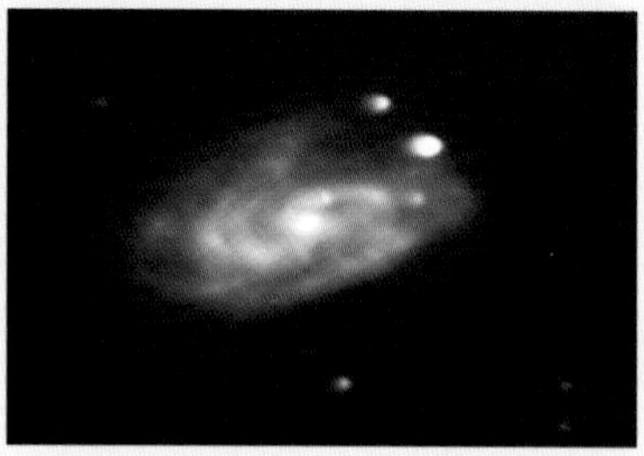

Telescope Aperture:	4" f/5	4" f/9	6" f/7	6" f/9	8" f/6.3	8" f/10	10" f/6.3	10" f/10	12" f/6.3	12" f/10
FOV(35mm film):	2.7° x 4.1°	1.50° x 2.26°	1.29° x 1.93°	1.0° x 1.50°	1.07° x 1.61°	0.68° x 1.02°	0.86° x 1.29°	0.54° x 0.81°	0.72° x 1.07°	0.45° x 0.68°

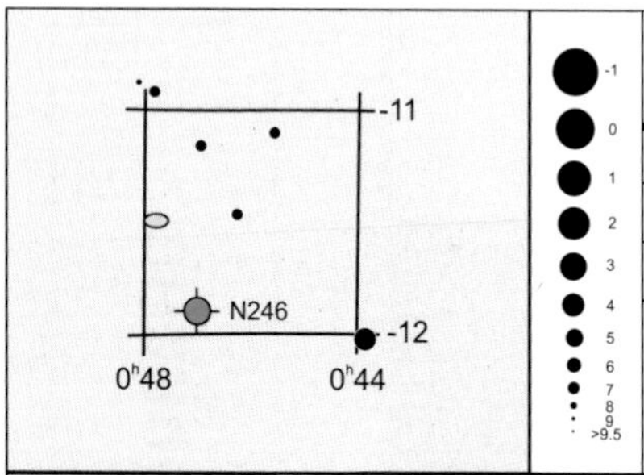

NGC 246

RA:	$00^h\ 47^m\ 5.2^s$	Con:	Cetus
Dec:	-11° 52' 24"	Type:	Planetary Nebula
Size:	3.8'	Mag:	8.0

NGC 246 is located a few degrees southeast of NGC 157. The object is one of the largest planetary nebulae.

Telescope Aperture:	4" f/5	4" f/9	6" f/7	6" f/9	8" f/6.3	8" f/10	10" f/6.3	10" f/10	12" f/6.3	12" f/10
FOV(35mm film):	2.7° x 4.1°	1.50° x 2.26°	1.29° x 1.93°	1.0° x 1.50°	1.07° x 1.61°	0.68° x 1.02°	0.86° x 1.29°	0.54° x 0.81°	0.72° x 1.07°	0.45° x 0.68°

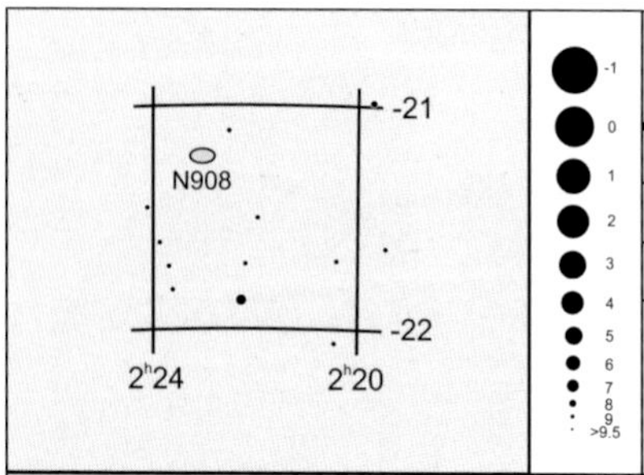

NGC 908

RA:	$02^h\ 23^m\ 8.9^s$	Con:	Cetus
Dec:	-21° 13' 31"	Type:	Spiral Galaxy
Size:	5.3' x 2.6'	Mag:	10.2

NGC 908 is found in the central region of the constellation. The object is a late type spiral galaxy which displays prominent and loose spiral arms.

Telescope Aperture:	4" f/5	4" f/9	6" f/7	6" f/9	8" f/6.3	8" f/10	10" f/6.3	10" f/10	12" f/6.3	12" f/10
FOV(35mm film):	2.7° x 4.1°	1.50° x 2.26°	1.29° x 1.93°	1.0° x 1.50°	1.07° x 1.61°	0.68° x 1.02°	0.86° x 1.29°	0.54° x 0.81°	0.72° x 1.07°	0.45° x 0.68°

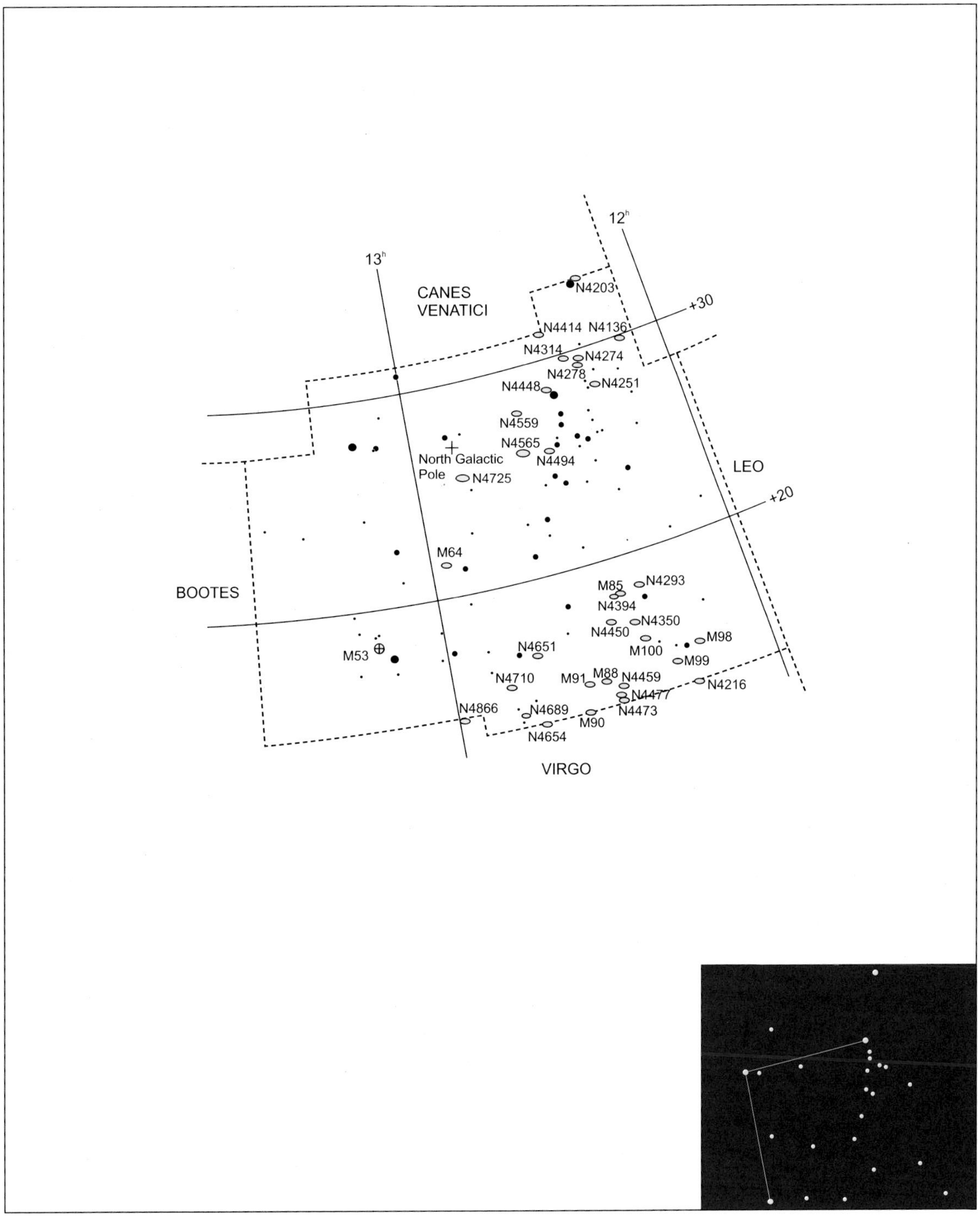

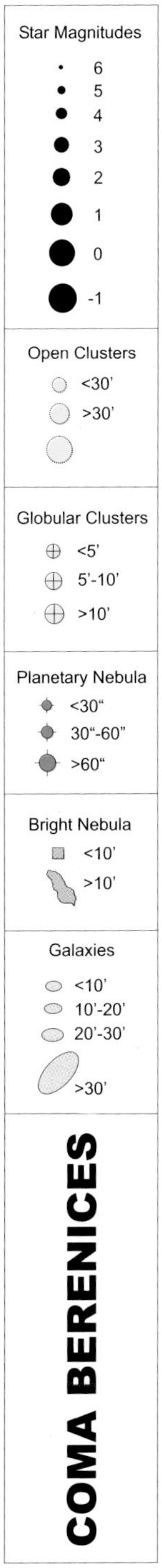

Constellation Facts:

Coma Berenices;

Coma Berenices, Berenices Hair.
Constellation covers 386 square degrees.

Constellation is visible from 90° N to 55° S. Partially visible from 55° S to 90° S.

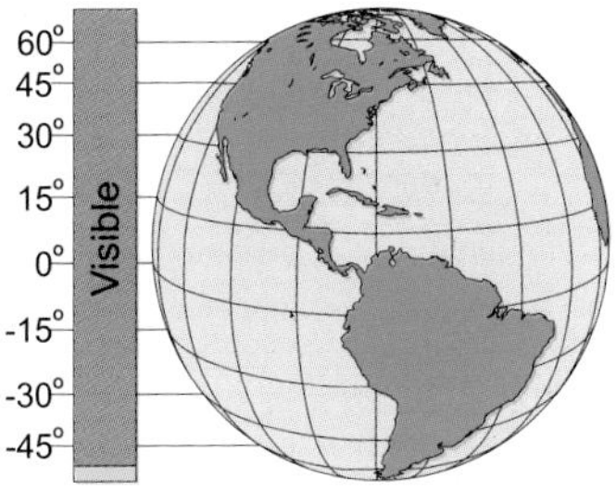

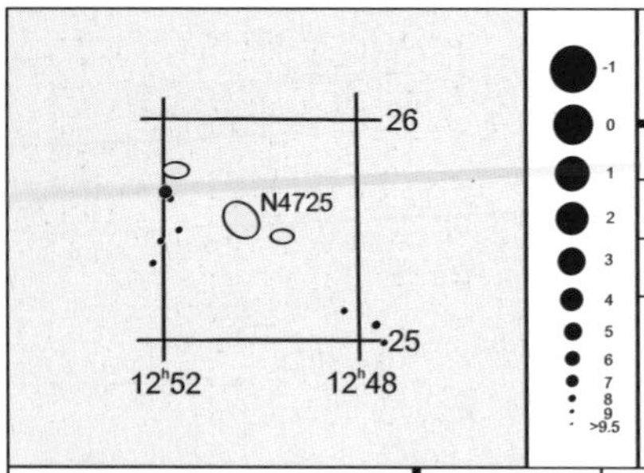

NGC 4725

RA:	$12^h\ 50^m\ 27.3^s$	Con:	Coma Berenices
Dec:	25° 29' 48"	Type:	Spiral Galaxy
Size:	11.0' x 6.0'	Mag:	9.2

NGC 4725 appears as an elongated galaxy with a bright core in the eyepiece.

Telescope Aperture:	4" f/5	4" f/9	6" f/7	6" f/9	8" f/6.3	8" f/10	10" f/6.3	10" f/10	12" f/6.3	12" f/10
FOV(35mm film):	2.7° x 4.1°	1.50° x 2.26°	1.29° x 1.93°	1.0° x 1.50°	1.07° x 1.61°	0.68° x 1.02°	0.86° x 1.29°	0.54° x 0.81°	0.72° x 1.07°	0.45° x 0.68°

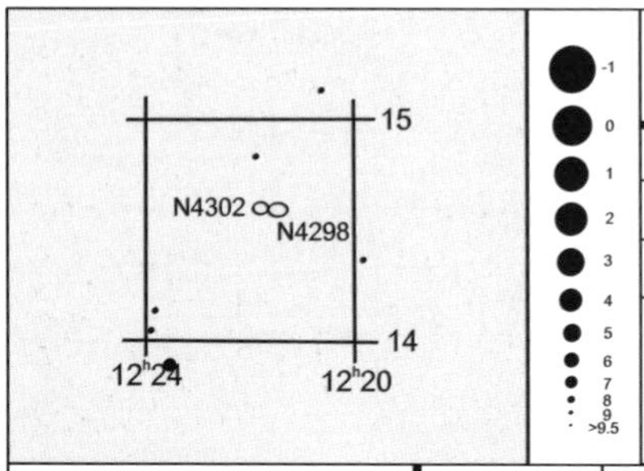

NGC 4298

RA:	$12^h\ 21^m\ 33.3^s$	Con:	Coma Berenices
Dec:	14° 35' 44"	Type:	Galaxy
Size:	3.0' x 1.5'	Mag:	11.4

NGC 4298 is a small bright galaxy. Object displays filamentary arms displaying dark lanes.

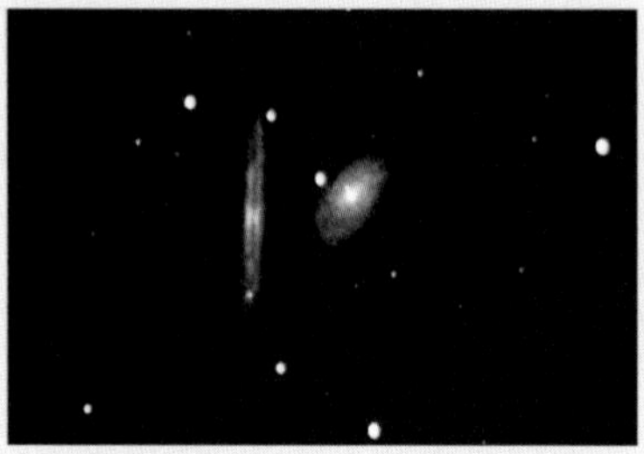

Telescope Aperture:	4" f/5	4" f/9	6" f/7	6" f/9	8" f/6.3	8" f/10	10" f/6.3	10" f/10	12" f/6.3	12" f/10
FOV(35mm film):	2.7° x 4.1°	1.50° x 2.26°	1.29° x 1.93°	1.0° x 1.50°	1.07° x 1.61°	0.68° x 1.02°	0.86° x 1.29°	0.54° x 0.81°	0.72° x 1.07°	0.45° x 0.68°

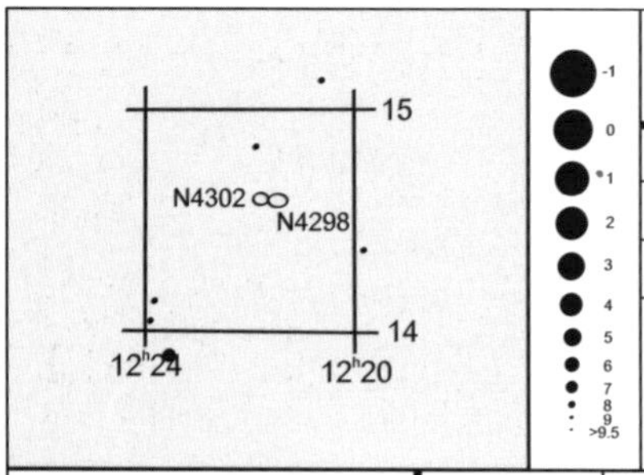

NGC 4302

RA:	$12^h\ 21^m\ 45.2^s$	Con:	Coma Berenices
Dec:	14° 35' 44"	Type:	Galaxy
Size:	5.0' x 0.6'	Mag:	11.6

NGC 4302 is one part of an interacting pair of galaxies.

Telescope Aperture:	4" f/5	4" f/9	6" f/7	6" f/9	8" f/6.3	8" f/10	10" f/6.3	10" f/10	12" f/6.3	12" f/10
FOV(35mm film):	2.7° x 4.1°	1.50° x 2.26°	1.29° x 1.93°	1.0° x 1.50°	1.07° x 1.61°	0.68° x 1.02°	0.86° x 1.29°	0.54° x 0.81°	0.72° x 1.07°	0.45° x 0.68°

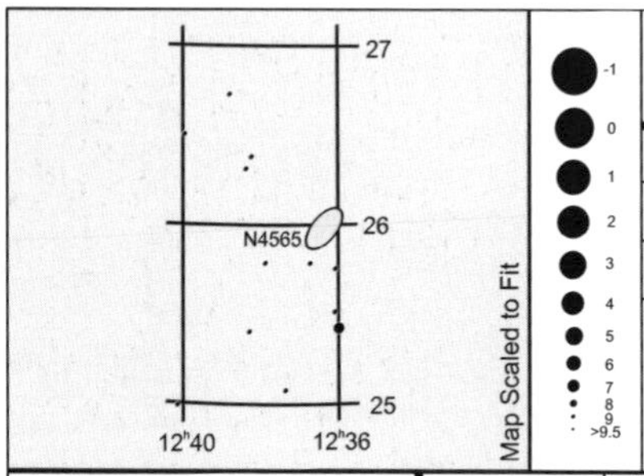

NGC 4565

RA:	$12^h\ 36^m\ 21.3^s$	Con:	Coma Berenices
Dec:	25° 58' 48"	Type:	Spiral Galaxy
Size:	12.0' x 1.5'	Mag:	9.6

NGC 4565 is located 3° from the north galactic pole. Object is the best example of an edge-on spiral galaxy.

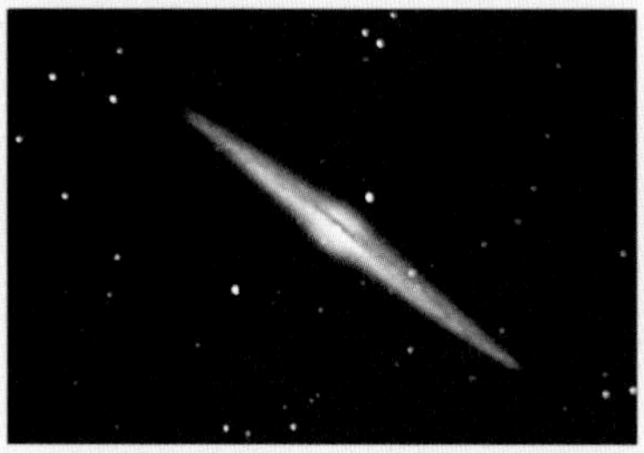

Telescope Aperture:	4" f/5	4" f/9	6" f/7	6" f/9	8" f/6.3	8" f/10	10" f/6.3	10" f/10	12" f/6.3	12" f/10
FOV(35mm film):	2.7° x 4.1°	1.50° x 2.26°	1.29° x 1.93°	1.0° x 1.50°	1.07° x 1.61°	0.68° x 1.02°	0.86° x 1.29°	0.54° x 0.81°	0.72° x 1.07°	0.45° x 0.68°

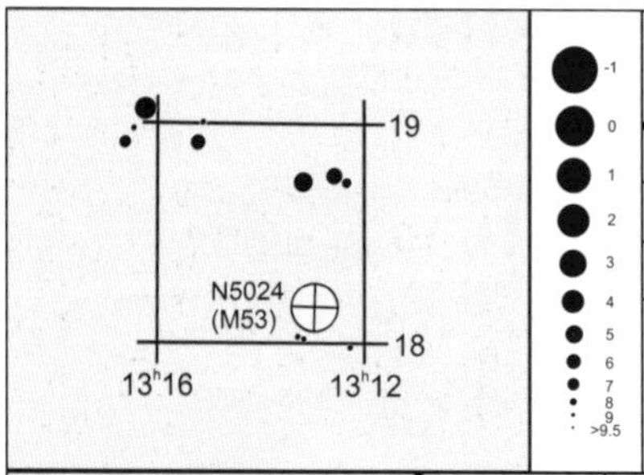

M53 (NGC 5024)

RA:	$13^h\ 12^m\ 57.5^s$	Con:	Coma Berenices
Dec:	18° 09' 46"	Type:	Globular Cluster
Size:	12.6'	Mag:	7.7

M53 is found in the eastern portion of the constellation. Object is 1° northwest of the binary star alpha Comae Berenices.

Telescope Aperture:	4" f/5	4" f/9	6" f/7	6" f/9	8" f/6.3	8" f/10	10" f/6.3	10" f/10	12" f/6.3	12" f/10
FOV(35mm film):	2.7° x 4.1°	1.50° x 2.26°	1.29° x 1.93°	1.0° x 1.50°	1.07° x 1.61°	0.68° x 1.02°	0.86° x 1.29°	0.54° x 0.81°	0.72° x 1.07°	0.45° x 0.68°

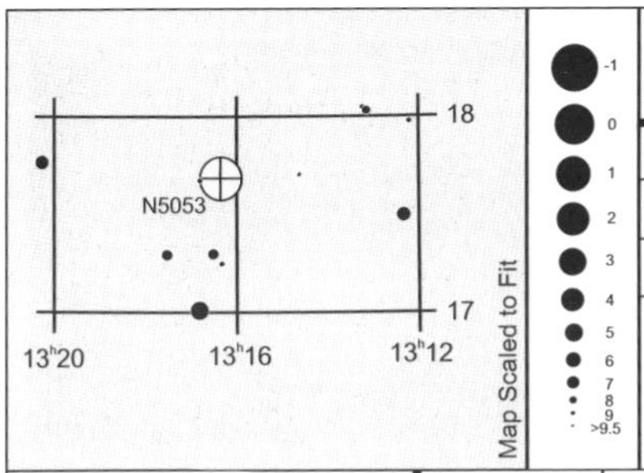

NGC 5053

RA:	$13^h 16^m 27.5^s$	Con:	Coma Berenices
Dec:	17° 41' 46"	Type:	Globular Cluster
Size:	10.5'	Mag:	9.8

NGC 5053 is located 1° southeast of M53 and is classified as a loose globular cluster. Object contains 3,400 stars.

Telescope Aperture:	4" f/5	4" f/9	6" f/7	6" f/9	8" f/6.3	8" f/10	10" f/6.3	10" f/10	12" f/6.3	12" f/10
FOV(35mm film):	2.7°x 4.1°	1.50°x 2.26°	1.29° x 1.93°	1.0° x 1.50°	1.07° x 1.61°	0.68°x 1.02°	0.86° x 1.29°	0.54° x 0.81°	0.72° x 1.07°	0.45°x 0.68°

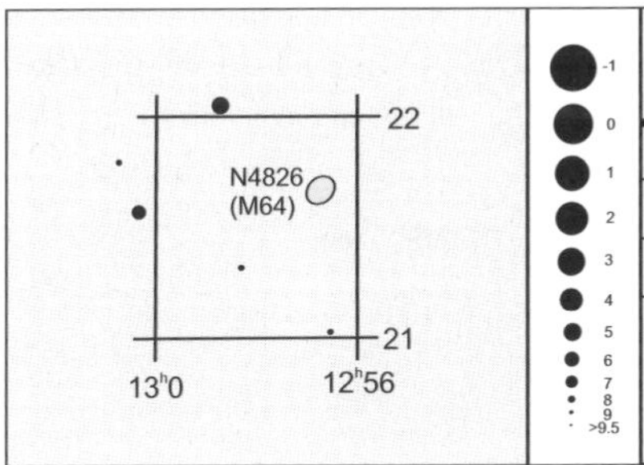

M64 (NGC 4826) "Black-eye Galaxy"

RA:	$12^h 56^m 45.4^s$	Con:	Coma Berenices
Dec:	21° 40' 47"	Type:	Spiral Galaxy
Size:	10.0' x 3.8'	Mag:	8.5

M64 received its name because of a dust patch superimposed across its face. Object is a type Sb galaxy.

Telescope Aperture:	4" f/5	4" f/9	6" f/7	6" f/9	8" f/6.3	8" f/10	10" f/6.3	10" f/10	12" f/6.3	12" f/10
FOV(35mm film):	2.7°x 4.1°	1.50°x 2.26°	1.29° x 1.93°	1.0° x 1.50°	1.07° x 1.61°	0.68°x 1.02°	0.86° x 1.29°	0.54° x 0.81°	0.72° x 1.07°	0.45°x 0.68°

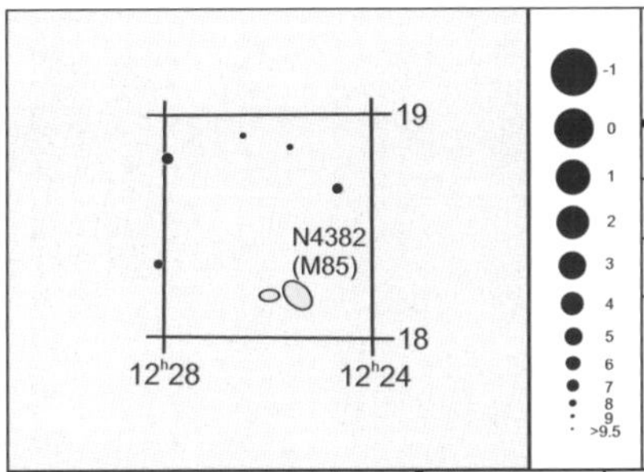

M85 (NGC 4382)

RA:	$12^h 25^m 27.3^s$	Con:	Coma Berenices
Dec:	18° 10' 45"	Type:	Lenticular Galaxy
Size:	7.5' x 6.5'	Mag:	9.2

M85 is located near the border with Virgo, about 5° north of the Virgo clusters heart. The object is a lenticular galaxy.

Telescope Aperture:	4" f/5	4" f/9	6" f/7	6" f/9	8" f/6.3	8" f/10	10" f/6.3	10" f/10	12" f/6.3	12" f/10
FOV(35mm film):	2.7°x 4.1°	1.50°x 2.26°	1.29° x 1.93°	1.0° x 1.50°	1.07° x 1.61°	0.68°x 1.02°	0.86° x 1.29°	0.54° x 0.81°	0.72° x 1.07°	0.45°x 0.68°

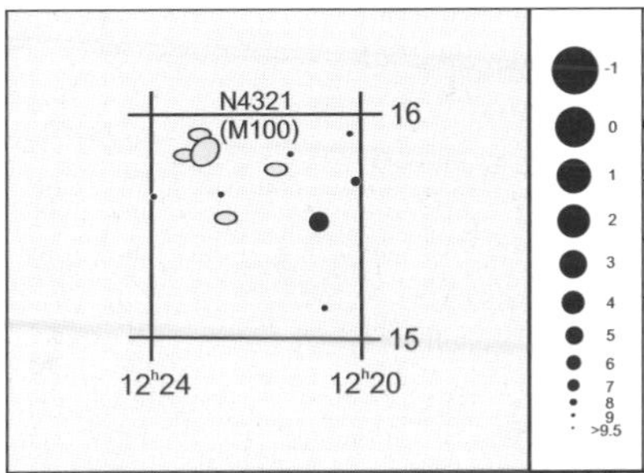

M100 (NGC 4321)

RA:	$12^h 22^m 57.3^s$	Con:	Coma Berenices
Dec:	15° 48' 44"	Type:	Spiral Galaxy
Size:	8.8' x 6.0'	Mag:	9.4

M100 is a loose Sc-type spiral galaxy. It is the largest spiral galaxy in the Virgo cluster.

Telescope Aperture:	4" f/5	4" f/9	6" f/7	6" f/9	8" f/6.3	8" f/10	10" f/6.3	10" f/10	12" f/6.3	12" f/10
FOV(35mm film):	2.7°x 4.1°	1.50°x 2.26°	1.29° x 1.93°	1.0° x 1.50°	1.07° x 1.61°	0.68°x 1.02°	0.86° x 1.29°	0.54° x 0.81°	0.72° x 1.07°	0.45°x 0.68°

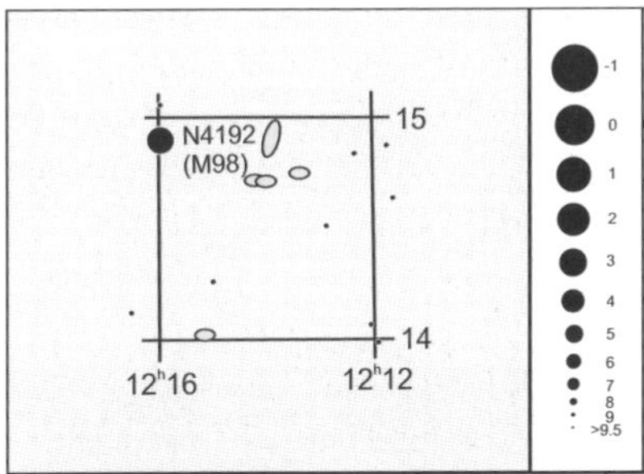

M98 (NGC 5024)

RA:	$12^h 13^m 51.3^s$	Con:	Coma Berenices
Dec:	14° 53' 44"	Type:	Spiral Galaxy
Size:	10.5' x 2.6'	Mag:	10.1

M98 is an elongated nearly edge-on Sb-type spiral galaxy.

Telescope Aperture:	4" f/5	4" f/9	6" f/7	6" f/9	8" f/6.3	8" f/10	10" f/6.3	10" f/10	12" f/6.3	12" f/10
FOV(35mm film):	2.7°x 4.1°	1.50°x 2.26°	1.29° x 1.93°	1.0° x 1.50°	1.07° x 1.61°	0.68°x 1.02°	0.86° x 1.29°	0.54° x 0.81°	0.72° x 1.07°	0.45°x 0.68°

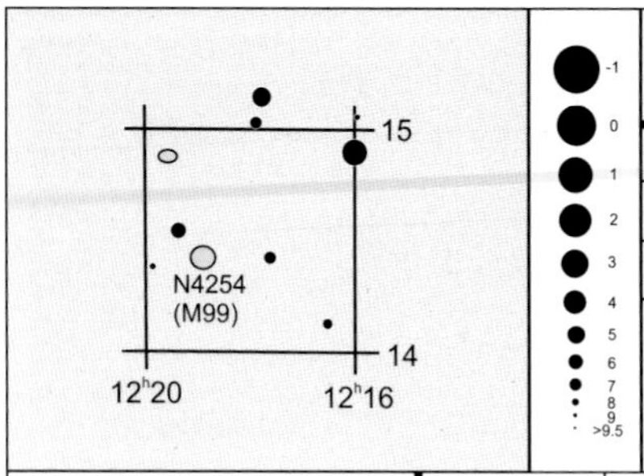

M99 (NGC 4254) "Pin-Wheel Galaxy"

RA:	12h 18m 51.3s	Con:	Coma Berenices
Dec:	14° 24' 44"	Type:	Spiral Galaxy
Size:	6.0' x 5.0'	Mag:	9.8

M99 is a face-on Sc-type spiral galaxy.

Telescope Aperture:	4" f/5	4" f/9	6" f/7	6" f/9	8" f/6.3	8" f/10	10" f/6.3	10" f/10	12" f/6.3	12" f/10
FOV(35mm film):	2.7° x 4.1°	1.50° x 2.26°	1.29° x 1.93°	1.0° x 1.50°	1.07° x 1.61°	0.68° x 1.02°	0.86° x 1.29°	0.54° x 0.81°	0.72° x 1.07°	0.45° x 0.68°

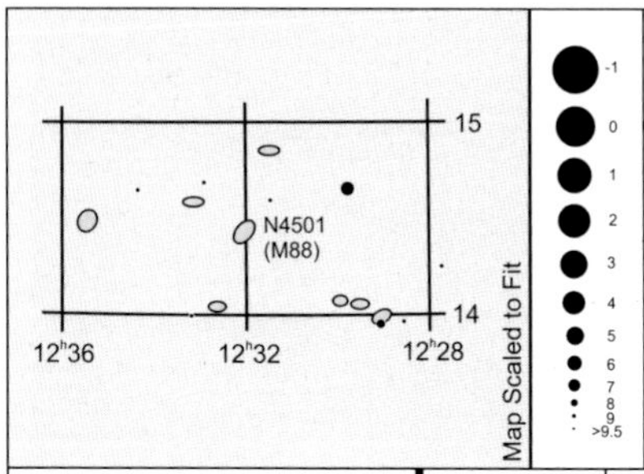

M88 (NGC 4501)

RA:	12h 32m 3.4s	Con:	Coma Berenices
Dec:	14° 24' 44"	Type:	Spiral Galaxy
Size:	6.0' x 3.0'	Mag:	9.5

M88 is a multiple arm spiral galaxy. Object is inclined 30° from edge-on to our line of sight.

Telescope Aperture:	4" f/5	4" f/9	6" f/7	6" f/9	8" f/6.3	8" f/10	10" f/6.3	10" f/10	12" f/6.3	12" f/10
FOV(35mm film):	2.7° x 4.1°	1.50° x 2.26°	1.29° x 1.93°	1.0° x 1.50°	1.07° x 1.61°	0.68° x 1.02°	0.86° x 1.29°	0.54° x 0.81°	0.72° x 1.07°	0.45° x 0.68°

Crosses Prime Meridian: April thru May

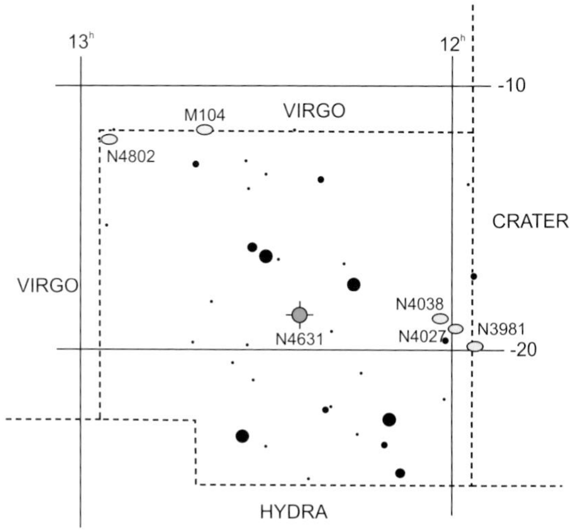

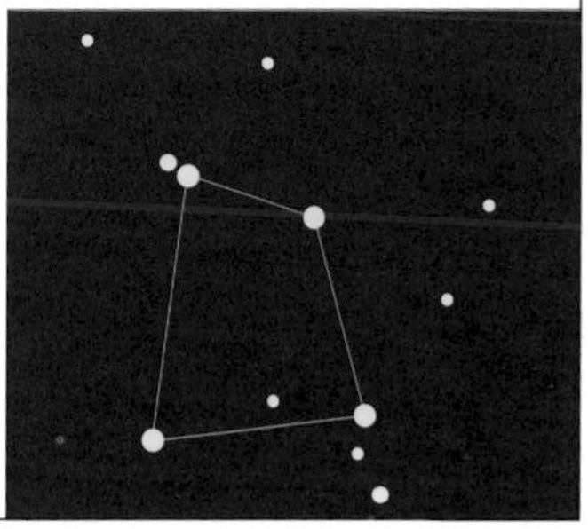

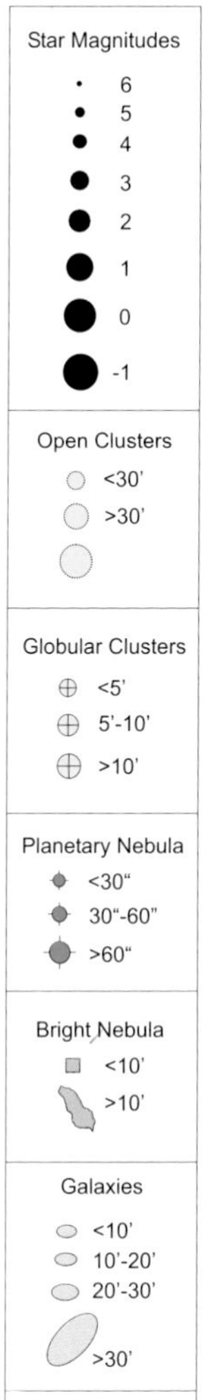

CORVUS

Constellation Facts:

Corvus; (CORE-VUSS)

Corvus, the Crow.
Corvus follows a southerly sky track, rising in the southeast and setting in the southwest.
Marked only by 3rd magnitude stars, the compact figure of Corvus is an interesting sight in an otherwise sparse region of the sky.
Corvus covers 184 square degrees.

Constellation is visible from 65° N to 90° S. Partially visible from 65° N to 80° N.

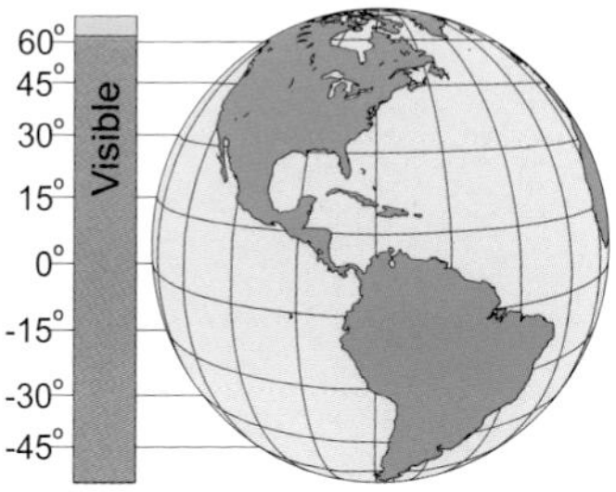

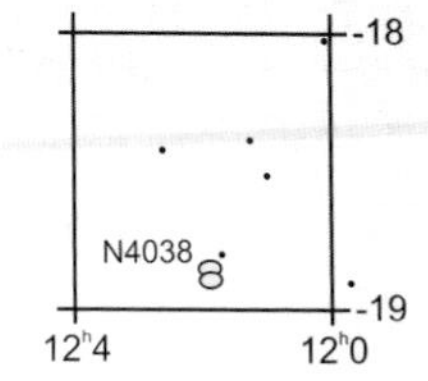

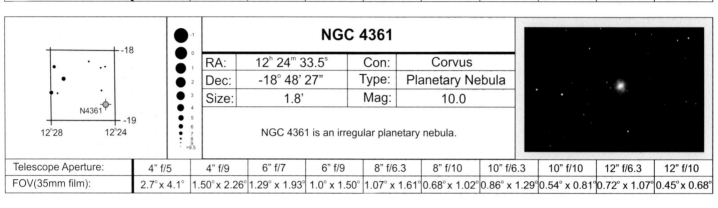

NGC 4038 "Antennae Galaxy"

RA:	12h 01m 57.3s	Con:	Corvus
Dec:	-18° 52' 27"	Type:	Interacting Galaxy
Size:	1.7' x 1.2'	Mag:	10.7

NGC 4038 is an interacting galaxy with NGC 4039.

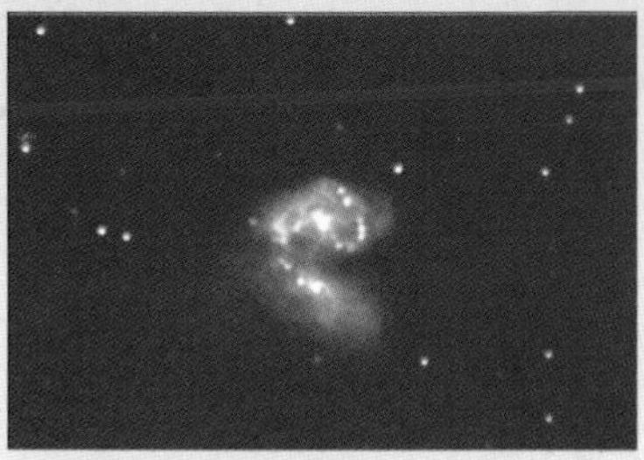

Telescope Aperture:	4" f/5	4" f/9	6" f/7	6" f/9	8" f/6.3	8" f/10	10" f/6.3	10" f/10	12" f/6.3	12" f/10
FOV(35mm film):	2.7° x 4.1°	1.50° x 2.26°	1.29° x 1.93°	1.0° x 1.50°	1.07° x 1.61°	0.68° x 1.02°	0.86° x 1.29°	0.54° x 0.81°	0.72° x 1.07°	0.45° x 0.68°

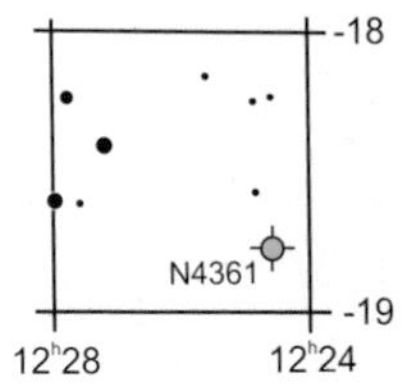

NGC 4361

RA:	12h 24m 33.5s	Con:	Corvus
Dec:	-18° 48' 27"	Type:	Planetary Nebula
Size:	1.8'	Mag:	10.0

NGC 4361 is an irregular planetary nebula.

Telescope Aperture:	4" f/5	4" f/9	6" f/7	6" f/9	8" f/6.3	8" f/10	10" f/6.3	10" f/10	12" f/6.3	12" f/10
FOV(35mm film):	2.7° x 4.1°	1.50° x 2.26°	1.29° x 1.93°	1.0° x 1.50°	1.07° x 1.61°	0.68° x 1.02°	0.86° x 1.29°	0.54° x 0.81°	0.72° x 1.07°	0.45° x 0.68°

Cyg

Crosses Prime Meridian: August thru September

SUMMER

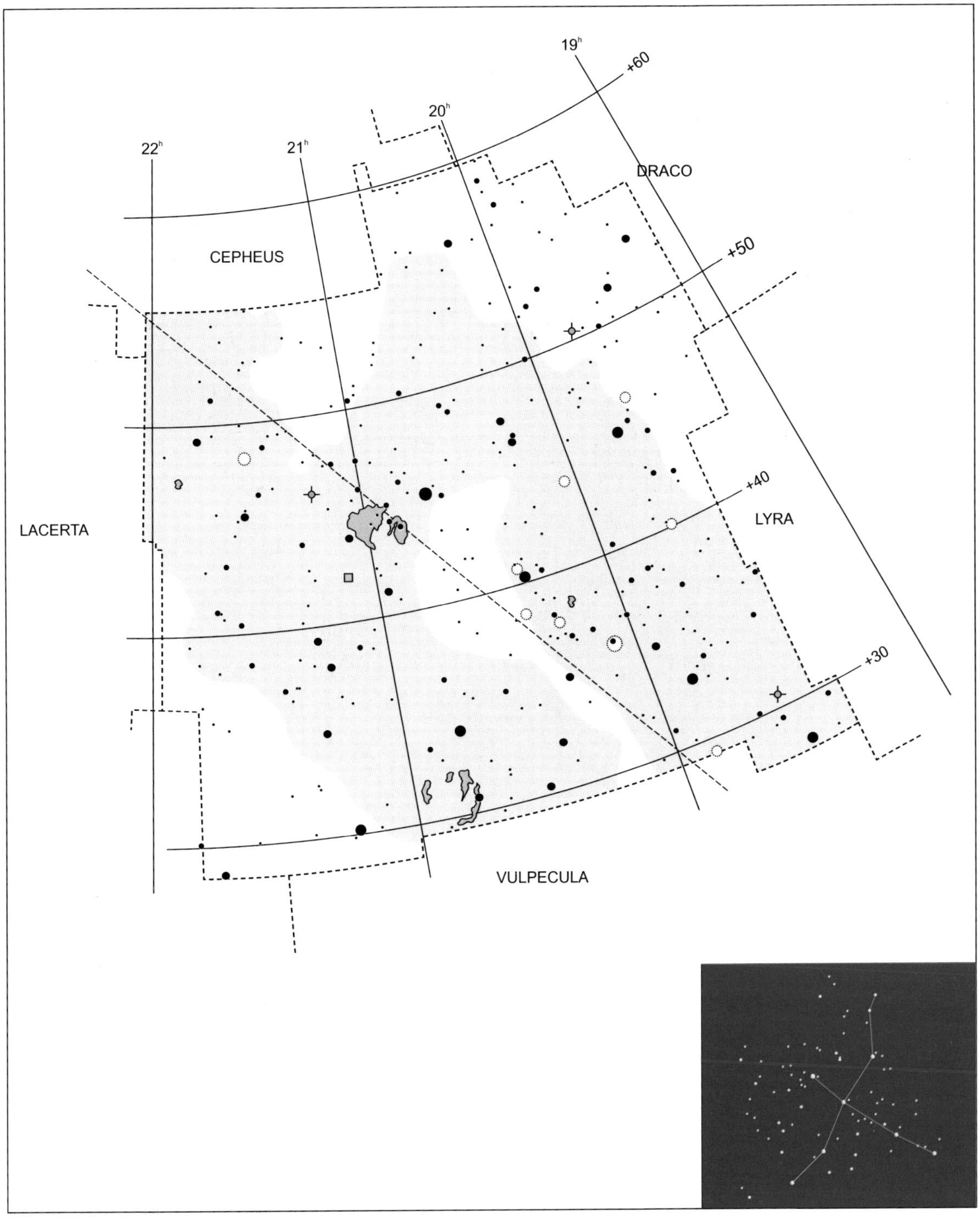

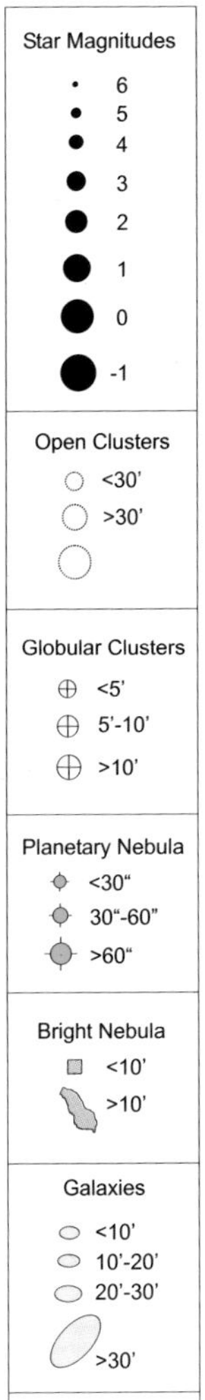

CYGNUS

Constellation Facts:

Cygnus; (SIG-nus)

Cygnus, the Swan;
is visible from June thru January.
The constellation rises in the northeast, crosses the meridian and sets towards the northwest.
The stars of Cygnus are part of an attractive asterism known as the Northern Cross. The bird is portrayed flying towards the southwest, directly along the centerline of the Milky Way.
The constellation covers 804 square degrees.

Constellation is visible from 90° N to 28° S. Partially visible from 28° S to 90° S.

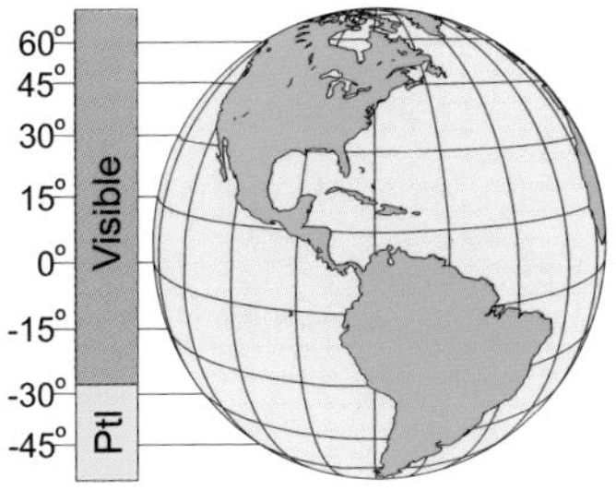

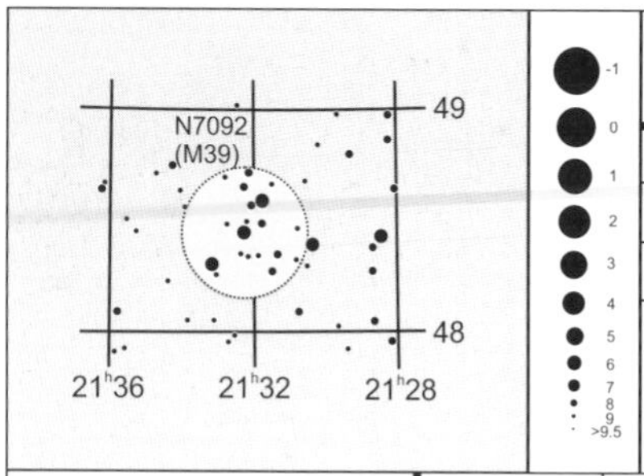

M39 (NGC 7092)

RA:	21h 32m 16.3s	Con:	Cygnus
Dec:	48° 26' 14"	Type:	Open Cluster
Size:	32.0'	Mag:	4.6

M39 (NGC 7092) is a large scattered open cluster. Object contains at least 28 stars.

Telescope Aperture:	4" f/5	4" f/9	6" f/7	6" f/9	8" f/6.3	8" f/10	10" f/6.3	10" f/10	12" f/6.3	12" f/10
FOV(35mm film):	2.7° x 4.1°	1.50° x 2.26°	1.29° x 1.93°	1.0° x 1.50°	1.07° x 1.61°	0.68° x 1.02°	0.86° x 1.29°	0.54° x 0.81°	0.72° x 1.07°	0.45° x 0.68°

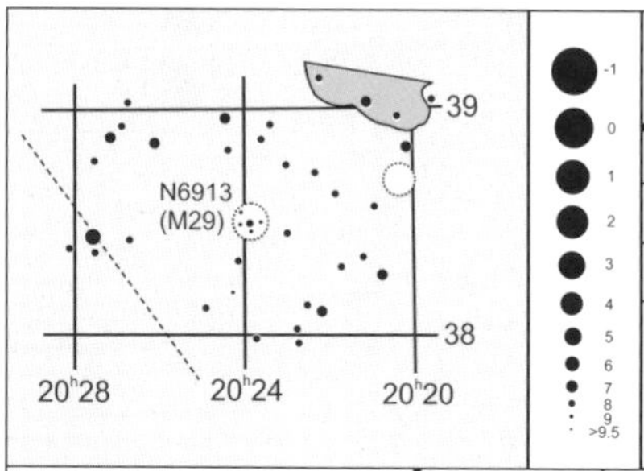

M29 (NGC 6913)

RA:	20h 23m 54.0s	Con:	Cygnus
Dec:	38° 32' 14"	Type:	Open Cluster
Size:	7.0'	Mag:	6.6

M29 (NGC 6913) is an unimpressive open cluster. Object contains 81 stars of 9th magnitude or fainter.

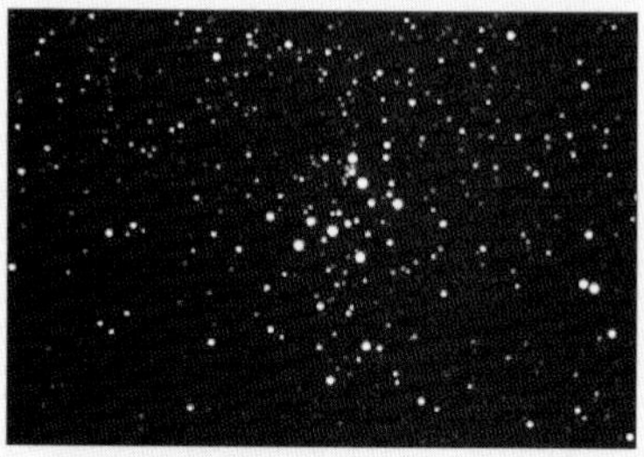

Telescope Aperture:	4" f/5	4" f/9	6" f/7	6" f/9	8" f/6.3	8" f/10	10" f/6.3	10" f/10	12" f/6.3	12" f/10
FOV(35mm film):	2.7° x 4.1°	1.50° x 2.26°	1.29° x 1.93°	1.0° x 1.50°	1.07° x 1.61°	0.68° x 1.02°	0.86° x 1.29°	0.54° x 0.81°	0.72° x 1.07°	0.45° x 0.68°

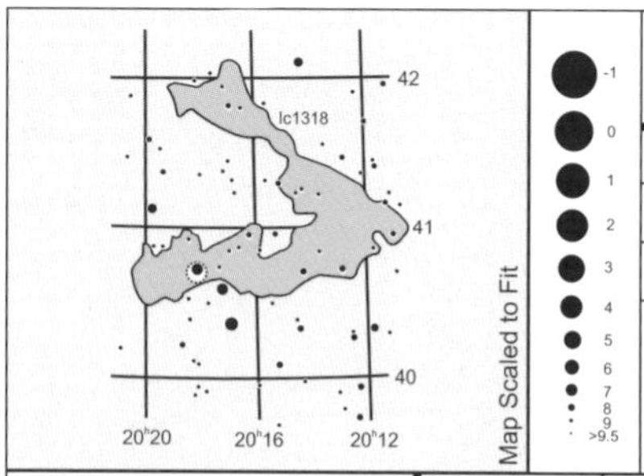

IC 1318

RA:	20h 22m 16.3s	Con:	Cygnus
Dec:	40° 15' 14"	Type:	Emission Nebula
Size:	240.0'	Mag:	

IC 1318 is a large region of emission nebula. Object is separated into five distinct regions by dark nebula.

Telescope Aperture:	4" f/5	4" f/9	6" f/7	6" f/9	8" f/6.3	8" f/10	10" f/6.3	10" f/10	12" f/6.3	12" f/10
FOV(35mm film):	2.7° x 4.1°	1.50° x 2.26°	1.29° x 1.93°	1.0° x 1.50°	1.07° x 1.61°	0.68° x 1.02°	0.86° x 1.29°	0.54° x 0.81°	0.72° x 1.07°	0.45° x 0.68°

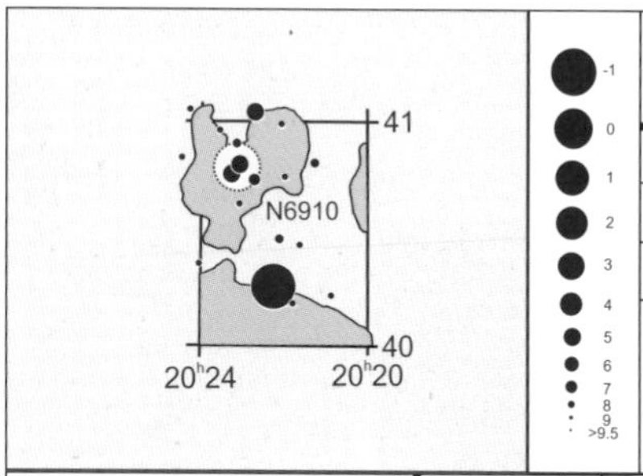

NGC 6910

RA:	20h 23m 10.3s	Con:	Cygnus
Dec:	40° 47' 14"	Type:	Open Cluster
Size:	8.0'	Mag:	7.4

NGC 6910 is a bright, scattered open cluster embedded in the nebula complex of IC 1318. Object consists of 66 stars.

Telescope Aperture:	4" f/5	4" f/9	6" f/7	6" f/9	8" f/6.3	8" f/10	10" f/6.3	10" f/10	12" f/6.3	12" f/10
FOV(35mm film):	2.7° x 4.1°	1.50° x 2.26°	1.29° x 1.93°	1.0° x 1.50°	1.07° x 1.61°	0.68° x 1.02°	0.86° x 1.29°	0.54° x 0.81°	0.72° x 1.07°	0.45° x 0.68°

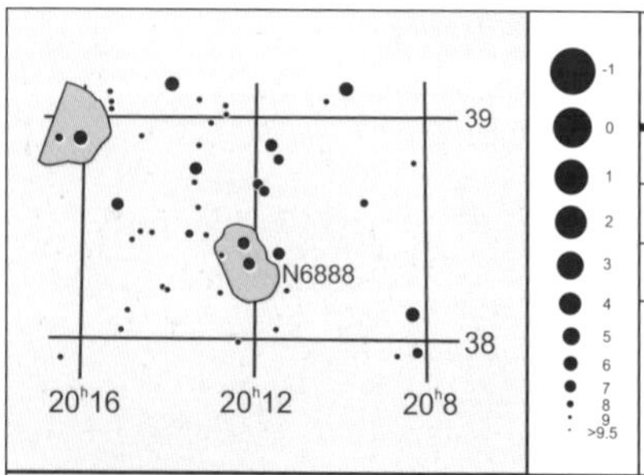

NGC 6888 "Crescent Nebula"

RA:	20h 12m 4.3s	Con:	Cygnus
Dec:	38° 21' 13"	Type:	Nebula
Size:	20.0'	Mag:	14.0

NGC 6888 the "Crescent Nebula" is a faint supernova remnant. It is a young shell of ionized gas that is expanding outward from a host central Wolf-Rayet star.

Telescope Aperture:	4" f/5	4" f/9	6" f/7	6" f/9	8" f/6.3	8" f/10	10" f/6.3	10" f/10	12" f/6.3	12" f/10
FOV(35mm film):	2.7° x 4.1°	1.50° x 2.26°	1.29° x 1.93°	1.0° x 1.50°	1.07° x 1.61°	0.68° x 1.02°	0.86° x 1.29°	0.54° x 0.81°	0.72° x 1.07°	0.45° x 0.68°

NGC 6871

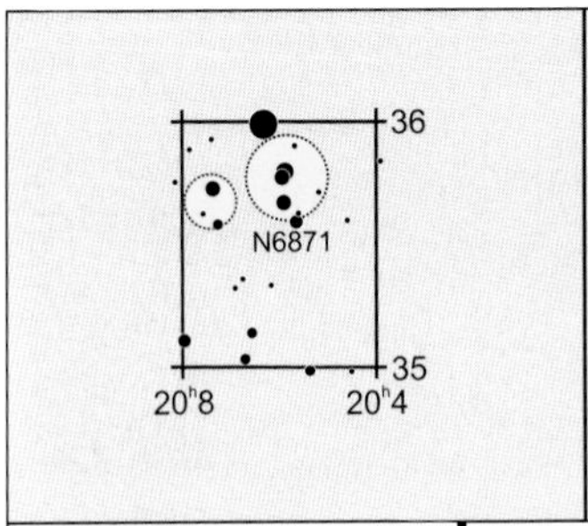

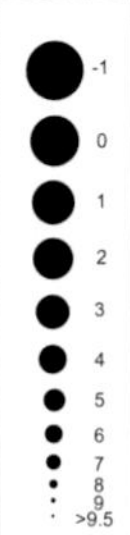

RA:	$20^h\ 05^m\ 58.3^s$	Con:	Cygnus
Dec:	35° 47' 13"	Type:	Open Cluster
Size:	20.0'	Mag:	5.2

NGC 6871 is a rich open cluster found in the central region of Cygnus.

Telescope Aperture:	4" f/5	4" f/9	6" f/7	6" f/9	8" f/6.3	8" f/10	10" f/6.3	10" f/10	12" f/6.3	12" f/10
FOV(35mm film):	2.7°x 4.1°	1.50°x 2.26°	1.29° x 1.93°	1.0° x 1.50°	1.07° x 1.61°	0.68°x 1.02°	0.86° x 1.29°	0.54° x 0.81°	0.72° x 1.07°	0.45°x 0.68°

NGC 7000 "North American Nebula"

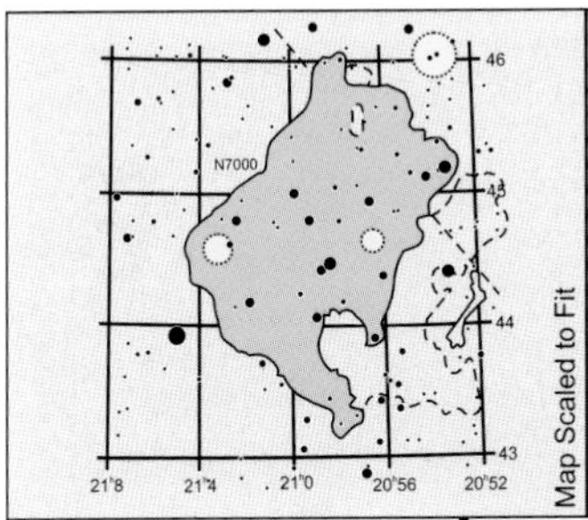

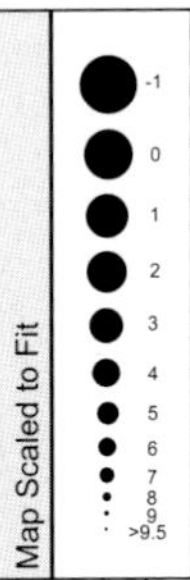

RA:	$20^h\ 58^m\ 52.3^s$	Con:	Cygnus
Dec:	44° 20' 14"	Type:	Emission Nebula
Size:	120.0'	Mag:	4.5

NGC 7000 the "North American Nebula" is a large emission nebula found just southeast of Deneb.

Telescope Aperture:	4" f/5	4" f/9	6" f/7	6" f/9	8" f/6.3	8" f/10	10" f/6.3	10" f/10	12" f/6.3	12" f/10
FOV(35mm film):	2.7°x 4.1°	1.50°x 2.26°	1.29° x 1.93°	1.0° x 1.50°	1.07° x 1.61°	0.68°x 1.02°	0.86° x 1.29°	0.54° x 0.81°	0.72° x 1.07°	0.45°x 0.68°

IC 5067/70 "Pelican Nebula"

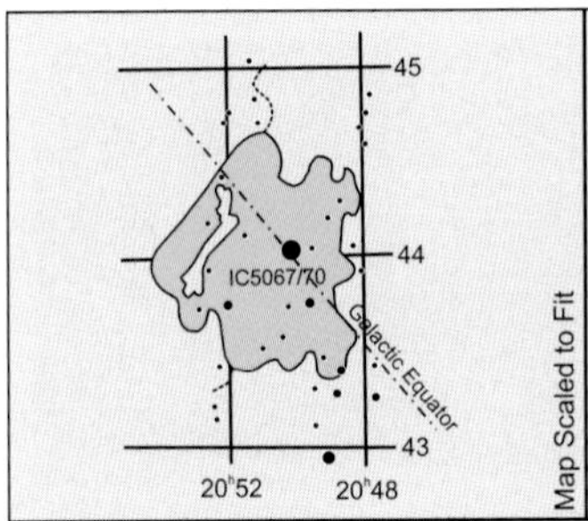

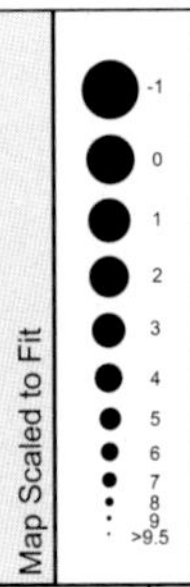

RA:	$20^h\ 50^m\ 52.3^s$	Con:	Cygnus
Dec:	44° 21' 14"	Type:	Emission Nebula
Size:	80.0'	Mag:	14.0

IC 5067/70 the "Pelican Nebula" is found due west of NGC 7000. Object is separated from NGC 7000 by a broad band of dust.

Telescope Aperture:	4" f/5	4" f/9	6" f/7	6" f/9	8" f/6.3	8" f/10	10" f/6.3	10" f/10	12" f/6.3	12" f/10
FOV(35mm film):	2.7°x 4.1°	1.50°x 2.26°	1.29° x 1.93°	1.0° x 1.50°	1.07° x 1.61°	0.68°x 1.02°	0.86° x 1.29°	0.54° x 0.81°	0.72° x 1.07°	0.45°x 0.68°

NGC 7027

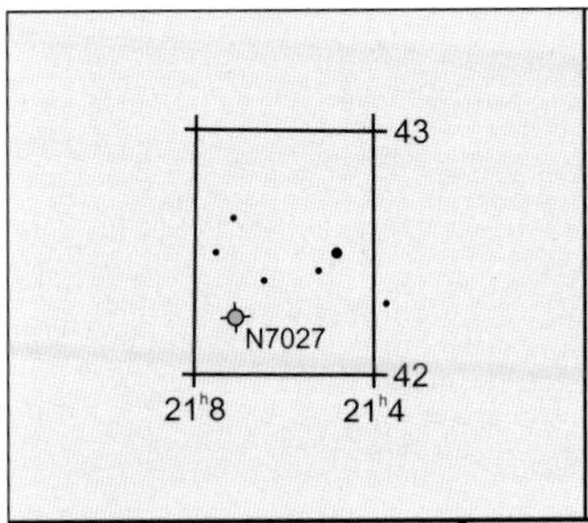

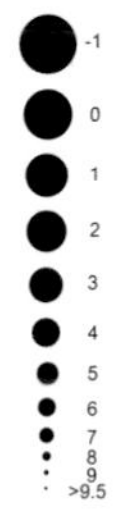

RA:	$21^h\ 07^m\ 10.3^s$	Con:	Cygnus
Dec:	42° 14' 14"	Type:	Planetary Nebula
Size:	0.3'	Mag:	10.0

NGC 7027 is a small irregular planetary nebula. Object is located near NGC 7000.

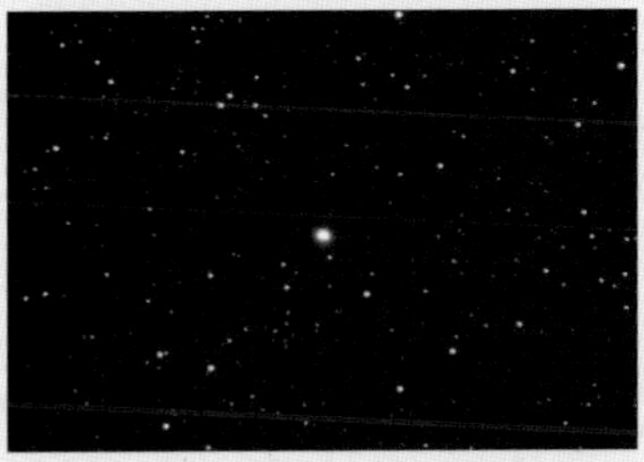

Telescope Aperture:	4" f/5	4" f/9	6" f/7	6" f/9	8" f/6.3	8" f/10	10" f/6.3	10" f/10	12" f/6.3	12" f/10
FOV(35mm film):	2.7°x 4.1°	1.50°x 2.26°	1.29° x 1.93°	1.0° x 1.50°	1.07° x 1.61°	0.68°x 1.02°	0.86° x 1.29°	0.54° x 0.81°	0.72° x 1.07°	0.45°x 0.68°

NGC 7026

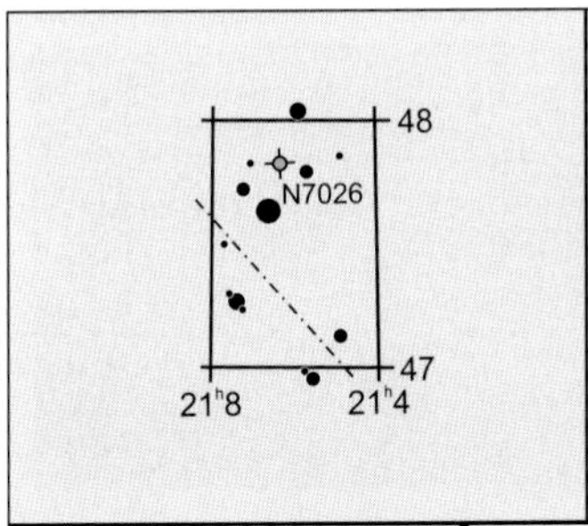

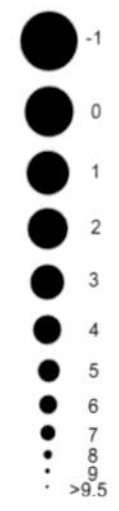

RA:	$21^h\ 06^m\ 22.3^s$	Con:	Cygnus
Dec:	47° 51' 13"	Type:	Planetary Nebula
Size:	0.4'	Mag:	13.0

NGC 7026 is another irregular planetary nebula located in Cygnus.

Telescope Aperture:	4" f/5	4" f/9	6" f/7	6" f/9	8" f/6.3	8" f/10	10" f/6.3	10" f/10	12" f/6.3	12" f/10
FOV(35mm film):	2.7°x 4.1°	1.50°x 2.26°	1.29° x 1.93°	1.0° x 1.50°	1.07° x 1.61°	0.68°x 1.02°	0.86° x 1.29°	0.54° x 0.81°	0.72° x 1.07°	0.45°x 0.68°

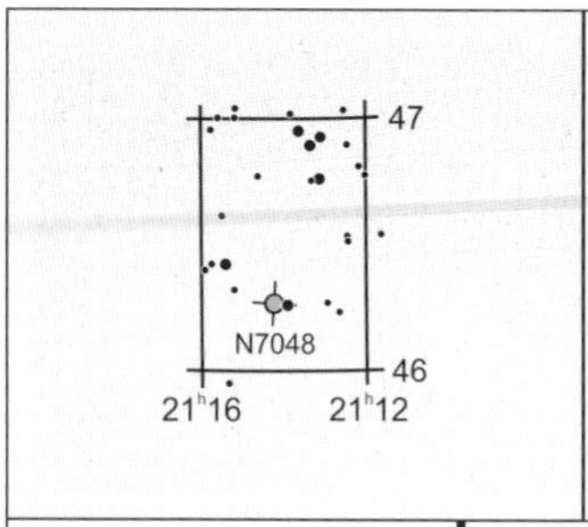

NGC 7048

RA:	$21^h\ 14^m\ 16.3^s$	Con:	Cygnus
Dec:	46° 16' 14"	Type:	Planetary Nebula
Size:	1.0'	Mag:	11.0

NGC 7048 is a planetary nebula.

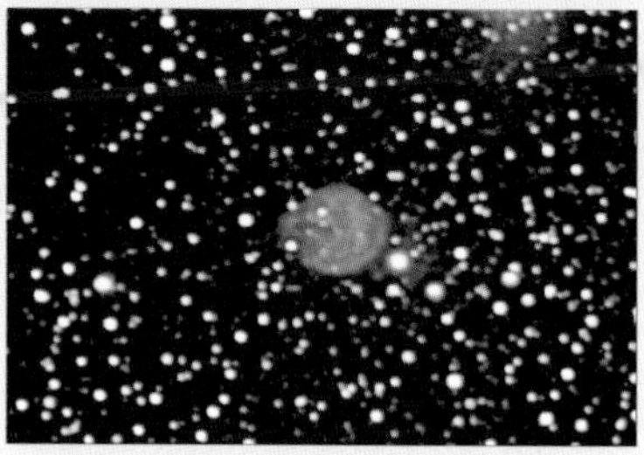

Telescope Aperture:	4" f/5	4" f/9	6" f/7	6" f/9	8" f/6.3	8" f/10	10" f/6.3	10" f/10	12" f/6.3	12" f/10
FOV(35mm film):	2.7° x 4.1°	1.50° x 2.26°	1.29° x 1.93°	1.0° x 1.50°	1.07° x 1.61°	0.68° x 1.02°	0.86° x 1.29°	0.54° x 0.81°	0.72° x 1.07°	0.45° x 0.68°

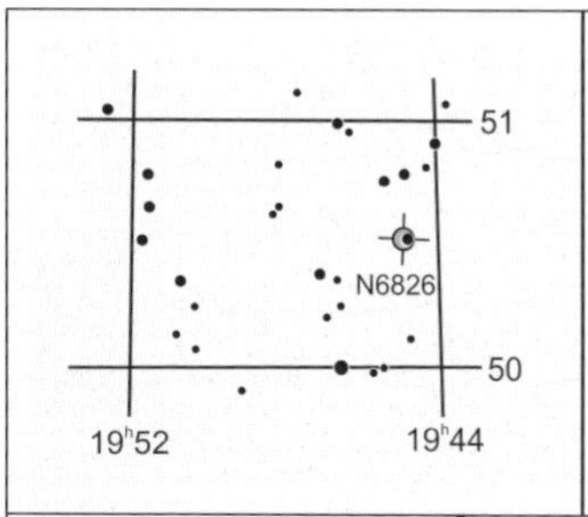

NGC 6826 "Blinking Planetary"

RA:	$19^h\ 44^m\ 52.1^s$	Con:	Cygnus
Dec:	50° 31' 12"	Type:	Planetary Nebula
Size:	2.3'	Mag:	10.0

NGC 6826; called the "Blinking Planetary". Visual intensity changes with direct and averted vision.

Telescope Aperture:	4" f/5	4" f/9	6" f/7	6" f/9	8" f/6.3	8" f/10	10" f/6.3	10" f/10	12" f/6.3	12" f/10
FOV(35mm film):	2.7° x 4.1°	1.50° x 2.26°	1.29° x 1.93°	1.0° x 1.50°	1.07° x 1.61°	0.68° x 1.02°	0.86° x 1.29°	0.54° x 0.81°	0.72° x 1.07°	0.45° x 0.68°

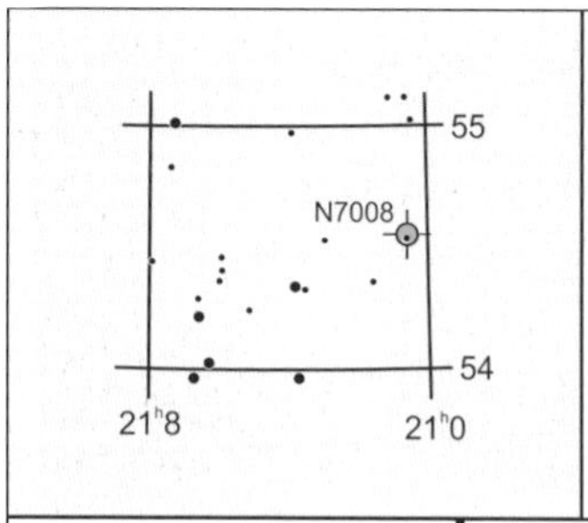

NGC 7008

RA:	$21^h\ 00^m\ 40.2^s$	Con:	Cygnus
Dec:	54° 33' 12"	Type:	Planetary Nebula
Size:	1.4'	Mag:	13.0

NGC 7008 is a planetary nebula ring with a central star.

Telescope Aperture:	4" f/5	4" f/9	6" f/7	6" f/9	8" f/6.3	8" f/10	10" f/6.3	10" f/10	12" f/6.3	12" f/10
FOV(35mm film):	2.7° x 4.1°	1.50° x 2.26°	1.29° x 1.93°	1.0° x 1.50°	1.07° x 1.61°	0.68° x 1.02°	0.86° x 1.29°	0.54° x 0.81°	0.72° x 1.07°	0.45° x 0.68°

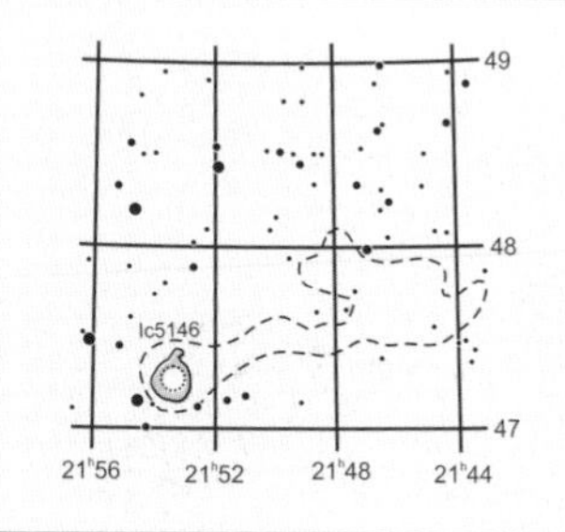

Map Scaled to Fit

IC 5146

RA:	$21^h\ 53^m\ 28.4^s$	Con:	Cygnus
Dec:	47° 16' 14"	Type:	Cluster & Nebula
Size:	12.0'	Mag:	7.0

IC 5146 is an open cluster and associated nebulosity.

Telescope Aperture:	4" f/5	4" f/9	6" f/7	6" f/9	8" f/6.3	8" f/10	10" f/6.3	10" f/10	12" f/6.3	12" f/10
FOV(35mm film):	2.7° x 4.1°	1.50° x 2.26°	1.29° x 1.93°	1.0° x 1.50°	1.07° x 1.61°	0.68° x 1.02°	0.86° x 1.29°	0.54° x 0.81°	0.72° x 1.07°	0.45° x 0.68°

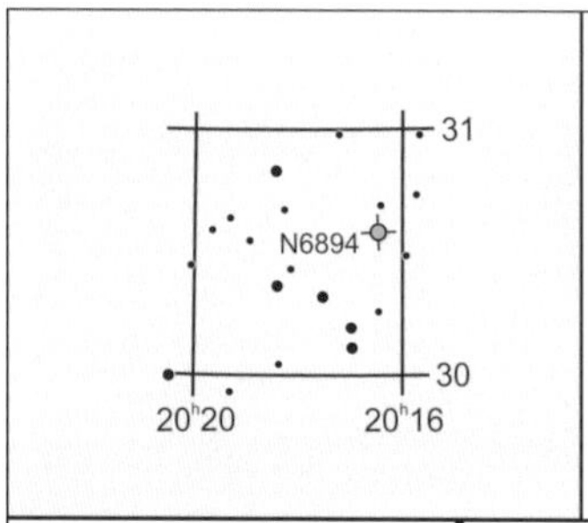

NGC 6894

RA:	$20^h\ 16^m\ 28.4^s$	Con:	Cygnus
Dec:	30° 34' 14"	Type:	Planetary Nebula
Size:	0.7'	Mag:	14.0

NGC 6894 is a planetary nebula ring.

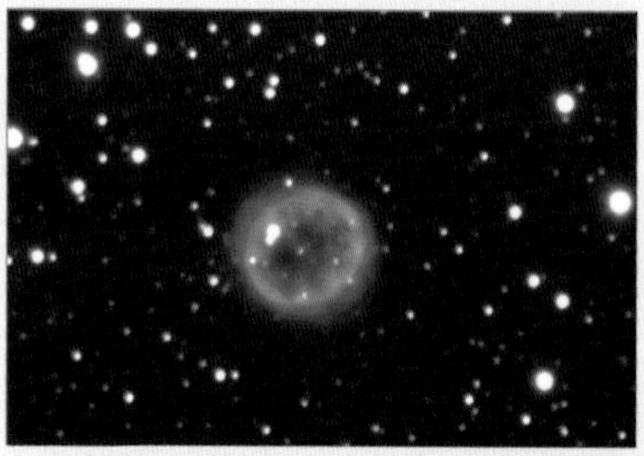

Telescope Aperture:	4" f/5	4" f/9	6" f/7	6" f/9	8" f/6.3	8" f/10	10" f/6.3	10" f/10	12" f/6.3	12" f/10
FOV(35mm film):	2.7° x 4.1°	1.50° x 2.26°	1.29° x 1.93°	1.0° x 1.50°	1.07° x 1.61°	0.68° x 1.02°	0.86° x 1.29°	0.54° x 0.81°	0.72° x 1.07°	0.45° x 0.68°

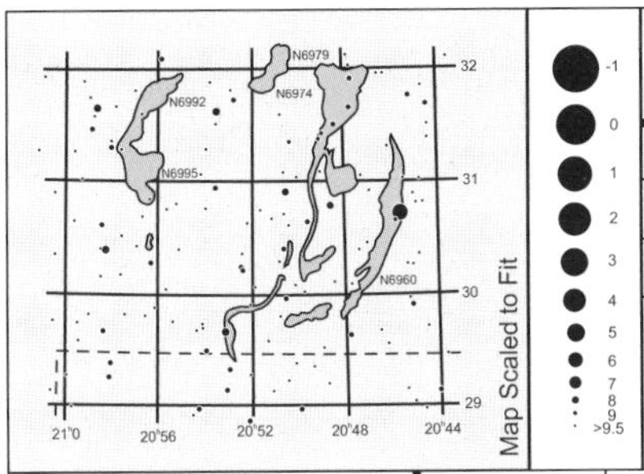

Veil Nebula

RA:		Con:	Cygnus
Dec:		Type:	Nebula
Size:		Mag:	

Veil Nebula consists of 4 distinct bright regions of nebulosity. NGC numbers of the regions are as follows: NGC 6960, NGC 6979, NGC 6992, NGC 6995

Telescope Aperture:	4" f/5	4" f/9	6" f/7	6" f/9	8" f/6.3	8" f/10	10" f/6.3	10" f/10	12" f/6.3	12" f/10
FOV(35mm film):	2.7° x 4.1°	1.50° x 2.26°	1.29° x 1.93°	1.0° x 1.50°	1.07° x 1.61°	0.68° x 1.02°	0.86° x 1.29°	0.54° x 0.81°	0.72° x 1.07°	0.45° x 0.68°

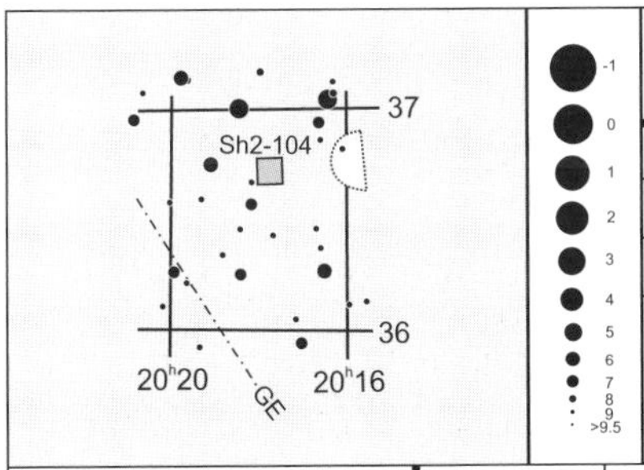

Sh2-104

RA:	19h 44m 52.1s	Con:	Cygnus
Dec:	50° 31' 12"	Type:	Emission Nebula
Size:	2.3'	Mag:	

Sh2-104 is an emission nebula.

Telescope Aperture:	4" f/5	4" f/9	6" f/7	6" f/9	8" f/6.3	8" f/10	10" f/6.3	10" f/10	12" f/6.3	12" f/10
FOV(35mm film):	2.7° x 4.1°	1.50° x 2.26°	1.29° x 1.93°	1.0° x 1.50°	1.07° x 1.61°	0.68° x 1.02°	0.86° x 1.29°	0.54° x 0.81°	0.72° x 1.07°	0.45° x 0.68°

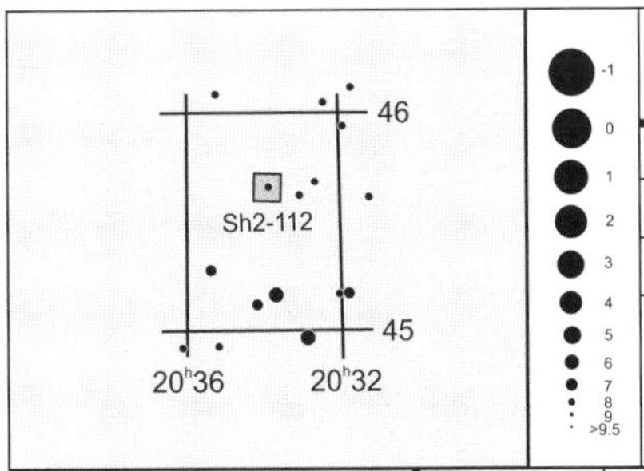

Sh2-112

RA:	21h 00m 40.2s	Con:	Cygnus
Dec:	54° 33' 12"	Type:	Emission Nebula
Size:	13.0' x 13.0'	Mag:	

Sh2-112 is an emission nebula.

Telescope Aperture:	4" f/5	4" f/9	6" f/7	6" f/9	8" f/6.3	8" f/10	10" f/6.3	10" f/10	12" f/6.3	12" f/10
FOV(35mm film):	2.7° x 4.1°	1.50° x 2.26°	1.29° x 1.93°	1.0° x 1.50°	1.07° x 1.61°	0.68° x 1.02°	0.86° x 1.29°	0.54° x 0.81°	0.72° x 1.07°	0.45° x 0.68°

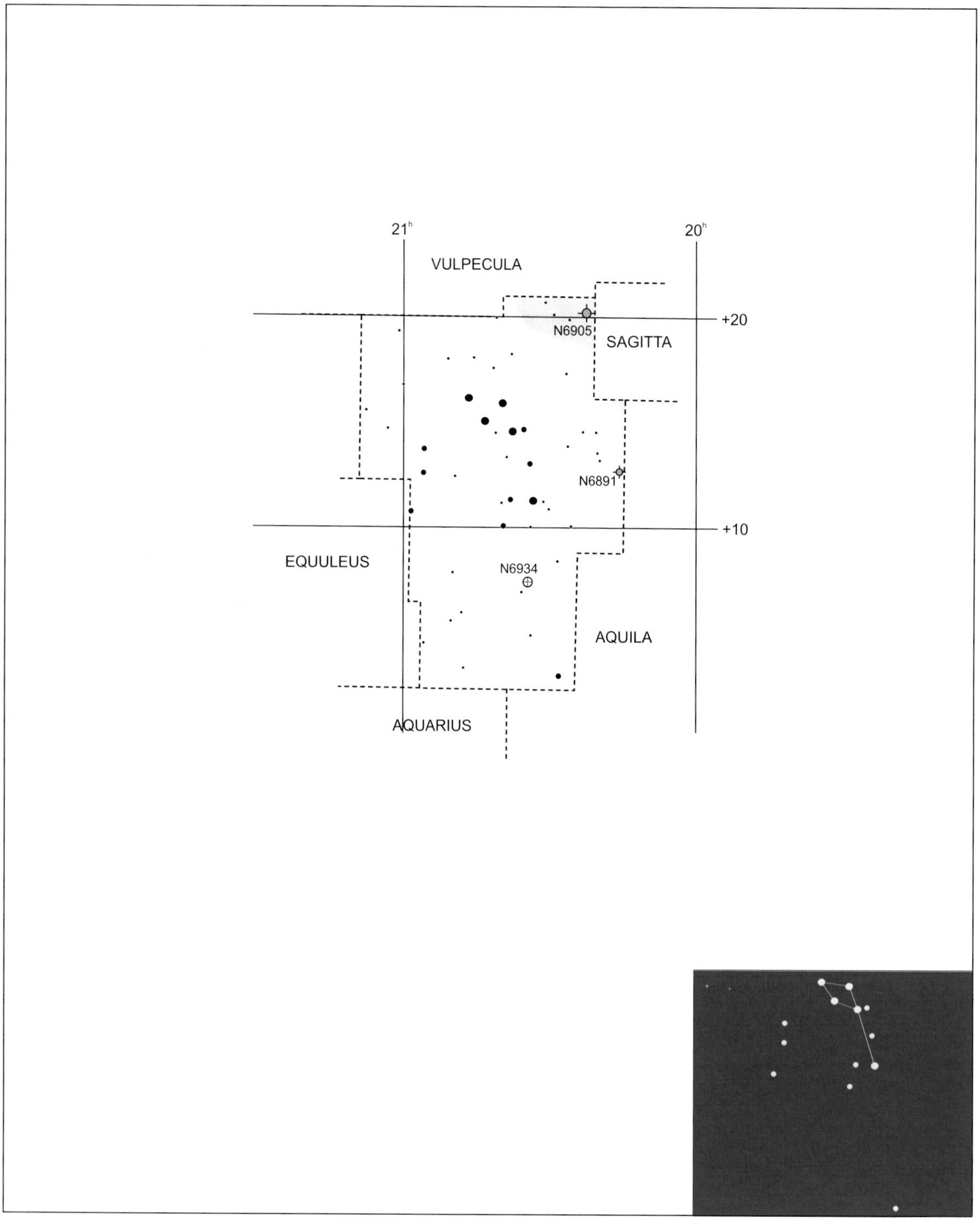

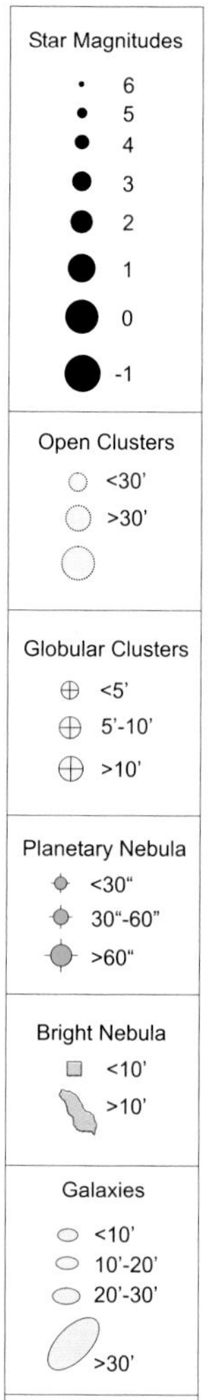

DELPHINUS

Constellation Facts:

Delphinus; (del-FEE-nus)

Delphinus, the Dolphin.
The constellation is tucked between Aquila and Pegasus. Constellation rises in the northeastern sky, passes overhead and sets towards the northwest.
Delphinus covers 189 square degrees.

Constellation is visible from 90° N to 69° S. Partially visible from 69° S to 90° S.

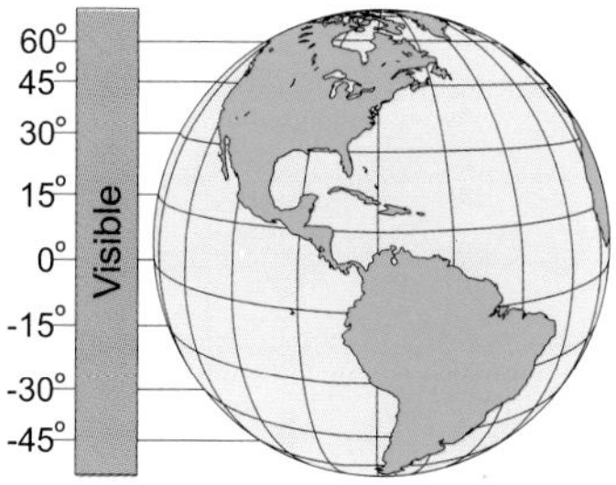

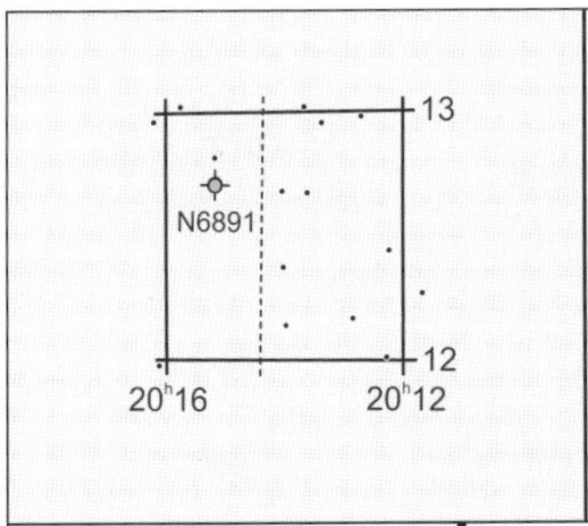

NGC 6891

RA:	$20^h\ 15^m\ 16.7^s$	Con:	Delphinus
Dec:	12° 42' 16"	Type:	Planetary Nebula
Size:	1.2'	Mag:	12.0

NGC 6891 is a planetary nebula found on the western edge of the constellation.

Telescope Aperture:	4" f/5	4" f/9	6" f/7	6" f/9	8" f/6.3	8" f/10	10" f/6.3	10" f/10	12" f/6.3	12" f/10
FOV(35mm film):	2.7° x 4.1°	1.50° x 2.26°	1.29° x 1.93°	1.0° x 1.50°	1.07° x 1.61°	0.68° x 1.02°	0.86° x 1.29°	0.54° x 0.81°	0.72° x 1.07°	0.45° x 0.68°

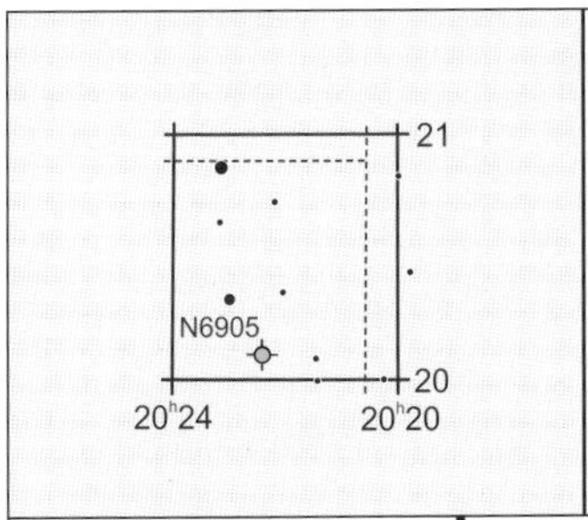

NGC 6905 "Blue Flash Nebula"

RA:	$20^h\ 22^m\ 28.6^s$	Con:	Delphinus
Dec:	20° 07' 16"	Type:	Planetary Nebula
Size:	1.7'	Mag:	12.0

NGC 6905 is a small bright planetary nebula found in the northern region of Delphinus.

Telescope Aperture:	4" f/5	4" f/9	6" f/7	6" f/9	8" f/6.3	8" f/10	10" f/6.3	10" f/10	12" f/6.3	12" f/10
FOV(35mm film):	2.7° x 4.1°	1.50° x 2.26°	1.29° x 1.93°	1.0° x 1.50°	1.07° x 1.61°	0.68° x 1.02°	0.86° x 1.29°	0.54° x 0.81°	0.72° x 1.07°	0.45° x 0.68°

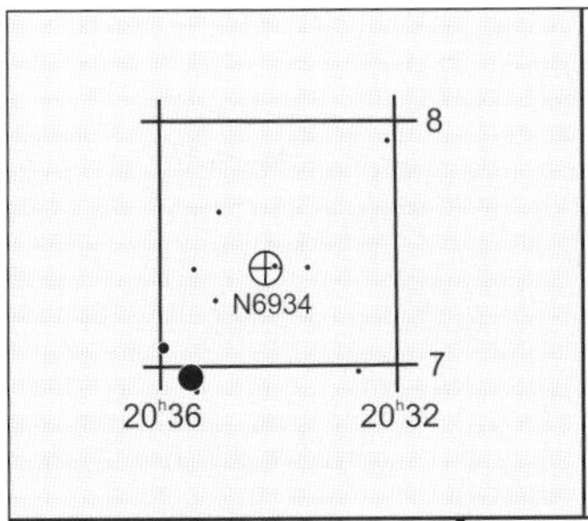

NGC 6934

RA:	$20^h 34^m\ 16.8^s$	Con:	Delphinus
Dec:	07° 24' 18"	Type:	Globular Cluster
Size:	5.9'	Mag:	8.9

NGC 6934 is found in the southern regions of the constellation.

Telescope Aperture:	4" f/5	4" f/9	6" f/7	6" f/9	8" f/6.3	8" f/10	10" f/6.3	10" f/10	12" f/6.3	12" f/10
FOV(35mm film):	2.7° x 4.1°	1.50° x 2.26°	1.29° x 1.93°	1.0° x 1.50°	1.07° x 1.61°	0.68° x 1.02°	0.86° x 1.29°	0.54° x 0.81°	0.72° x 1.07°	0.45° x 0.68°

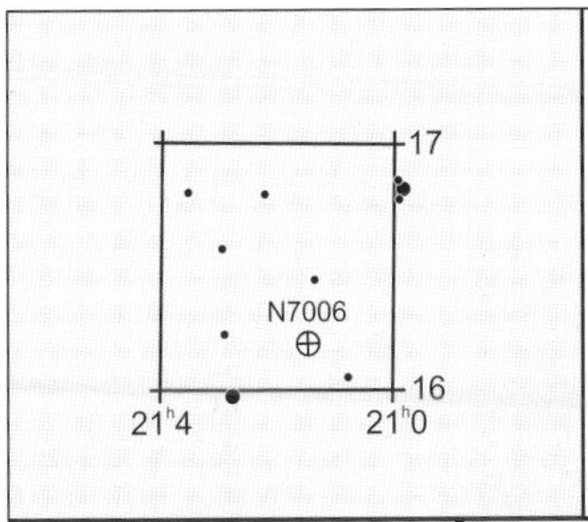

NGC 7006

RA:	$21^h 01^m\ 34.6^s$	Con:	Delphinus
Dec:	16° 11' 18"	Type:	Globular Cluster
Size:	2.8'	Mag:	10.6

NGC 7006 is a mottled globular cluster found 10° from NGC 6934 in the northeast portion of the constellation.

Telescope Aperture:	4" f/5	4" f/9	6" f/7	6" f/9	8" f/6.3	8" f/10	10" f/6.3	10" f/10	12" f/6.3	12" f/10
FOV(35mm film):	2.7° x 4.1°	1.50° x 2.26°	1.29° x 1.93°	1.0° x 1.50°	1.07° x 1.61°	0.68° x 1.02°	0.86° x 1.29°	0.54° x 0.81°	0.72° x 1.07°	0.45° x 0.68°

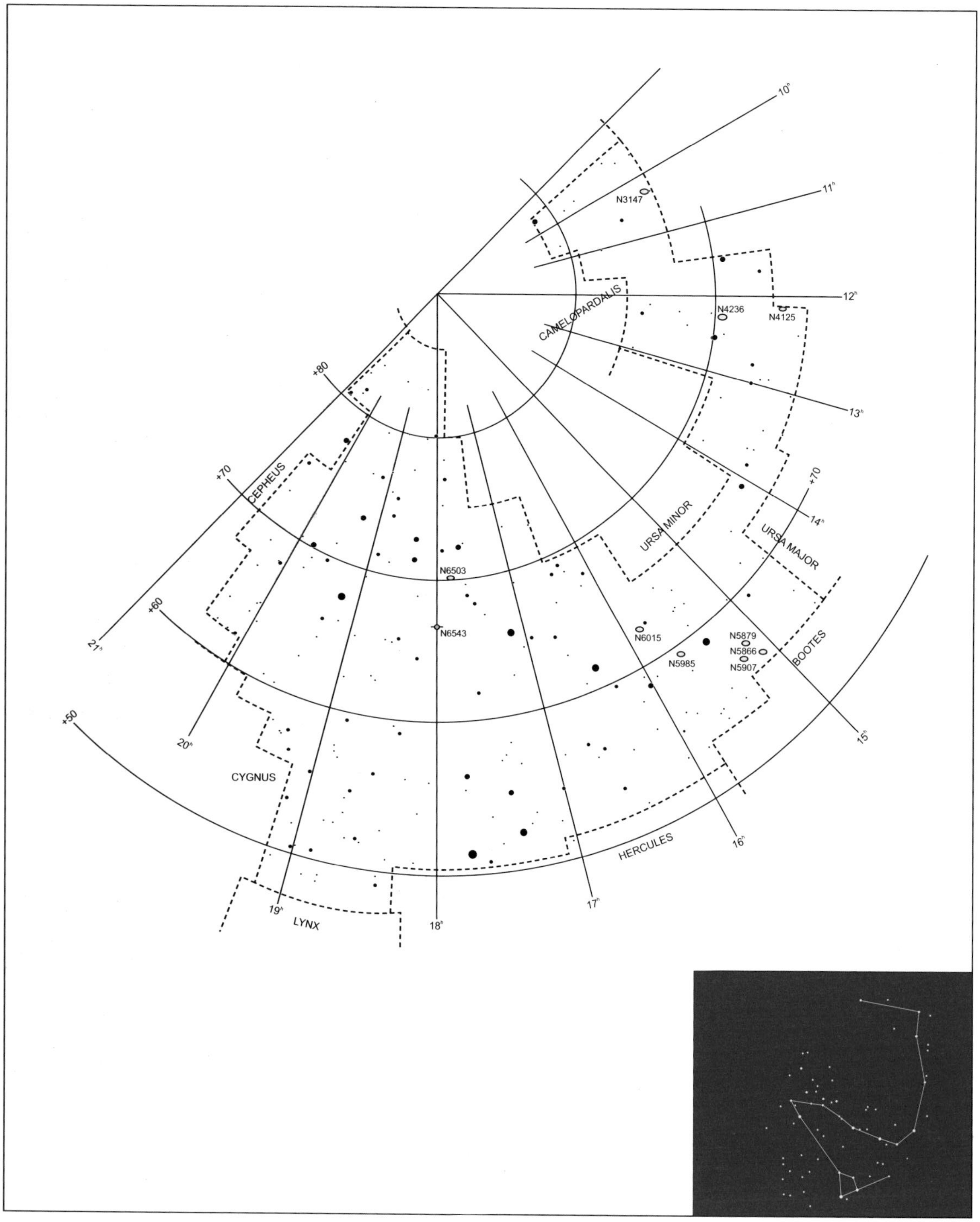

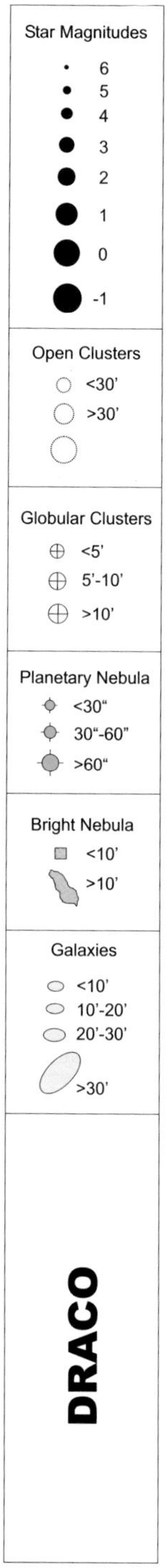

Constellation Facts:

Draco; (DRAY-ko)

Is a northern circumpolar constellation, and is visible throughout the year.
The constellation covers 1083 square degrees.

Constellation is visible from 90° N to 4° S. Partially visible from 4° S to 30° S.

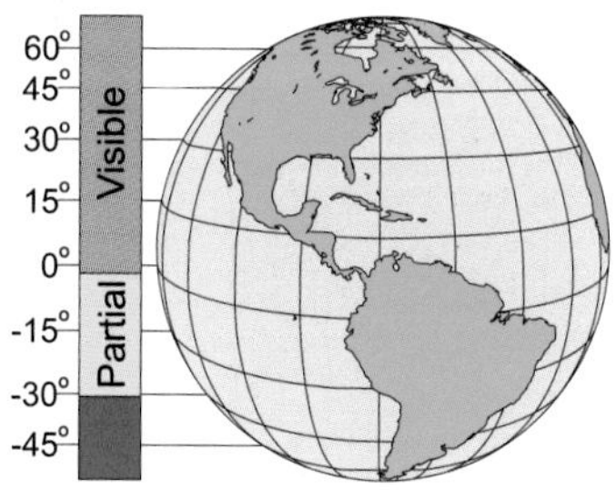

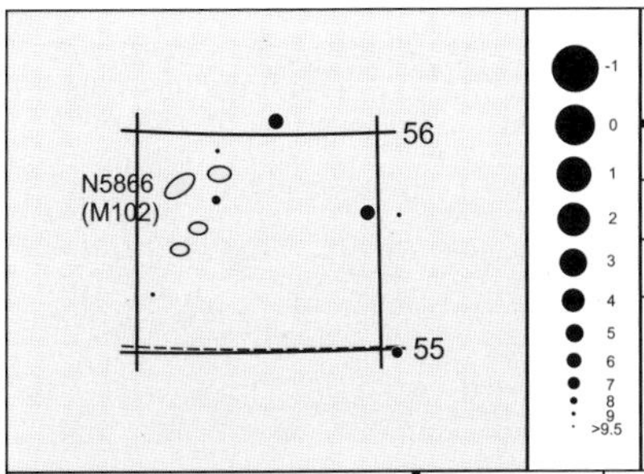

M102 (NGC 5866)

RA:	$15^h\ 06^m\ 33.0^s$	Con:	Draco
Dec:	55° 46′ 01″	Type:	Lenticular Galaxy
Size:	2.8′ x 1.2′	Mag:	10.0

M102 (NGC 5866) is a bright lenticular galaxy found at the southern edge of the constellation, 4° southwest of the star Iota Draconis.

Telescope Aperture:	4″ f/5	4″ f/9	6″ f/7	6″ f/9	8″ f/6.3	8″ f/10	10″ f/6.3	10″ f/10	12″ f/6.3	12″ f/10
FOV(35mm film):	2.7° x 4.1°	1.50° x 2.26°	1.29° x 1.93°	1.0° x 1.50°	1.07° x 1.61°	0.68° x 1.02°	0.86° x 1.29°	0.54° x 0.81°	0.72° x 1.07°	0.45° x 0.68°

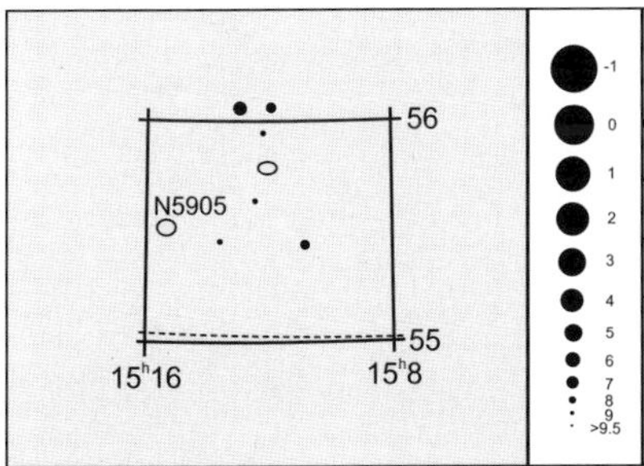

NGC 5905

RA:	$15^h\ 15^m\ 27.0^s$	Con:	Draco
Dec:	55° 31′ 01″	Type:	Barred Spiral
Size:	3.2′ x 3.0′	Mag:	12.0

NGC 5905 is a barred spiral galaxy located 2° due east of M102.

Telescope Aperture:	4″ f/5	4″ f/9	6″ f/7	6″ f/9	8″ f/6.3	8″ f/10	10″ f/6.3	10″ f/10	12″ f/6.3	12″ f/10
FOV(35mm film):	2.7° x 4.1°	1.50° x 2.26°	1.29° x 1.93°	1.0° x 1.50°	1.07° x 1.61°	0.68° x 1.02°	0.86° x 1.29°	0.54° x 0.81°	0.72° x 1.07°	0.45° x 0.68°

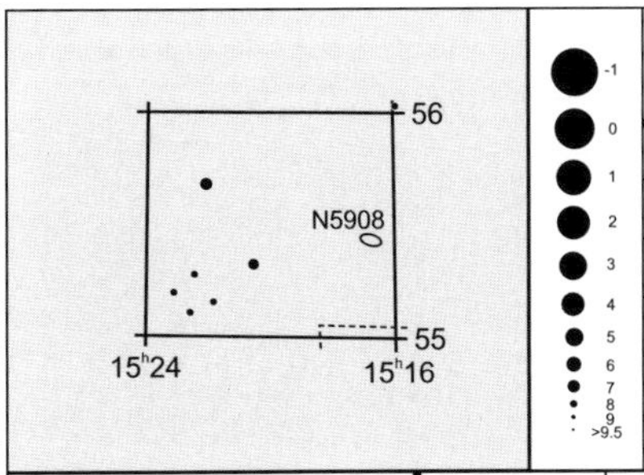

NGC 5908

RA:	$15^h\ 16^m\ 45.0^s$	Con:	Draco
Dec:	55° 25′ 01″	Type:	Spiral Galaxy
Size:	3.0′ x 1.0′	Mag:	11.9

NGC 5908 is also located 2° due east of M102. NGC 5908 and NGC 5905 are separated by 12.5′.

Telescope Aperture:	4″ f/5	4″ f/9	6″ f/7	6″ f/9	8″ f/6.3	8″ f/10	10″ f/6.3	10″ f/10	12″ f/6.3	12″ f/10
FOV(35mm film):	2.7° x 4.1°	1.50° x 2.26°	1.29° x 1.93°	1.0° x 1.50°	1.07° x 1.61°	0.68° x 1.02°	0.86° x 1.29°	0.54° x 0.81°	0.72° x 1.07°	0.45° x 0.68°

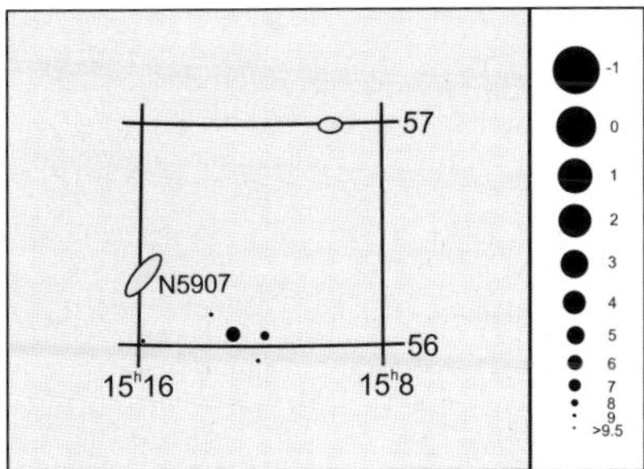

NGC 5907

RA:	$15^h\ 15^m\ 57.0^s$	Con:	Draco
Dec:	56° 19′ 01″	Type:	Edge-on Spiral
Size:	11.0′ x 1.0′	Mag:	10.4

NGC 5907 is a large edge-on Sb-type spiral galaxy.

Telescope Aperture:	4″ f/5	4″ f/9	6″ f/7	6″ f/9	8″ f/6.3	8″ f/10	10″ f/6.3	10″ f/10	12″ f/6.3	12″ f/10
FOV(35mm film):	2.7° x 4.1°	1.50° x 2.26°	1.29° x 1.93°	1.0° x 1.50°	1.07° x 1.61°	0.68° x 1.02°	0.86° x 1.29°	0.54° x 0.81°	0.72° x 1.07°	0.45° x 0.68°

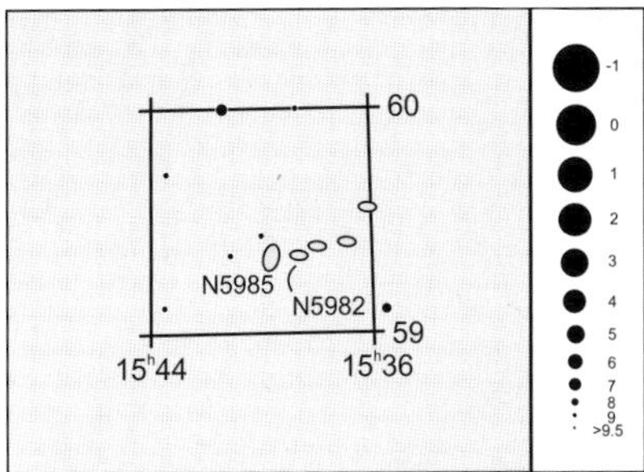

NGC 5985

RA:	$15^h\ 39^m\ 38.9^s$	Con:	Draco
Dec:	59° 20′ 02″	Type:	Spiral Galaxy
Size:	5.4′ x 2.3′	Mag:	11.0

NGC 5985 is an Sb-type spiral galaxy located 4.5° northeast of NGC 5879.

Telescope Aperture:	4″ f/5	4″ f/9	6″ f/7	6″ f/9	8″ f/6.3	8″ f/10	10″ f/6.3	10″ f/10	12″ f/6.3	12″ f/10
FOV(35mm film):	2.7° x 4.1°	1.50° x 2.26°	1.29° x 1.93°	1.0° x 1.50°	1.07° x 1.61°	0.68° x 1.02°	0.86° x 1.29°	0.54° x 0.81°	0.72° x 1.07°	0.45° x 0.68°

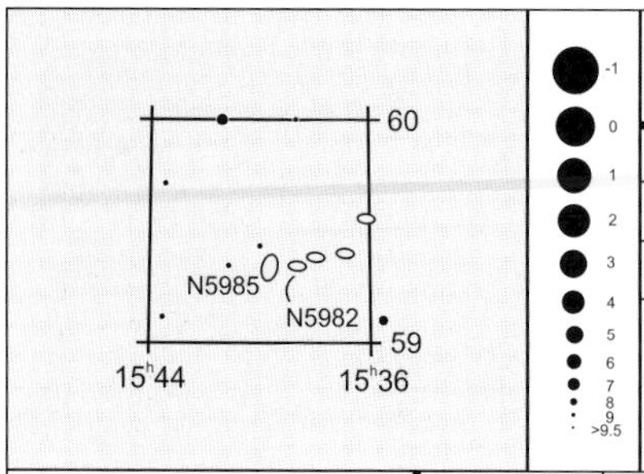

NGC 5982

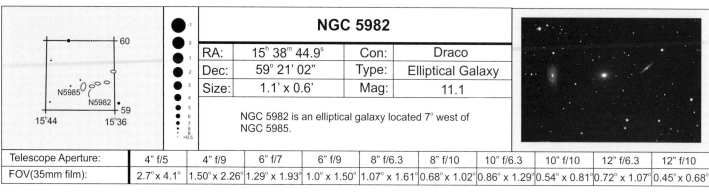

RA:	$15^h\ 38^m\ 44.9^s$	Con:	Draco
Dec:	59° 21’ 02”	Type:	Elliptical Galaxy
Size:	1.1’ x 0.6’	Mag:	11.1

NGC 5982 is an elliptical galaxy located 7° west of NGC 5985.

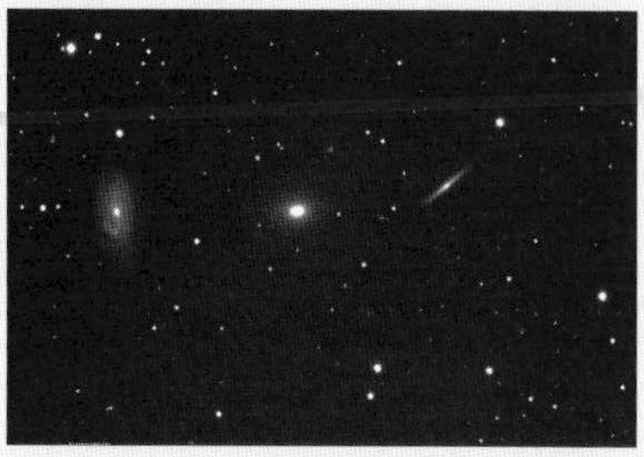

Telescope Aperture:	4” f/5	4” f/9	6” f/7	6” f/9	8” f/6.3	8” f/10	10” f/6.3	10” f/10	12” f/6.3	12” f/10
FOV(35mm film):	2.7° x 4.1°	1.50° x 2.26°	1.29° x 1.93°	1.0° x 1.50°	1.07° x 1.61°	0.68° x 1.02°	0.86° x 1.29°	0.54° x 0.81°	0.72° x 1.07°	0.45° x 0.68°

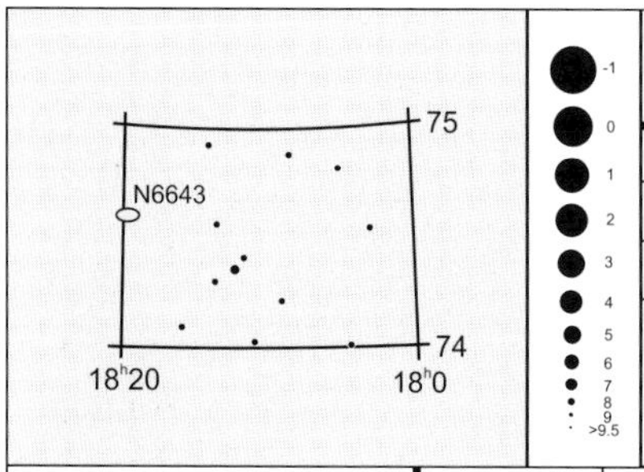

NGC 6643

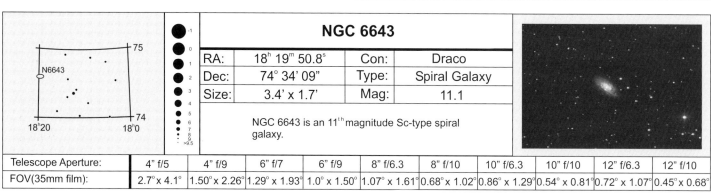

RA:	$18^h\ 19^m\ 50.8^s$	Con:	Draco
Dec:	74° 34’ 09”	Type:	Spiral Galaxy
Size:	3.4’ x 1.7’	Mag:	11.1

NGC 6643 is an 11th magnitude Sc-type spiral galaxy.

Telescope Aperture:	4” f/5	4” f/9	6” f/7	6” f/9	8” f/6.3	8” f/10	10” f/6.3	10” f/10	12” f/6.3	12” f/10
FOV(35mm film):	2.7° x 4.1°	1.50° x 2.26°	1.29° x 1.93°	1.0° x 1.50°	1.07° x 1.61°	0.68° x 1.02°	0.86° x 1.29°	0.54° x 0.81°	0.72° x 1.07°	0.45° x 0.68°

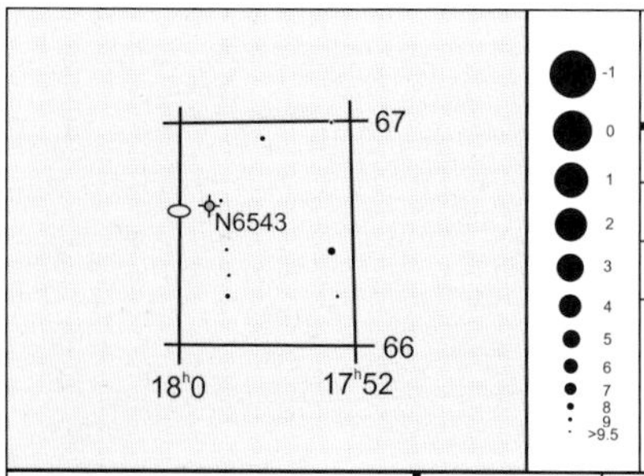

NGC 6543

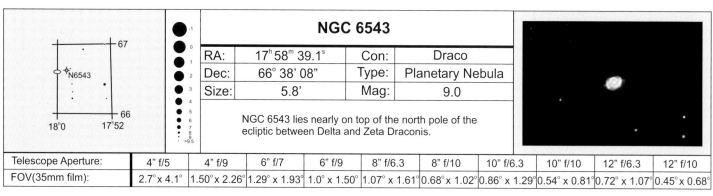

RA:	$17^h\ 58^m\ 39.1^s$	Con:	Draco
Dec:	66° 38’ 08”	Type:	Planetary Nebula
Size:	5.8’	Mag:	9.0

NGC 6543 lies nearly on top of the north pole of the ecliptic between Delta and Zeta Draconis.

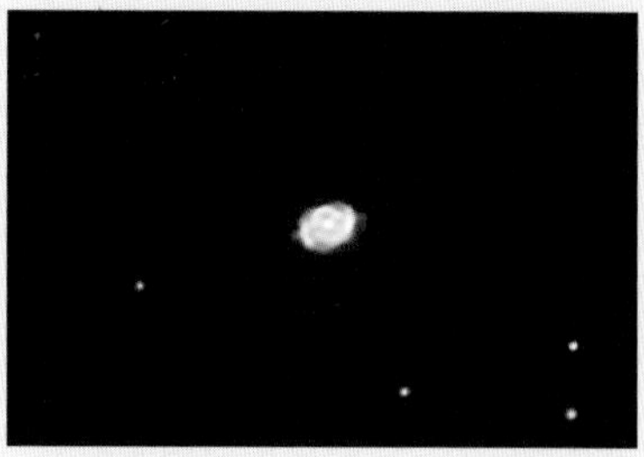

Telescope Aperture:	4” f/5	4” f/9	6” f/7	6” f/9	8” f/6.3	8” f/10	10” f/6.3	10” f/10	12” f/6.3	12” f/10
FOV(35mm film):	2.7° x 4.1°	1.50° x 2.26°	1.29° x 1.93°	1.0° x 1.50°	1.07° x 1.61°	0.68° x 1.02°	0.86° x 1.29°	0.54° x 0.81°	0.72° x 1.07°	0.45° x 0.68°

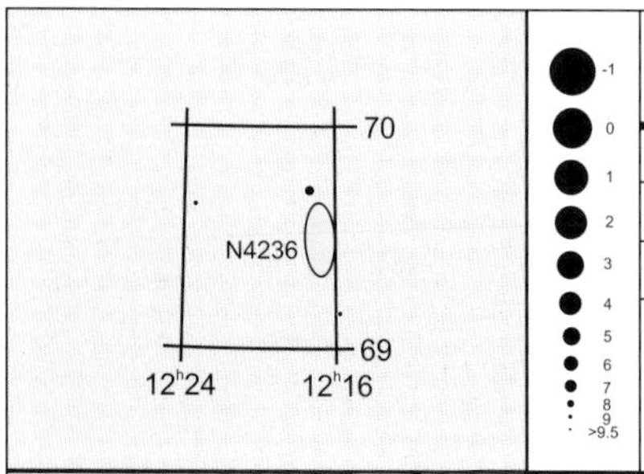

NGC 4236

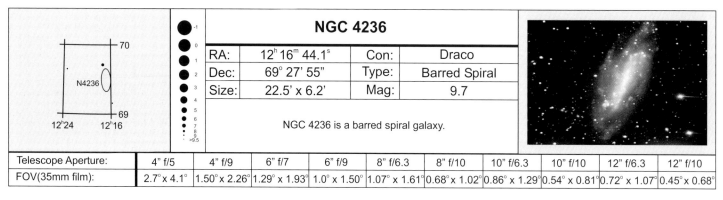

RA:	$12^h\ 16^m\ 44.1^s$	Con:	Draco
Dec:	69° 27’ 55”	Type:	Barred Spiral
Size:	22.5’ x 6.2’	Mag:	9.7

NGC 4236 is a barred spiral galaxy.

Telescope Aperture:	4” f/5	4” f/9	6” f/7	6” f/9	8” f/6.3	8” f/10	10” f/6.3	10” f/10	12” f/6.3	12” f/10
FOV(35mm film):	2.7° x 4.1°	1.50° x 2.26°	1.29° x 1.93°	1.0° x 1.50°	1.07° x 1.61°	0.68° x 1.02°	0.86° x 1.29°	0.54° x 0.81°	0.72° x 1.07°	0.45° x 0.68°

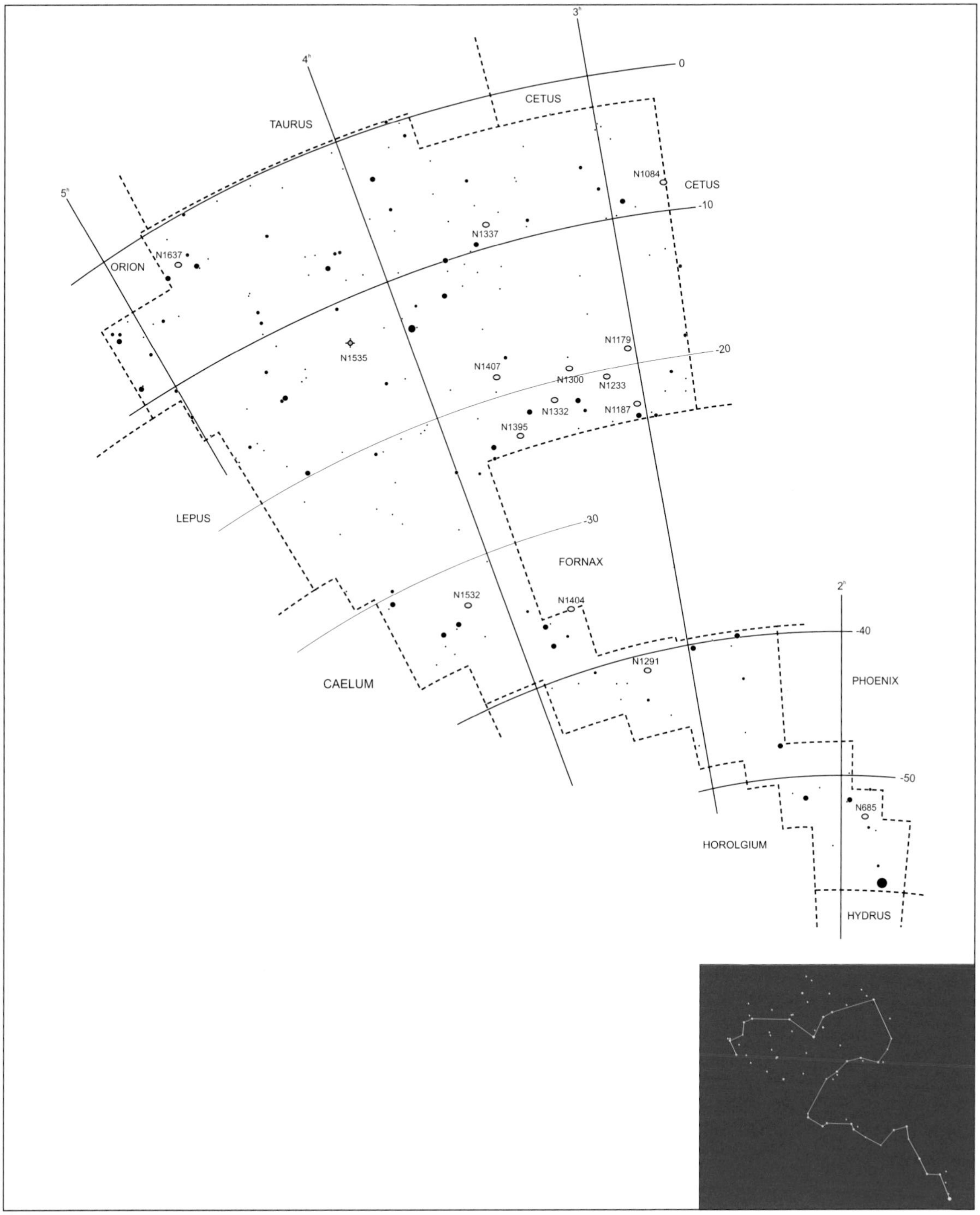

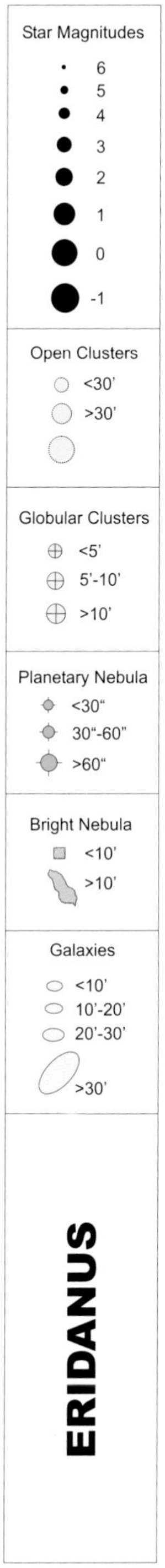

Constellation Facts:

Eridanus; (eh-RID-uh-nuss)

Eridanus, the River.
Constellation follows a southern track, rising in the southeast, and setting in the southwest.
Eridanus is one of the longest and faintest of the constellations. The stars extend towards the west, then track south and east, disappearing below the horizon.
Eridanus covers 1138 square degrees.

Constellation is visible from 32° N to 89° S. Partially visible from 32° N to 90° N.

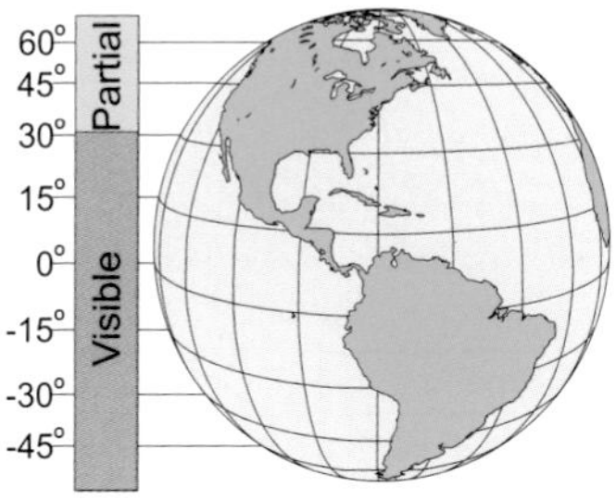

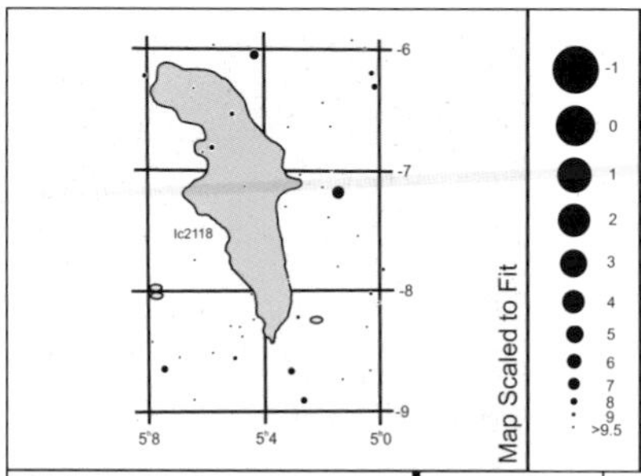

IC 2118 "Witchead Nebula"

RA:	$05^h\ 06^m\ 56.4^s$	Con:	Eridanus
Dec:	$-07^\circ\ 12'\ 51''$	Type:	Nebula
Size:	180.0'	Mag:	

IC 2118 is a large faint nebula that appears as a profile view of characterized witches.

Telescope Aperture:	4" f/5	4" f/9	6" f/7	6" f/9	8" f/6.3	8" f/10	10" f/6.3	10" f/10	12" f/6.3	12" f/10
FOV(35mm film):	2.7° x 4.1°	1.50° x 2.26°	1.29° x 1.93°	1.0° x 1.50°	1.07° x 1.61°	0.68° x 1.02°	0.86° x 1.29°	0.54° x 0.81°	0.72° x 1.07°	0.45° x 0.68°

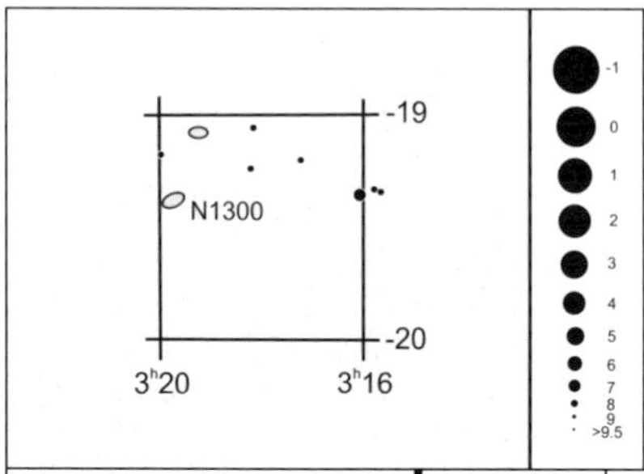

NGC 1300

RA:	$03^h\ 19^m\ 44.5^s$	Con:	Eridanus
Dec:	$-19^\circ\ 24'\ 37''$	Type:	Barred Spiral
Size:	6.0' x 3.5'	Mag:	10.4

NGC 1300 is located near the center of the constellation. Object is one of the finest face-on barred spiral galaxies in the night sky.

Telescope Aperture:	4" f/5	4" f/9	6" f/7	6" f/9	8" f/6.3	8" f/10	10" f/6.3	10" f/10	12" f/6.3	12" f/10
FOV(35mm film):	2.7° x 4.1°	1.50° x 2.26°	1.29° x 1.93°	1.0° x 1.50°	1.07° x 1.61°	0.68° x 1.02°	0.86° x 1.29°	0.54° x 0.81°	0.72° x 1.07°	0.45° x 0.68°

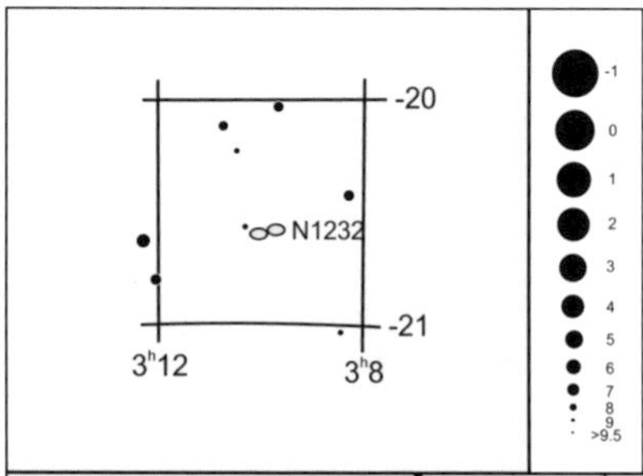

NGC 1232

RA:	$03^h\ 09^m\ 50.6^s$	Con:	Eridanus
Dec:	$-20^\circ\ 34'\ 36''$	Type:	Spiral Galaxy
Size:	8.5' x 7.5'	Mag:	9.9

NGC 1232 is located southwest of NGC 1300. Object is an Sc-type galaxy that is face-on, displaying many loose arms.

Telescope Aperture:	4" f/5	4" f/9	6" f/7	6" f/9	8" f/6.3	8" f/10	10" f/6.3	10" f/10	12" f/6.3	12" f/10
FOV(35mm film):	2.7° x 4.1°	1.50° x 2.26°	1.29° x 1.93°	1.0° x 1.50°	1.07° x 1.61°	0.68° x 1.02°	0.86° x 1.29°	0.54° x 0.81°	0.72° x 1.07°	0.45° x 0.68°

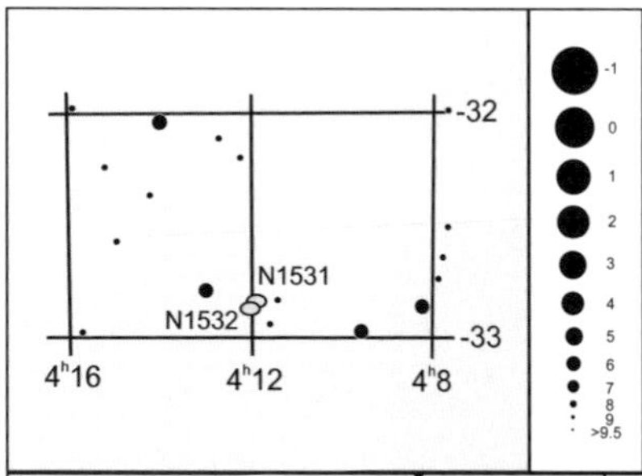

NGC 1531

RA:	$04^h\ 12^m\ 1.7^s$	Con:	Eridanus
Dec:	$-32^\circ\ 50'\ 40''$	Type:	Elliptical Galaxy
Size:	0.6' x 0.4'	Mag:	12.1

NGC 1531 is located on the eastern fringe of the constellation.

Telescope Aperture:	4" f/5	4" f/9	6" f/7	6" f/9	8" f/6.3	8" f/10	10" f/6.3	10" f/10	12" f/6.3	12" f/10
FOV(35mm film):	2.7° x 4.1°	1.50° x 2.26°	1.29° x 1.93°	1.0° x 1.50°	1.07° x 1.61°	0.68° x 1.02°	0.86° x 1.29°	0.54° x 0.81°	0.72° x 1.07°	0.45° x 0.68°

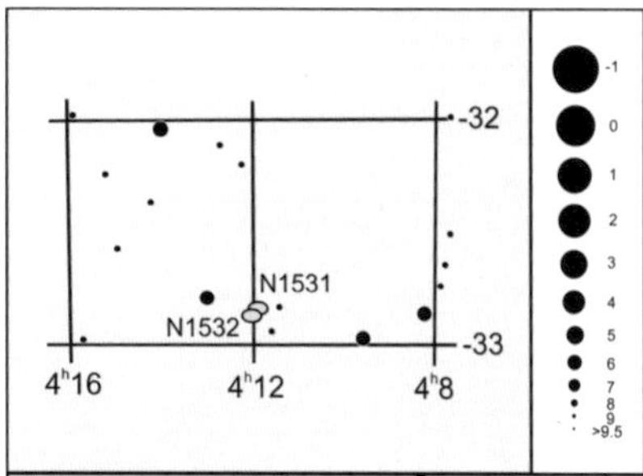

NGC 1532

RA:	$04^h\ 12^m\ 7.6^s$	Con:	Eridanus
Dec:	$-32^\circ\ 51'\ 40''$	Type:	Spiral Galaxy
Size:	4.0' x 1.3'	Mag:	11.0

NGC 1532 like NGC 1531 is found on the eastern fringes of the constellation. Object is nearly edge-on to our line of sight.

Telescope Aperture:	4" f/5	4" f/9	6" f/7	6" f/9	8" f/6.3	8" f/10	10" f/6.3	10" f/10	12" f/6.3	12" f/10
FOV(35mm film):	2.7° x 4.1°	1.50° x 2.26°	1.29° x 1.93°	1.0° x 1.50°	1.07° x 1.61°	0.68° x 1.02°	0.86° x 1.29°	0.54° x 0.81°	0.72° x 1.07°	0.45° x 0.68°

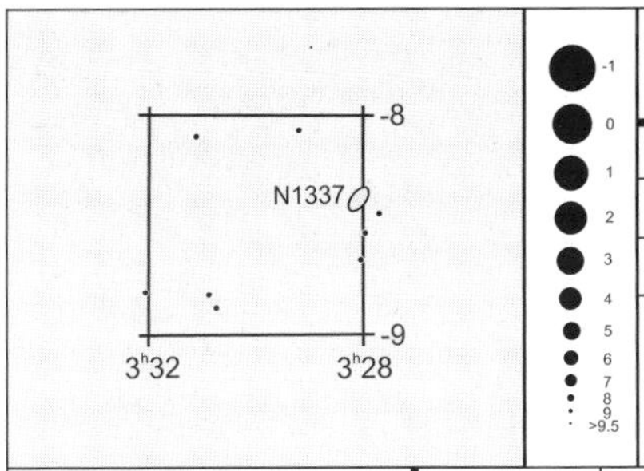

NGC 1337

RA:	$03^h\ 28^m\ 8.8^s$	Con:	Eridanus
Dec:	-08° 22' 41"	Type:	Spiral Galaxy
Size:	7.0' x 1.5'	Mag:	11.7

NGC 1337 is located in the northern part of the constellation. Object is an Sc-type edge-on spiral galaxy.

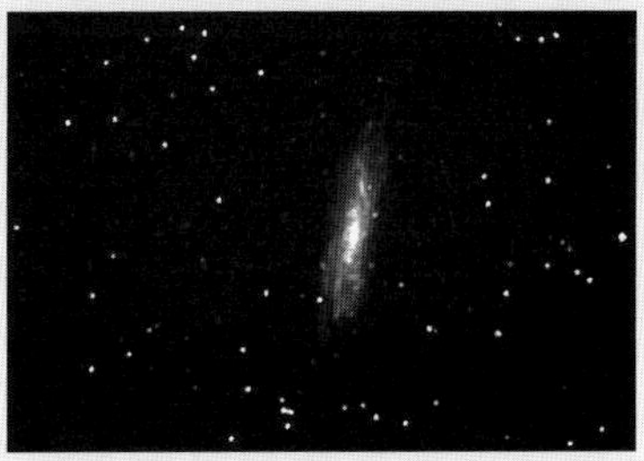

Telescope Aperture:	4" f/5	4" f/9	6" f/7	6" f/9	8" f/6.3	8" f/10	10" f/6.3	10" f/10	12" f/6.3	12" f/10
FOV(35mm film):	2.7° x 4.1°	1.50° x 2.26°	1.29° x 1.93°	1.0° x 1.50°	1.07° x 1.61°	0.68° x 1.02°	0.86° x 1.29°	0.54° x 0.81°	0.72° x 1.07°	0.45° x 0.68°

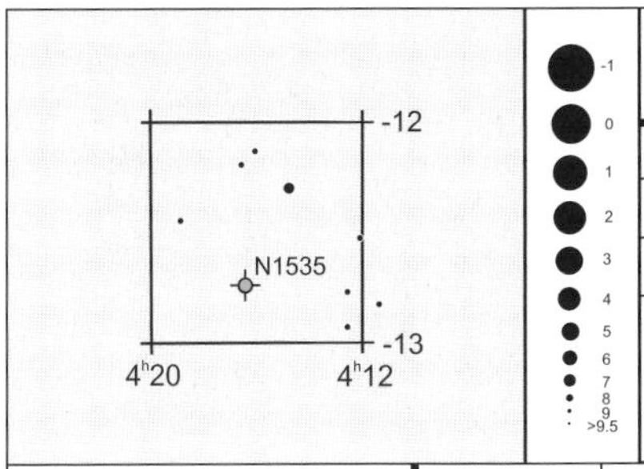

NGC 1535

RA:	$04^h\ 14^m\ 14.4^s$	Con:	Eridanus
Dec:	-12° 43' 44"	Type:	Planetary Nebula
Size:	0.7'	Mag:	10.0

NGC 1535 is a small planetary nebula.

Telescope Aperture:	4" f/5	4" f/9	6" f/7	6" f/9	8" f/6.3	8" f/10	10" f/6.3	10" f/10	12" f/6.3	12" f/10
FOV(35mm film):	2.7° x 4.1°	1.50° x 2.26°	1.29° x 1.93°	1.0° x 1.50°	1.07° x 1.61°	0.68° x 1.02°	0.86° x 1.29°	0.54° x 0.81°	0.72° x 1.07°	0.45° x 0.68°

Gem

Crosses Prime Meridian:
January thru February

WINTER

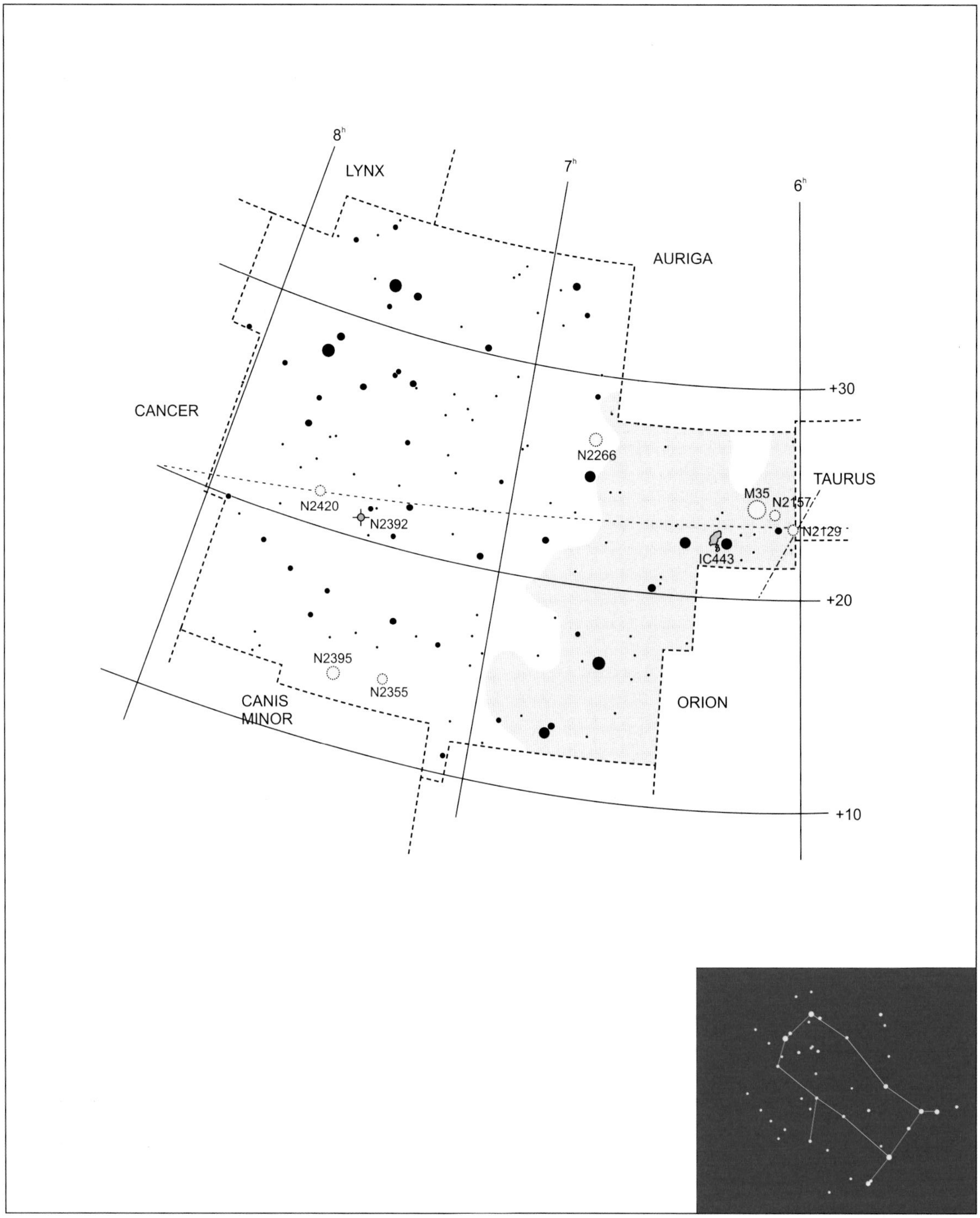

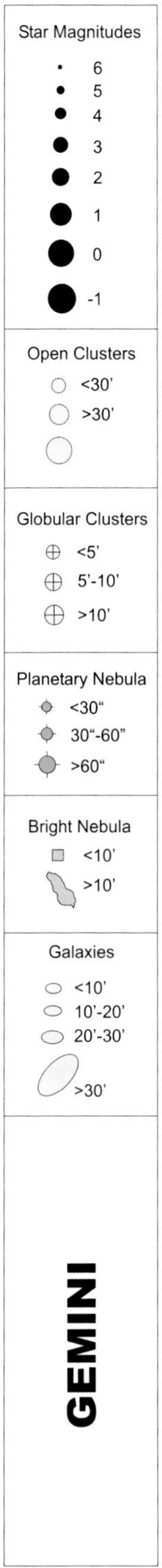

Constellation Facts:

Gemini; (GEM-in-eye)

Gemini, the Twins, rises in the northeast, transits the meridian high in the southern sky, and sets in the northwest during the evenings of late spring. The stars of Gemini are on the eastern edge of the Milky Way.
Gemini covers 514 square degrees.

Constellation is visible from 90° N to 55° S. Partially visible from 55° S to 80° S.

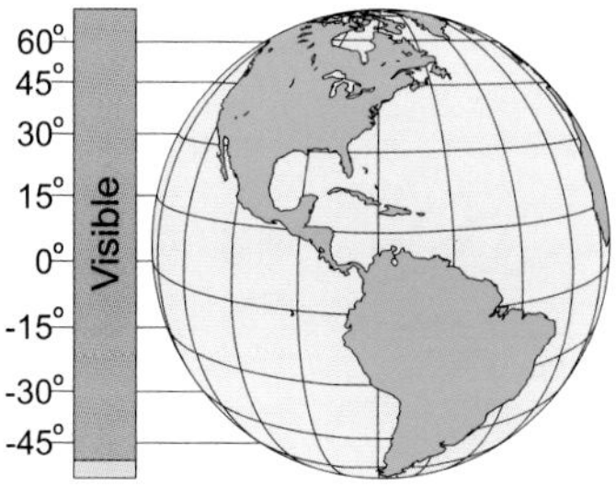

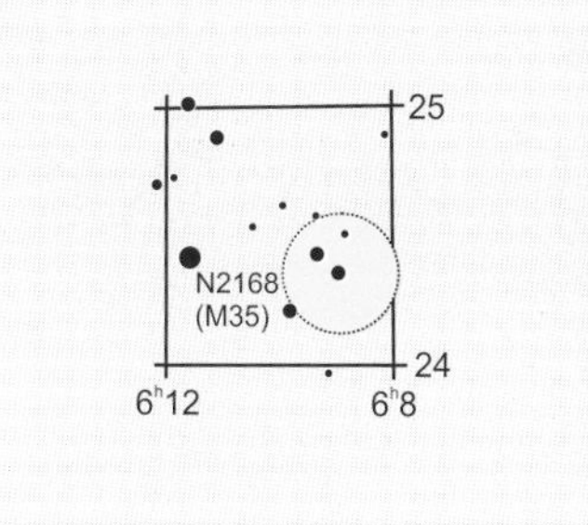

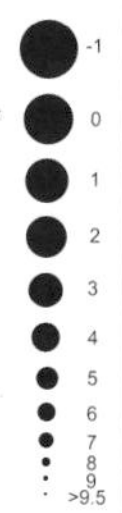

M35 (NGC 2168)

RA:	$06^h\ 08^m\ 57.0^s$	Con:	Gemini
Dec:	24° 19' 58"	Type:	Open Cluster
Size:	28.0'	Mag:	5.1

M35 is a rich open cluster of 5th magnitude. It is a group of approximately 120 stars evenly spread out across its diameter.

Telescope Aperture:	4" f/5	4" f/9	6" f/7	6" f/9	8" f/6.3	8" f/10	10" f/6.3	10" f/10	12" f/6.3	12" f/10
FOV(35mm film):	2.7° x 4.1°	1.50° x 2.26°	1.29° x 1.93°	1.0° x 1.50°	1.07° x 1.61°	0.68° x 1.02°	0.86° x 1.29°	0.54° x 0.81°	0.72° x 1.07°	0.45° x 0.68°

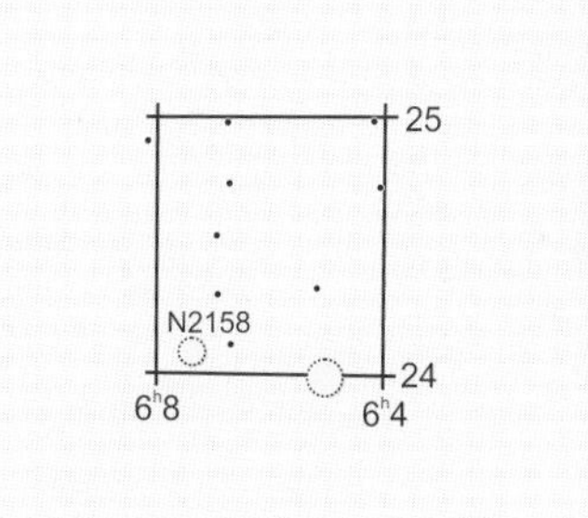

NGC 2158

RA:	$06^h\ 07^m\ 33.0^s$	Con:	Gemini
Dec:	24° 05' 58"	Type:	Open Cluster
Size:	5.0'	Mag:	8.6

NGC 2158 is a rich open cluster comprised of 150 stars.

Telescope Aperture:	4" f/5	4" f/9	6" f/7	6" f/9	8" f/6.3	8" f/10	10" f/6.3	10" f/10	12" f/6.3	12" f/10
FOV(35mm film):	2.7° x 4.1°	1.50° x 2.26°	1.29° x 1.93°	1.0° x 1.50°	1.07° x 1.61°	0.68° x 1.02°	0.86° x 1.29°	0.54° x 0.81°	0.72° x 1.07°	0.45° x 0.68°

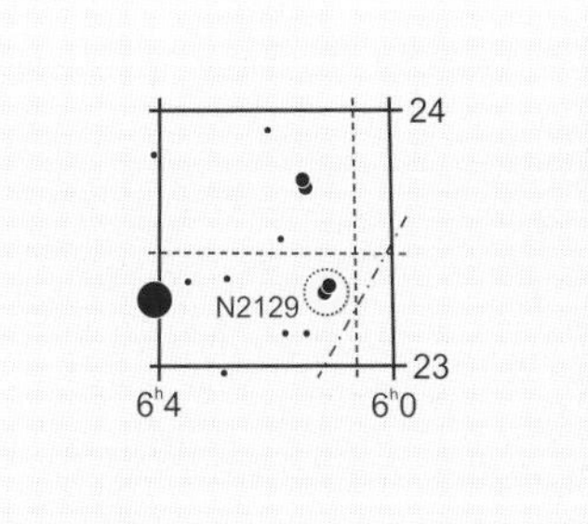

NGC 2129

RA:	$06^h\ 51^m\ 3.0^s$	Con:	Gemini
Dec:	23° 17' 59"	Type:	Open Cluster
Size:	7.0'	Mag:	6.7

NGC 2129 is a rich open cluster comprised of 50 stars. Object shines at 7th magnitude due to a triangle of 8th and 9th magnitude stars at the center.

Telescope Aperture:	4" f/5	4" f/9	6" f/7	6" f/9	8" f/6.3	8" f/10	10" f/6.3	10" f/10	12" f/6.3	12" f/10
FOV(35mm film):	2.7° x 4.1°	1.50° x 2.26°	1.29° x 1.93°	1.0° x 1.50°	1.07° x 1.61°	0.68° x 1.02°	0.86° x 1.29°	0.54° x 0.81°	0.72° x 1.07°	0.45° x 0.68°

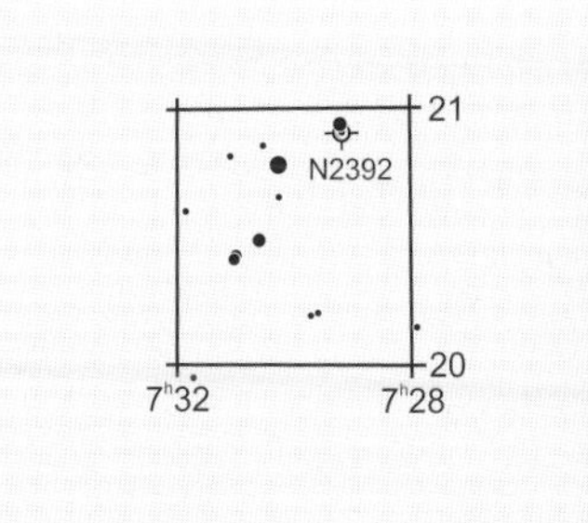

NGC 2392 "Eskimo Nebula"

RA:	$07^h\ 29^m\ 14.8^s$	Con:	Gemini
Dec:	20° 54' 53"	Type:	Planetary Nebula
Size:	0.7'	Mag:	10.0

NGC 2392 the "Eskimo Nebula" is located midway between Kappa and Lambda Geminorum. Object is one of the youngest planetary nebulae in the sky.

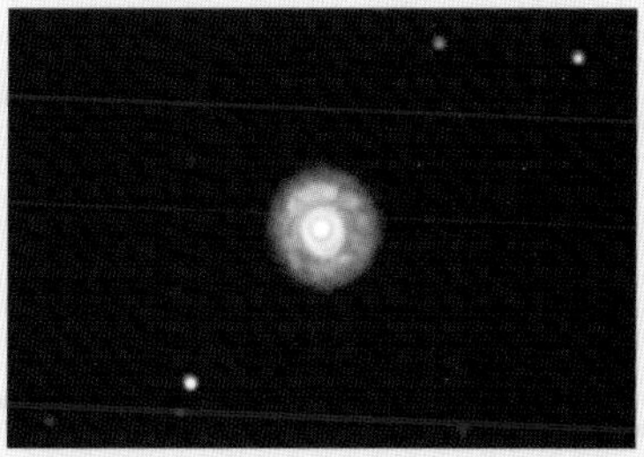

Telescope Aperture:	4" f/5	4" f/9	6" f/7	6" f/9	8" f/6.3	8" f/10	10" f/6.3	10" f/10	12" f/6.3	12" f/10
FOV(35mm film):	2.7° x 4.1°	1.50° x 2.26°	1.29° x 1.93°	1.0° x 1.50°	1.07° x 1.61°	0.68° x 1.02°	0.86° x 1.29°	0.54° x 0.81°	0.72° x 1.07°	0.45° x 0.68°

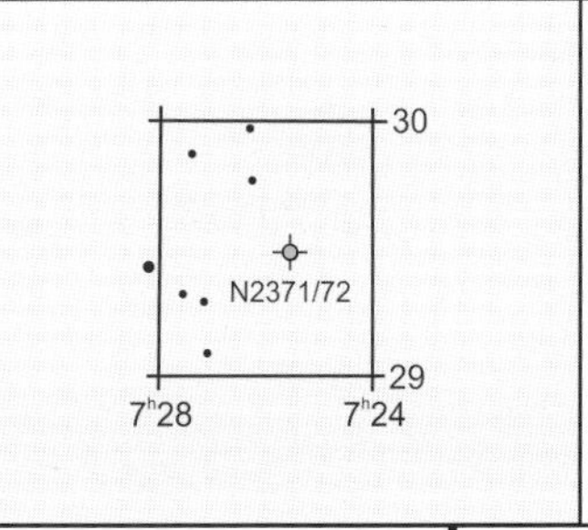

NGC 2371/72

RA:	$07^h\ 25^m\ 39.0^s$	Con:	Gemini
Dec:	29° 28' 53"	Type:	Planetary Nebula
Size:	0.9'	Mag:	13.0

NGC 2371/72 is an irregular planetary nebula larger than the Eskimo Nebula. Object has a double-lobed structure.

Telescope Aperture:	4" f/5	4" f/9	6" f/7	6" f/9	8" f/6.3	8" f/10	10" f/6.3	10" f/10	12" f/6.3	12" f/10
FOV(35mm film):	2.7° x 4.1°	1.50° x 2.26°	1.29° x 1.93°	1.0° x 1.50°	1.07° x 1.61°	0.68° x 1.02°	0.86° x 1.29°	0.54° x 0.81°	0.72° x 1.07°	0.45° x 0.68°

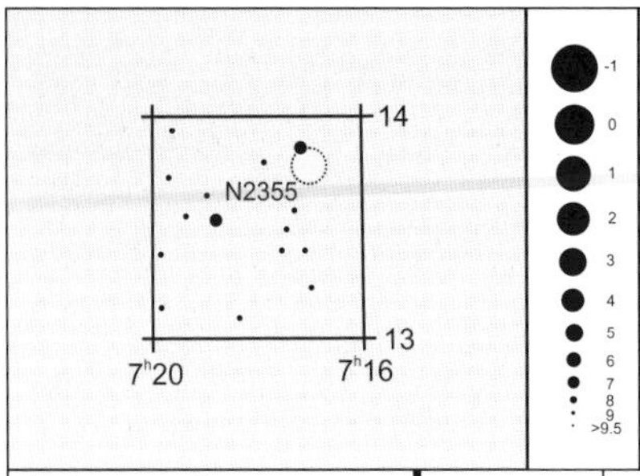

NGC 2355

RA:	07h 16m 56.7s	Con:	Gemini
Dec:	13° 46' 54"	Type:	Open Cluster
Size:	9.0'	Mag:	10.0

NGC 2355 is a dense open cluster.

Telescope Aperture:	4" f/5	4" f/9	6" f/7	6" f/9	8" f/6.3	8" f/10	10" f/6.3	10" f/10	12" f/6.3	12" f/10
FOV(35mm film):	2.7° x 4.1°	1.50° x 2.26°	1.29° x 1.93°	1.0° x 1.50°	1.07° x 1.61°	0.68° x 1.02°	0.86° x 1.29°	0.54° x 0.81°	0.72° x 1.07°	0.45° x 0.68°

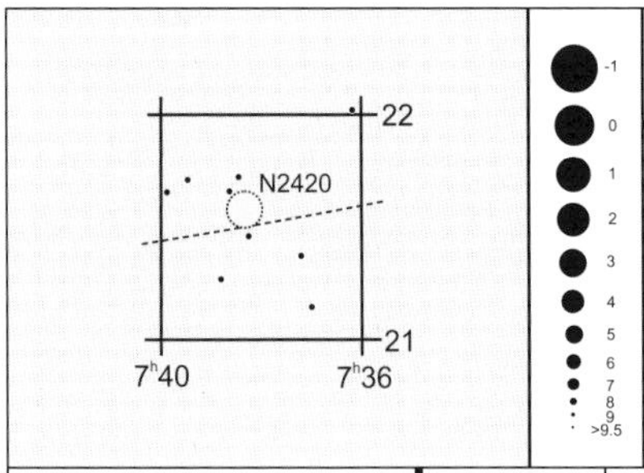

NGC 2420

RA:	07h 38m 32.8s	Con:	Gemini
Dec:	21° 33' 52"	Type:	Open Cluster
Size:	10.0'	Mag:	8.3

NGC 2420 is a rich open cluster comprised of 30 stars.

Telescope Aperture:	4" f/5	4" f/9	6" f/7	6" f/9	8" f/6.3	8" f/10	10" f/6.3	10" f/10	12" f/6.3	12" f/10
FOV(35mm film):	2.7° x 4.1°	1.50° x 2.26°	1.29° x 1.93°	1.0° x 1.50°	1.07° x 1.61°	0.68° x 1.02°	0.86° x 1.29°	0.54° x 0.81°	0.72° x 1.07°	0.45° x 0.68°

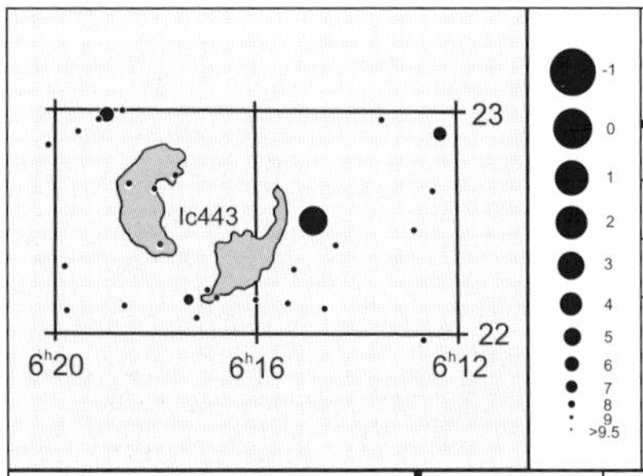

IC 443

RA:	06h 16m 57.0s	Con:	Gemini
Dec:	22° 46' 58"	Type:	Emission Nebula
Size:	50.0'	Mag:	

IC 443 displays a crescent shape with a sharply defined convex side. Object appears very filamentary.

Telescope Aperture:	4" f/5	4" f/9	6" f/7	6" f/9	8" f/6.3	8" f/10	10" f/6.3	10" f/10	12" f/6.3	12" f/10
FOV(35mm film):	2.7° x 4.1°	1.50° x 2.26°	1.29° x 1.93°	1.0° x 1.50°	1.07° x 1.61°	0.68° x 1.02°	0.86° x 1.29°	0.54° x 0.81°	0.72° x 1.07°	0.45° x 0.68°

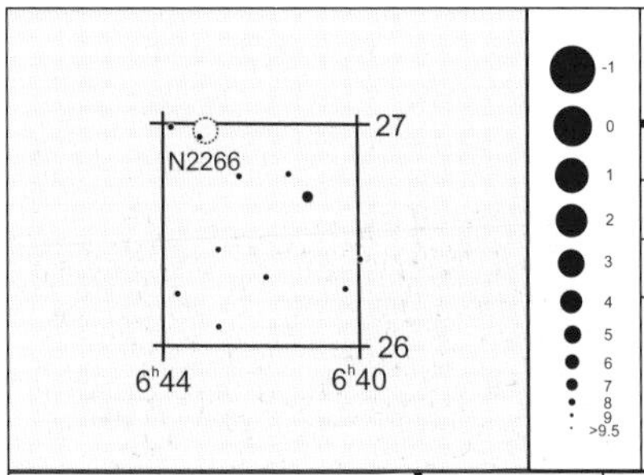

NGC 2266

RA:	06h 43m 15.0s	Con:	Gemini
Dec:	26° 57' 56"	Type:	Open Cluster
Size:	7.0'	Mag:	10.0

NGC 2266 is a dense open cluster comprised of 35 stars.

Telescope Aperture:	4" f/5	4" f/9	6" f/7	6" f/9	8" f/6.3	8" f/10	10" f/6.3	10" f/10	12" f/6.3	12" f/10
FOV(35mm film):	2.7° x 4.1°	1.50° x 2.26°	1.29° x 1.93°	1.0° x 1.50°	1.07° x 1.61°	0.68° x 1.02°	0.86° x 1.29°	0.54° x 0.81°	0.72° x 1.07°	0.45° x 0.68°

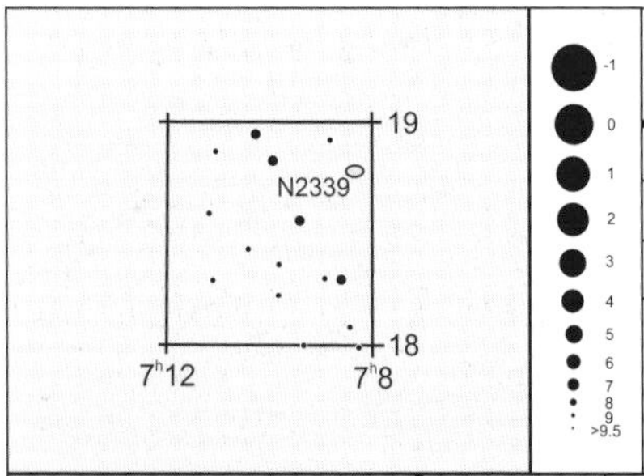

NGC 2339

RA:	07h 08m 20.8s	Con:	Gemini
Dec:	18° 46' 54"	Type:	Spiral Galaxy
Size:	2.8' x 2.0'	Mag:	11.6

NGC 2339 is an Sc-type spiral galaxy found within the plane of the Milky Way. Object is found southwest of the Eskimo Nebula.

Telescope Aperture:	4" f/5	4" f/9	6" f/7	6" f/9	8" f/6.3	8" f/10	10" f/6.3	10" f/10	12" f/6.3	12" f/10
FOV(35mm film):	2.7° x 4.1°	1.50° x 2.26°	1.29° x 1.93°	1.0° x 1.50°	1.07° x 1.61°	0.68° x 1.02°	0.86° x 1.29°	0.54° x 0.81°	0.72° x 1.07°	0.45° x 0.68°

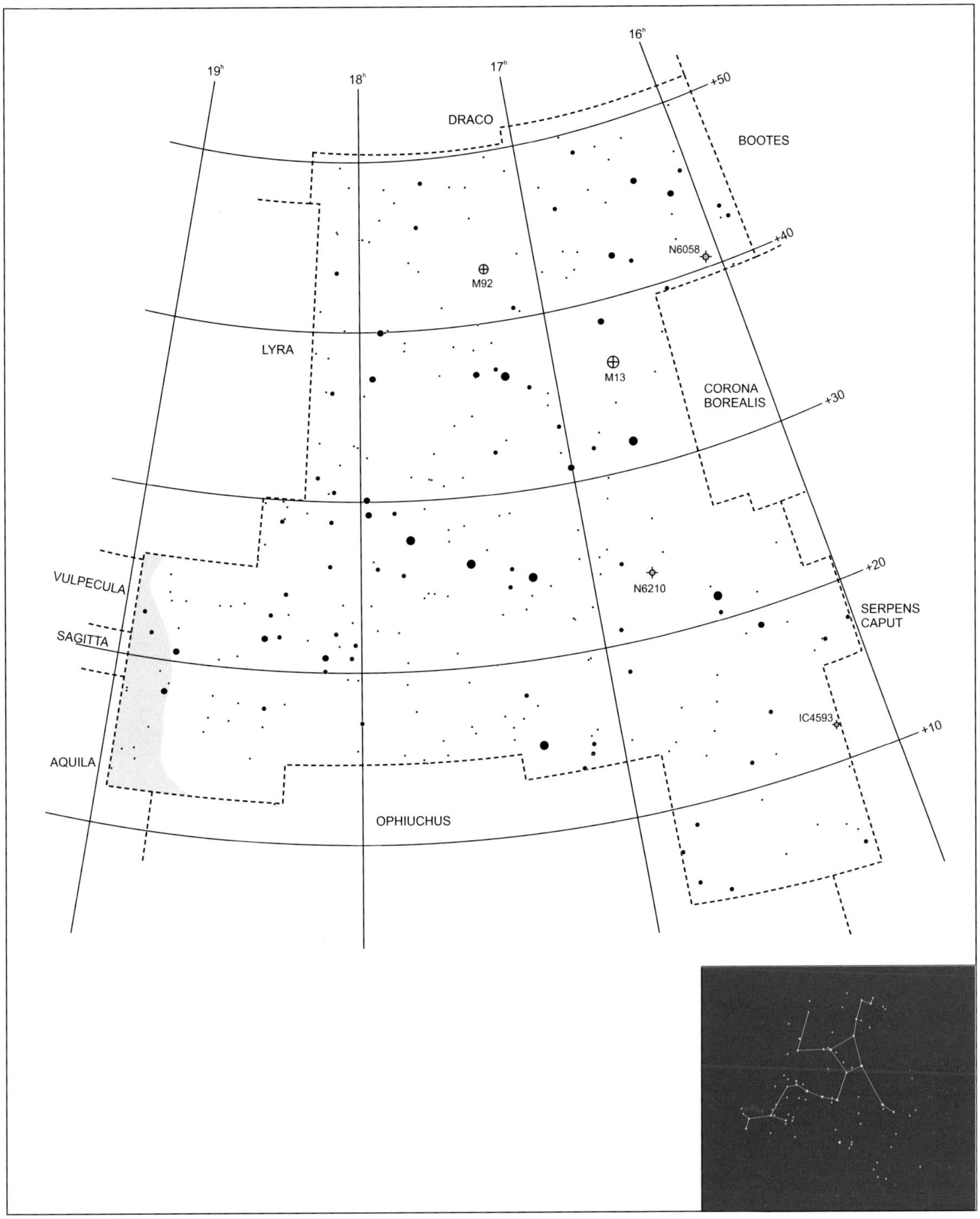

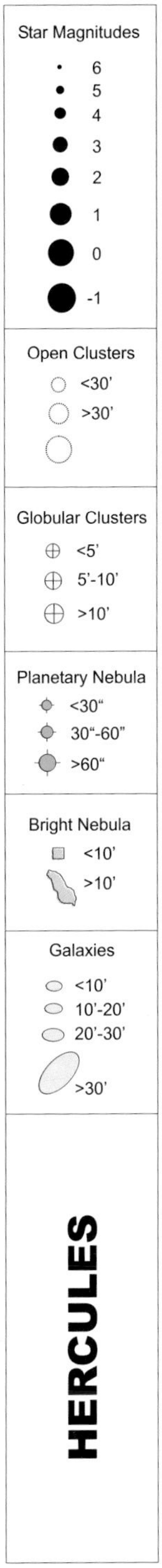

Constellation Facts:

Hercules; (HER-kyou-Leez)

Hercules, the Hero.
The stars of Hercules rise in the northeast, pass overhead, and set towards the northwest.
Central feature of the constellation is a keystone asterism, halfway between Vega and Alphecca.
Constellation covers 1225 square degrees.

Constellation is visible from 90° N to 38° S. Partially visible from 38° S to 90° S.

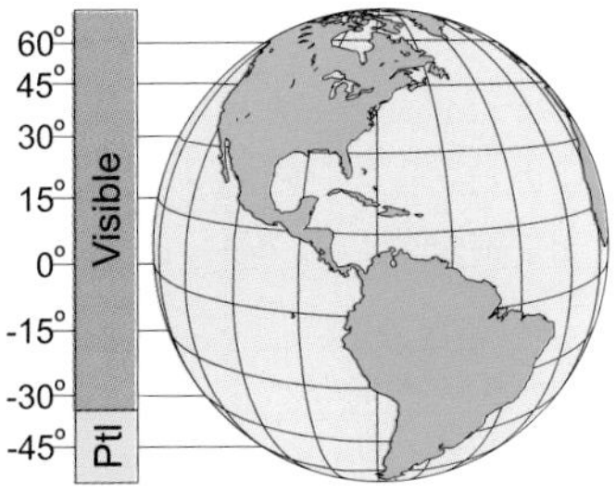

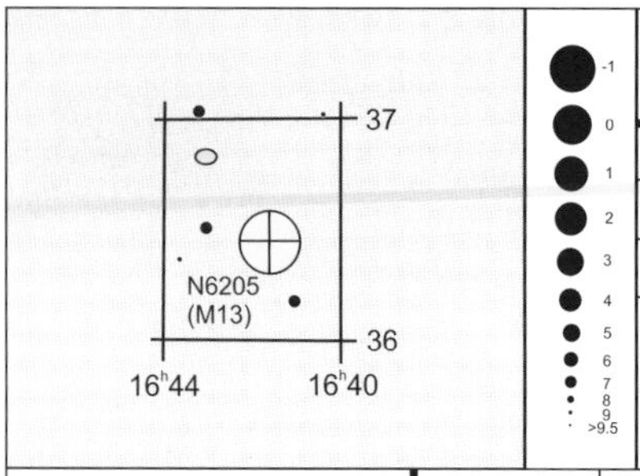

M13 (NGC 6205)

RA:	$16^h\ 41^m\ 45.8^s$	Con:	Hercules
Dec:	36° 28' 03"	Type:	Globular Cluster
Size:	16.6'	Mag:	5.9

M13 (NGC 6205) is a large well defined globular cluster. Object lies 2.5° south of Eta Herculis, one of the keystone stars.

Telescope Aperture:	4" f/5	4" f/9	6" f/7	6" f/9	8" f/6.3	8" f/10	10" f/6.3	10" f/10	12" f/6.3	12" f/10
FOV(35mm film):	2.7° x 4.1°	1.50° x 2.26°	1.29° x 1.93°	1.0° x 1.50°	1.07° x 1.61°	0.68° x 1.02°	0.86° x 1.29°	0.54° x 0.81°	0.72° x 1.07°	0.45° x 0.68°

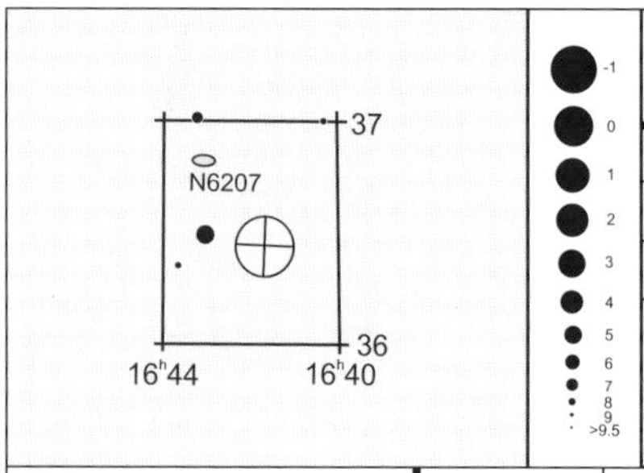

NGC 6207

RA:	$16^h\ 43^m\ 9.8^s$	Con:	Hercules
Dec:	36° 50' 03"	Type:	Spiral Galaxy
Size:	2.0' x 1.0'	Mag:	11.6

NGC 6207 is a small Sc-type spiral galaxy tilted 45° to our line of site. Galaxy is found 1.5° north and slightly east of M13.

Telescope Aperture:	4" f/5	4" f/9	6" f/7	6" f/9	8" f/6.3	8" f/10	10" f/6.3	10" f/10	12" f/6.3	12" f/10
FOV(35mm film):	2.7° x 4.1°	1.50° x 2.26°	1.29° x 1.93°	1.0° x 1.50°	1.07° x 1.61°	0.68° x 1.02°	0.86° x 1.29°	0.54° x 0.81°	0.72° x 1.07°	0.45° x 0.68°

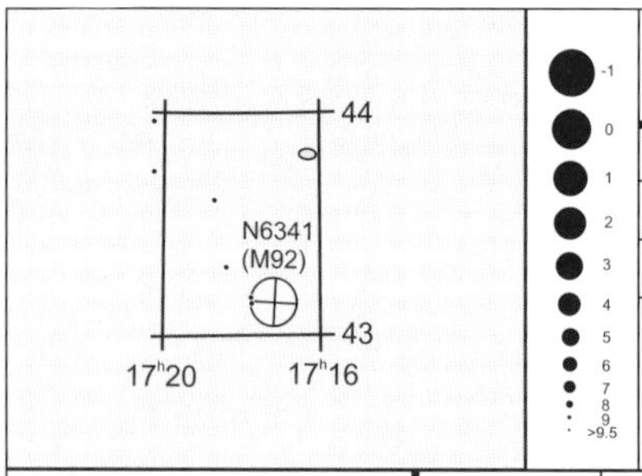

M92 (NGC 6341)

RA:	$17^h\ 17^m\ 9.7^s$	Con:	Hercules
Dec:	43° 08' 06"	Type:	Globular Cluster
Size:	11.2'	Mag:	6.5

M92 (NGC 6341) is located 10° northeast of M13. Object is nearly as large and bright as M13, but because it is more compact, it is harder to resolve into individual stars.

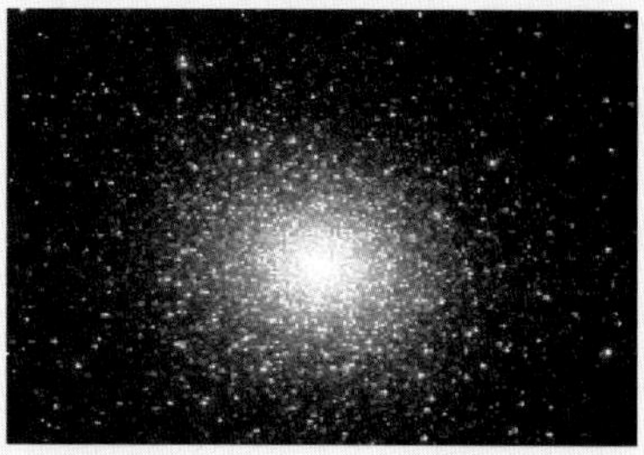

Telescope Aperture:	4" f/5	4" f/9	6" f/7	6" f/9	8" f/6.3	8" f/10	10" f/6.3	10" f/10	12" f/6.3	12" f/10
FOV(35mm film):	2.7° x 4.1°	1.50° x 2.26°	1.29° x 1.93°	1.0° x 1.50°	1.07° x 1.61°	0.68° x 1.02°	0.86° x 1.29°	0.54° x 0.81°	0.72° x 1.07°	0.45° x 0.68°

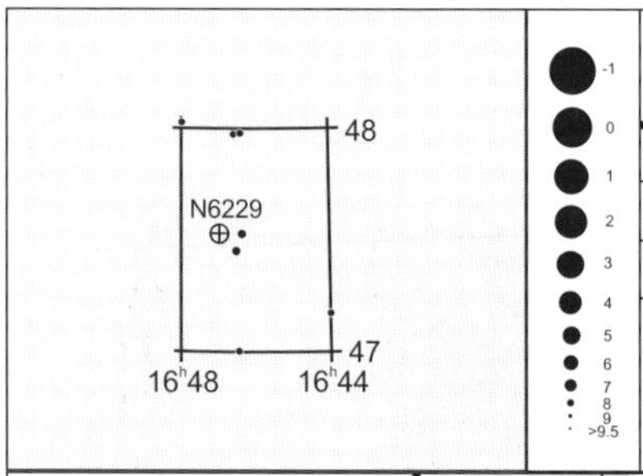

NGC 6229

RA:	$16^h\ 47^m\ 3.6^s$	Con:	Hercules
Dec:	47° 32' 05"	Type:	Globular Cluster
Size:	4.5'	Mag:	9.4

NGC 6229 is the third globular cluster found in Hercules. Object is located 7° northwest of M92.

Telescope Aperture:	4" f/5	4" f/9	6" f/7	6" f/9	8" f/6.3	8" f/10	10" f/6.3	10" f/10	12" f/6.3	12" f/10
FOV(35mm film):	2.7° x 4.1°	1.50° x 2.26°	1.29° x 1.93°	1.0° x 1.50°	1.07° x 1.61°	0.68° x 1.02°	0.86° x 1.29°	0.54° x 0.81°	0.72° x 1.07°	0.45° x 0.68°

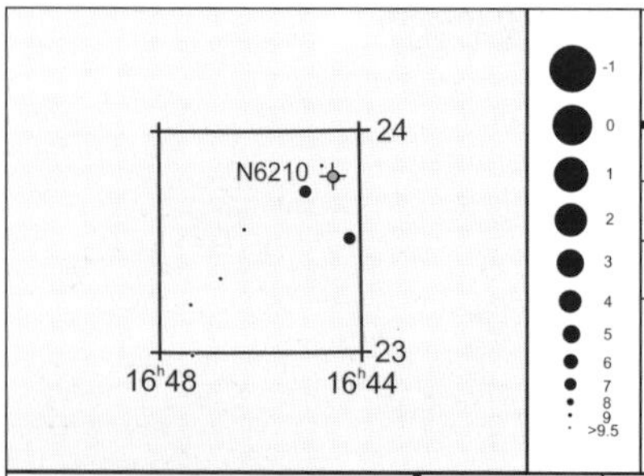

NGC 6210

RA:	$16^h\ 44^m\ 34.1^s$	Con:	Hercules
Dec:	23° 49' 01"	Type:	Planetary Nebula
Size:	0.2'	Mag:	9.0

NGC 6210 is a bright planetary nebula. Object appears as a blueish disk.

Telescope Aperture:	4" f/5	4" f/9	6" f/7	6" f/9	8" f/6.3	8" f/10	10" f/6.3	10" f/10	12" f/6.3	12" f/10
FOV(35mm film):	2.7° x 4.1°	1.50° x 2.26°	1.29° x 1.93°	1.0° x 1.50°	1.07° x 1.61°	0.68° x 1.02°	0.86° x 1.29°	0.54° x 0.81°	0.72° x 1.07°	0.45° x 0.68°

Leo

Crosses Prime Meridian: April thru March

WINTER

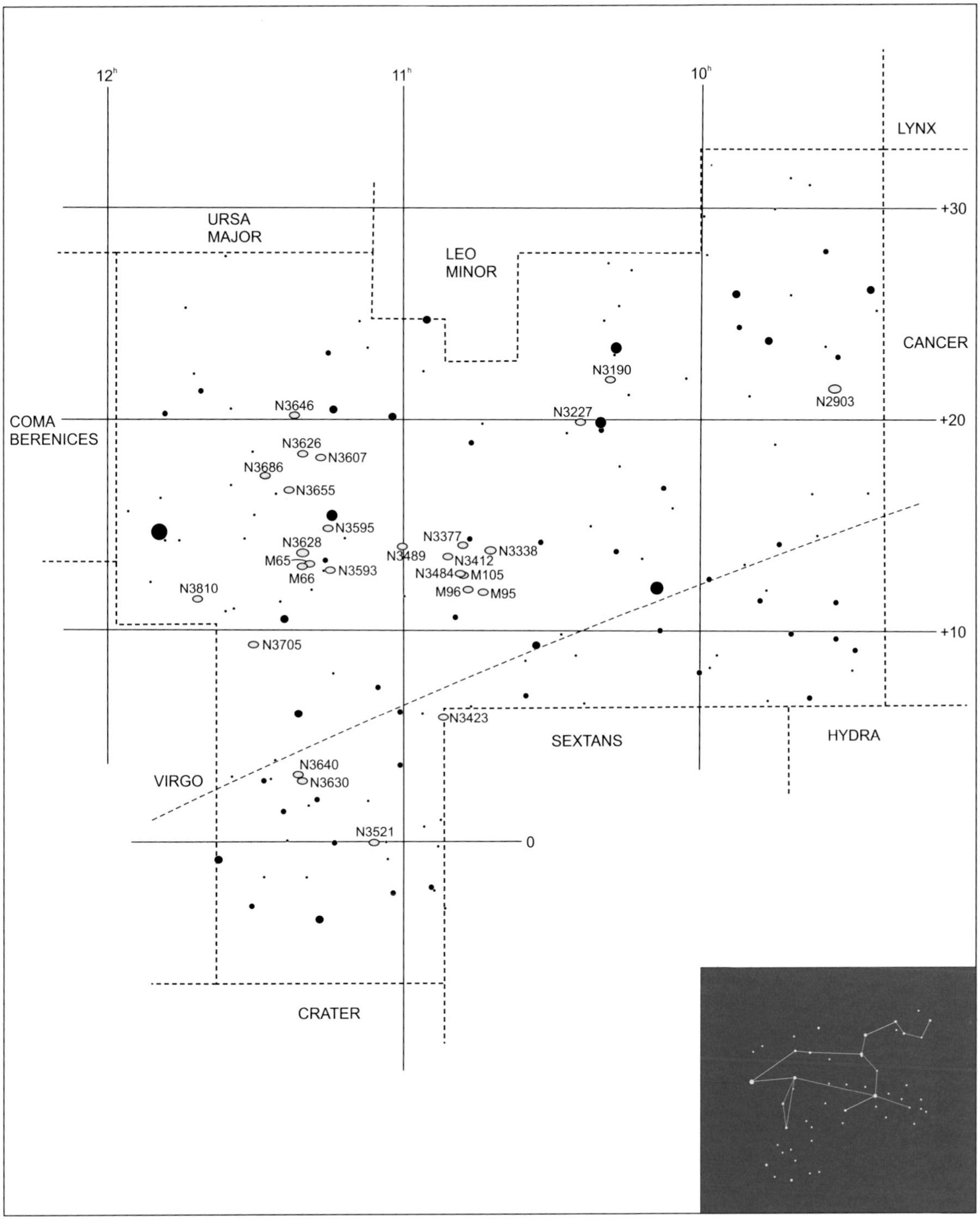

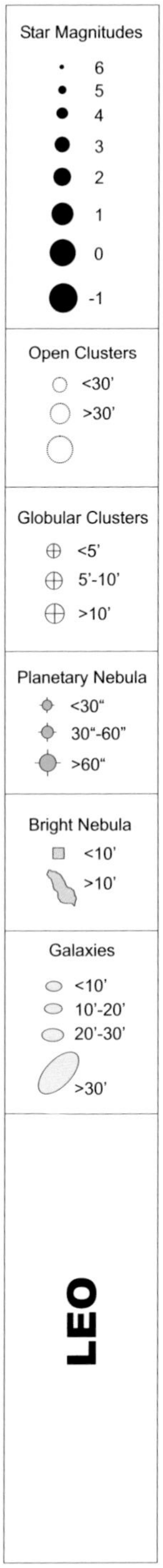

Constellation Facts:

Leo; (LEE-oh)

Leo, the Lion, because it is a zodiacal constellation, the stars rise in the east, pass the meridian high in the south, and sets towards the west.
The constellation covers 947 square degrees.

Constellation is visible from 82° N to 57° S. Partially visible from 57° S to 90° S.

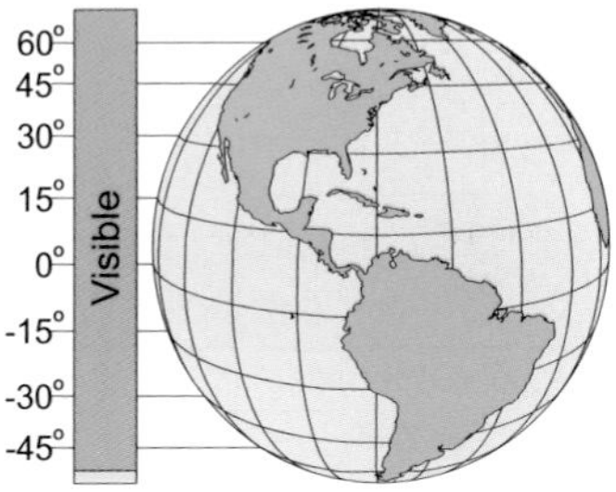

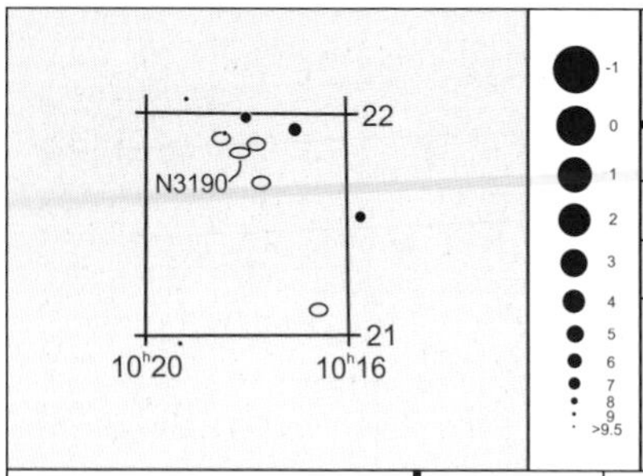

NGC 3190

RA:	$10^h\ 18^m\ 8.9^s$	Con:	Leo
Dec:	21° 49' 46"	Type:	Spiral Galaxy
Size:	3.5' x 1.3'	Mag:	11.0

NGC 3190 appears as a very elongated galaxy that is the brightest galaxy in its local group.

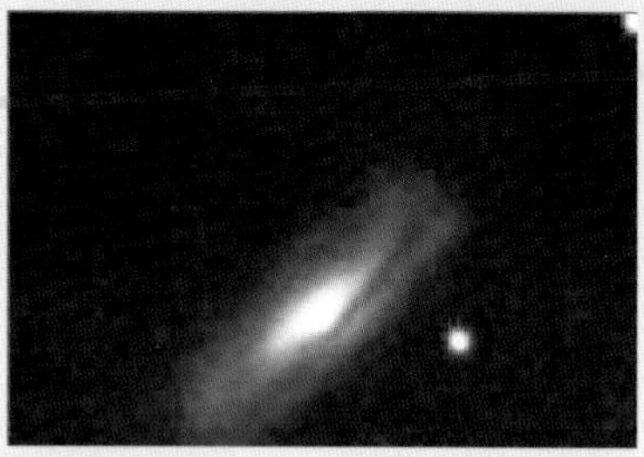

Telescope Aperture:	4" f/5	4" f/9	6" f/7	6" f/9	8" f/6.3	8" f/10	10" f/6.3	10" f/10	12" f/6.3	12" f/10
FOV(35mm film):	2.7° x 4.1°	1.50° x 2.26°	1.29° x 1.93°	1.0° x 1.50°	1.07° x 1.61°	0.68° x 1.02°	0.86° x 1.29°	0.54° x 0.81°	0.72° x 1.07°	0.45° x 0.68°

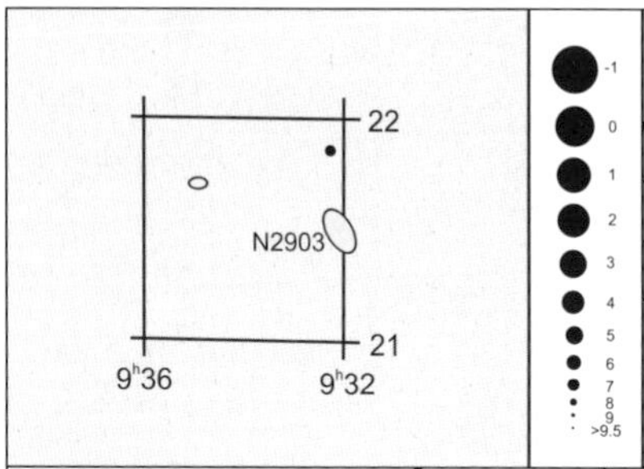

NGC 2903

RA:	$09^h\ 32^m\ 14.9^s$	Con:	Leo
Dec:	21° 29' 47"	Type:	Spiral Galaxy
Size:	13.0' x 5.0'	Mag:	8.9

NGC 2903 is located 1.5° south of Lambda Leonis. Object is a loosely wound spiral galaxy.

Telescope Aperture:	4" f/5	4" f/9	6" f/7	6" f/9	8" f/6.3	8" f/10	10" f/6.3	10" f/10	12" f/6.3	12" f/10
FOV(35mm film):	2.7° x 4.1°	1.50° x 2.26°	1.29° x 1.93°	1.0° x 1.50°	1.07° x 1.61°	0.68° x 1.02°	0.86° x 1.29°	0.54° x 0.81°	0.72° x 1.07°	0.45° x 0.68°

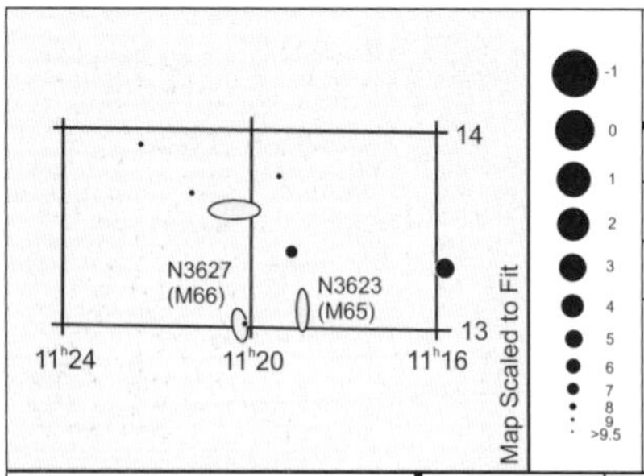

M66 (NGC 3627)

RA:	$11^h\ 20^m\ 15.0^s$	Con:	Leo
Dec:	12° 58' 43"	Type:	Spiral Galaxy
Size:	7.6' x 3.3'	Mag:	9.0

M66 (NGC 3627) is an Sb-type spiral galaxy, and is the largest galaxy in a group of three bright galaxies. Found midway between the stars Theta and Iota Leonis.

Telescope Aperture:	4" f/5	4" f/9	6" f/7	6" f/9	8" f/6.3	8" f/10	10" f/6.3	10" f/10	12" f/6.3	12" f/10
FOV(35mm film):	2.7° x 4.1°	1.50° x 2.26°	1.29° x 1.93°	1.0° x 1.50°	1.07° x 1.61°	0.68° x 1.02°	0.86° x 1.29°	0.54° x 0.81°	0.72° x 1.07°	0.45° x 0.68°

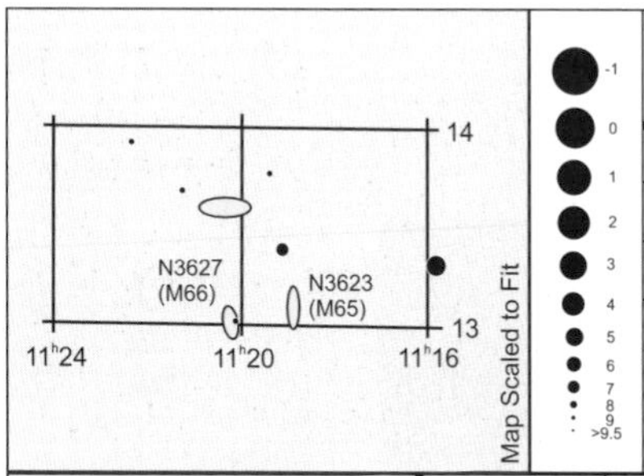

M65 (NGC 3623)

RA:	$11^h\ 18^m\ 58.3$	Con:	Leo
Dec:	13° 05' 18"	Type:	Spiral Galaxy
Size:	9.7' x 2.8'	Mag:	10.1

M65 (NGC 3623) is a type Sa/Sb spiral galaxy, much flatter than M-66. Objects M66 and M65 lie 21' apart.

Telescope Aperture:	4" f/5	4" f/9	6" f/7	6" f/9	8" f/6.3	8" f/10	10" f/6.3	10" f/10	12" f/6.3	12" f/10
FOV(35mm film):	2.7° x 4.1°	1.50° x 2.26°	1.29° x 1.93°	1.0° x 1.50°	1.07° x 1.61°	0.68° x 1.02°	0.86° x 1.29°	0.54° x 0.81°	0.72° x 1.07°	0.45° x 0.68°

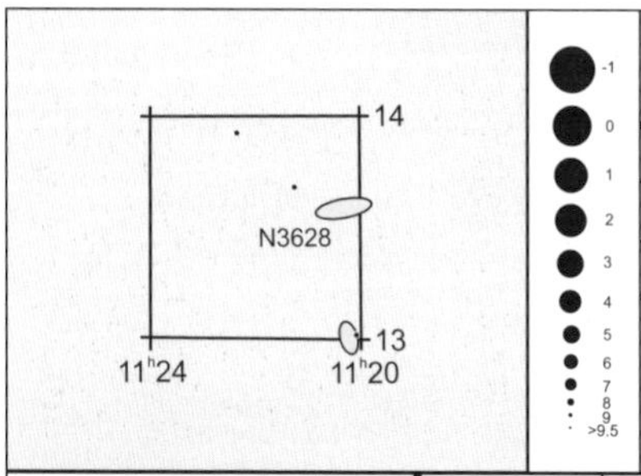

NGC 3628

RA:	$11^h\ 20^m\ 21.1^s$	Con:	Leo
Dec:	13° 35' 43"	Type:	Spiral Galaxy
Size:	10.0' x 3.3'	Mag:	9.5

NGC 3628 is a large faint galaxy aligned edge-on to our line-of-site. Object found 35' north of M66.

Telescope Aperture:	4" f/5	4" f/9	6" f/7	6" f/9	8" f/6.3	8" f/10	10" f/6.3	10" f/10	12" f/6.3	12" f/10
FOV(35mm film):	2.7° x 4.1°	1.50° x 2.26°	1.29° x 1.93°	1.0° x 1.50°	1.07° x 1.61°	0.68° x 1.02°	0.86° x 1.29°	0.54° x 0.81°	0.72° x 1.07°	0.45° x 0.68°

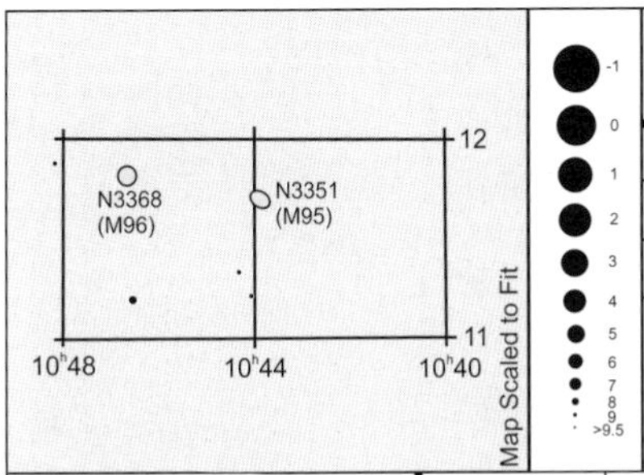

M95 (NGC 3351)

RA:	$10^h\ 44^m\ 2.9^s$	Con:	Leo
Dec:	11° 41' 43"	Type:	Barred Spiral
Size:	7.0' x 4.0'	Mag:	9.7

M95 (NGC 3351) is located 42' west of M96. Object comprises a second trio of galaxies found in Leo. M95 is a small barred spiral galaxy.

Telescope Aperture:	4" f/5	4" f/9	6" f/7	6" f/9	8" f/6.3	8" f/10	10" f/6.3	10" f/10	12" f/6.3	12" f/10
FOV(35mm film):	2.7° x 4.1°	1.50° x 2.26°	1.29° x 1.93°	1.0° x 1.50°	1.07° x 1.61°	0.68° x 1.02°	0.86° x 1.29°	0.54° x 0.81°	0.72° x 1.07°	0.45° x 0.68°

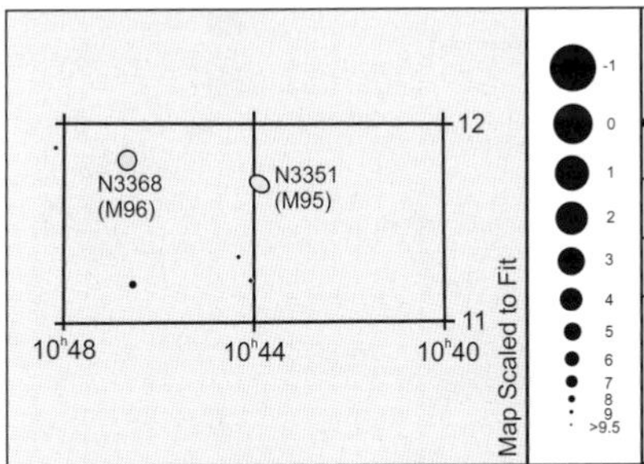

M96 (NGC 3368)

RA:	$10^h\ 46^m\ 50.9^s$	Con:	Leo
Dec:	11° 48' 43"	Type:	Spiral Galaxy
Size:	7.5' x 5.0'	Mag:	9.2

M96 (NGC 3368) is the brightest in the galactic trio.

Telescope Aperture:	4" f/5	4" f/9	6" f/7	6" f/9	8" f/6.3	8" f/10	10" f/6.3	10" f/10	12" f/6.3	12" f/10
FOV(35mm film):	2.7° x 4.1°	1.50° x 2.26°	1.29° x 1.93°	1.0° x 1.50°	1.07° x 1.61°	0.68° x 1.02°	0.86° x 1.29°	0.54° x 0.81°	0.72° x 1.07°	0.45° x 0.68°

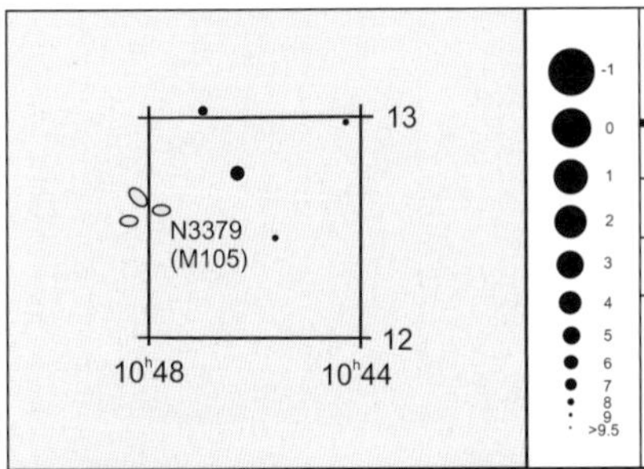

M105 (NGC 3379)

RA:	$10^h\ 47^m\ 51.0^s$	Con:	Leo
Dec:	12° 34' 43"	Type:	Elliptical Galaxy
Size:	4.6' x 4.0'	Mag:	9.3

M105 (NGC 3379) is located roughly 48' north-northwest of M96. Object displays reasonable surface brightness, but little detail is revealed.

Telescope Aperture:	4" f/5	4" f/9	6" f/7	6" f/9	8" f/6.3	8" f/10	10" f/6.3	10" f/10	12" f/6.3	12" f/10
FOV(35mm film):	2.7° x 4.1°	1.50° x 2.26°	1.29° x 1.93°	1.0° x 1.50°	1.07° x 1.61°	0.68° x 1.02°	0.86° x 1.29°	0.54° x 0.81°	0.72° x 1.07°	0.45° x 0.68°

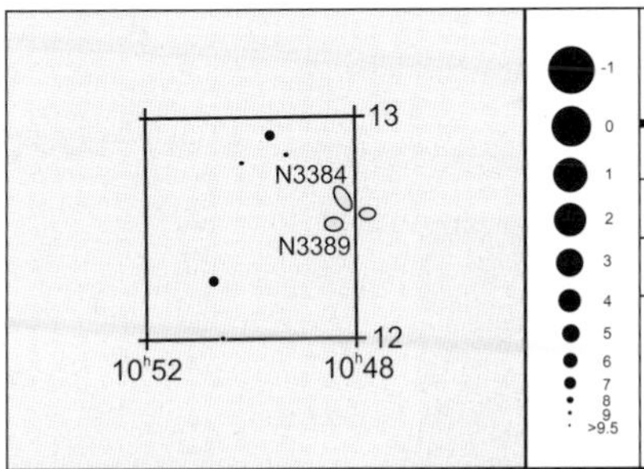

NGC 3384

RA:	$10^h\ 48^m\ 21.0^s$	Con:	Leo
Dec:	12° 37' 43"	Type:	Spiral Galaxy
Size:	1.6' x 1.1'	Mag:	10.0

NGC 3384 forms a small triangle with M105 and NGC 3389 that measures approximately 8' across.

Telescope Aperture:	4" f/5	4" f/9	6" f/7	6" f/9	8" f/6.3	8" f/10	10" f/6.3	10" f/10	12" f/6.3	12" f/10
FOV(35mm film):	2.7° x 4.1°	1.50° x 2.26°	1.29° x 1.93°	1.0° x 1.50°	1.07° x 1.61°	0.68° x 1.02°	0.86° x 1.29°	0.54° x 0.81°	0.72° x 1.07°	0.45° x 0.68°

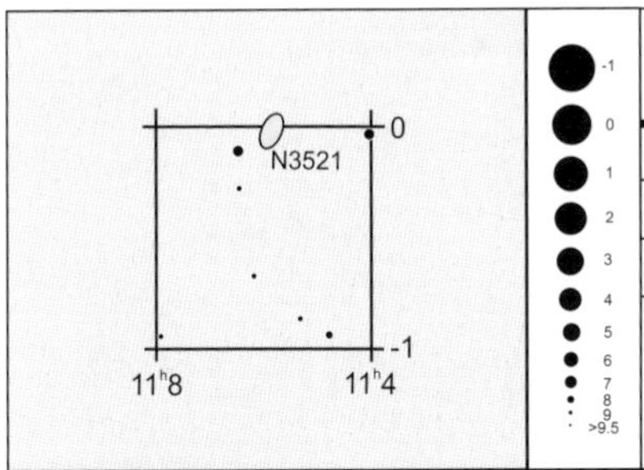

NGC 3521

RA:	$11^h\ 05^m\ 51.0^s$	Con:	Leo
Dec:	-00° 02' 21"	Type:	Spiral Galaxy
Size:	7.0' x 4.0'	Mag:	8.9

NGC 3521 is a large bright spiral galaxy found in the southern region of the constellation. Object is a tightly wound multi-arm Sb-type spiral galaxy.

Telescope Aperture:	4" f/5	4" f/9	6" f/7	6" f/9	8" f/6.3	8" f/10	10" f/6.3	10" f/10	12" f/6.3	12" f/10
FOV(35mm film):	2.7° x 4.1°	1.50° x 2.26°	1.29° x 1.93°	1.0° x 1.50°	1.07° x 1.61°	0.68° x 1.02°	0.86° x 1.29°	0.54° x 0.81°	0.72° x 1.07°	0.45° x 0.68°

LMi

Crosses Prime Meridian: March thru April

WINTER

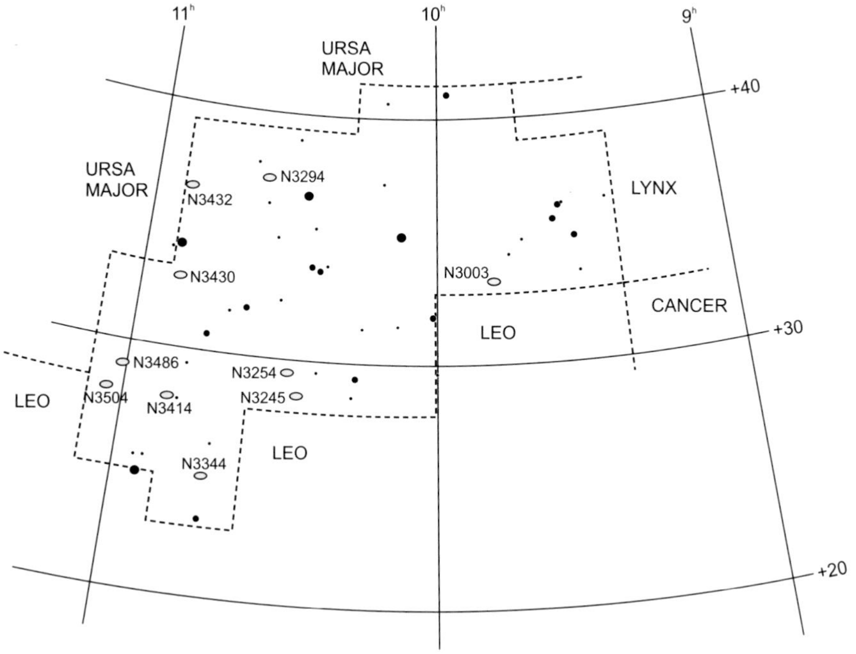

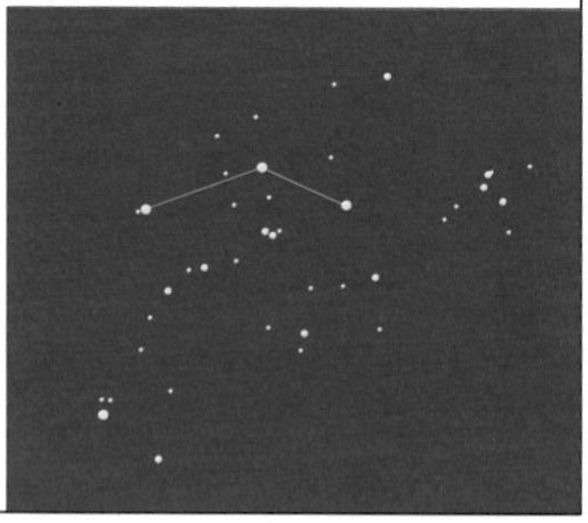

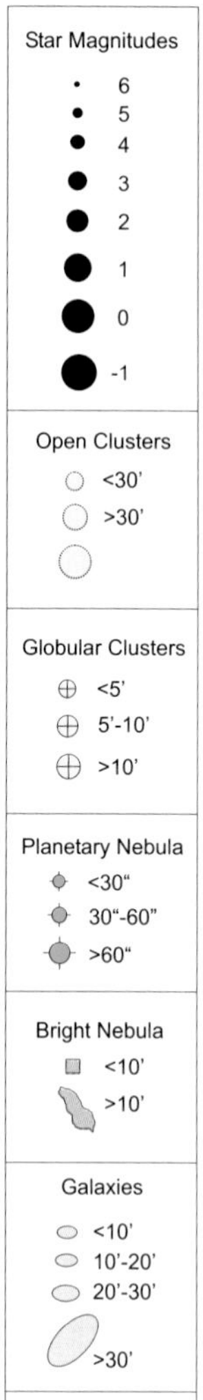

LEO MINOR

Constellation Facts:

Leo Minor; (LEE-oh-MY-nor))

Leo Minor, The Little Lion rises in the northeast, passes the meridian overhead and sets towards the northwest.

The constellation covers 232 square degrees.

Constellation is visible from 90° N to 48° S. Partially visible from 48° S to 90° S.

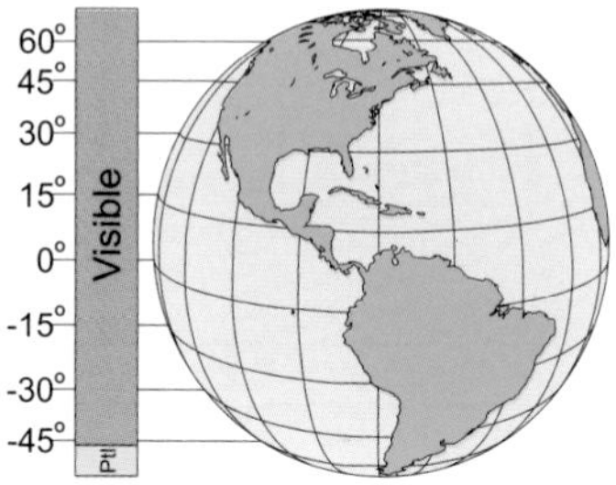

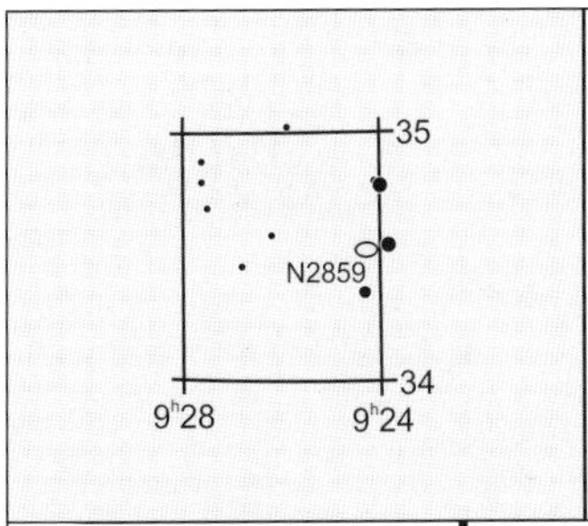

NGC 2859

RA:	$09^h\ 24^m\ 20.9^s$	Con:	Leo Minor
Dec:	34° 30' 49"	Type:	Spiral Galaxy
Size:	4.0' x 3.5'	Mag:	10.7

NGC 2859 appears as a round knot with a bright core, in medium to large telescopes.

Telescope Aperture:	4" f/5	4" f/9	6" f/7	6" f/9	8" f/6.3	8" f/10	10" f/6.3	10" f/10	12" f/6.3	12" f/10
FOV(35mm film):	2.7° x 4.1°	1.50° x 2.26°	1.29° x 1.93°	1.0° x 1.50°	1.07° x 1.61°	0.68° x 1.02°	0.86° x 1.29°	0.54° x 0.81°	0.72° x 1.07°	0.45° x 0.68°

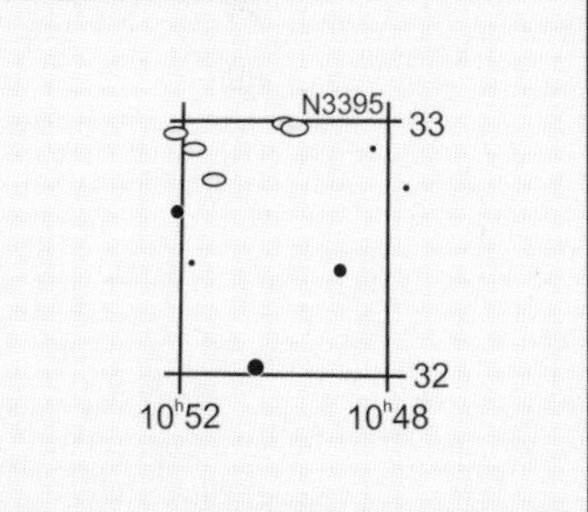

NGC 3395

RA:	$10^h\ 49^m\ 51.0^s$	Con:	Leo Minor
Dec:	32° 58' 48"	Type:	Spiral Galaxy
Size:	1.5' x 0.8'	Mag:	12.1

NGC 3395 is visible as an elongated galaxy with a visible close companion.

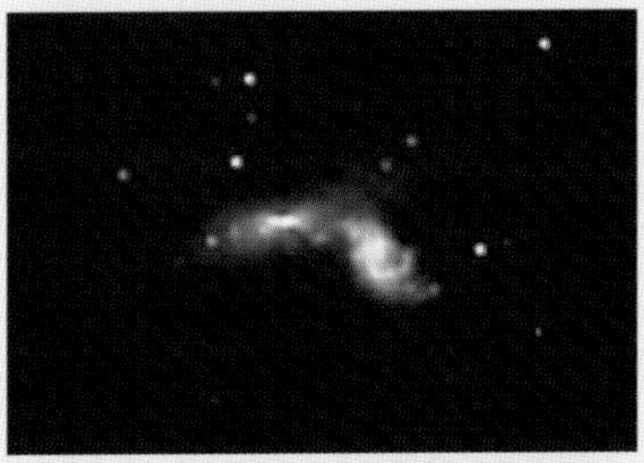

Telescope Aperture:	4" f/5	4" f/9	6" f/7	6" f/9	8" f/6.3	8" f/10	10" f/6.3	10" f/10	12" f/6.3	12" f/10
FOV(35mm film):	2.7° x 4.1°	1.50° x 2.26°	1.29° x 1.93°	1.0° x 1.50°	1.07° x 1.61°	0.68° x 1.02°	0.86° x 1.29°	0.54° x 0.81°	0.72° x 1.07°	0.45° x 0.68°

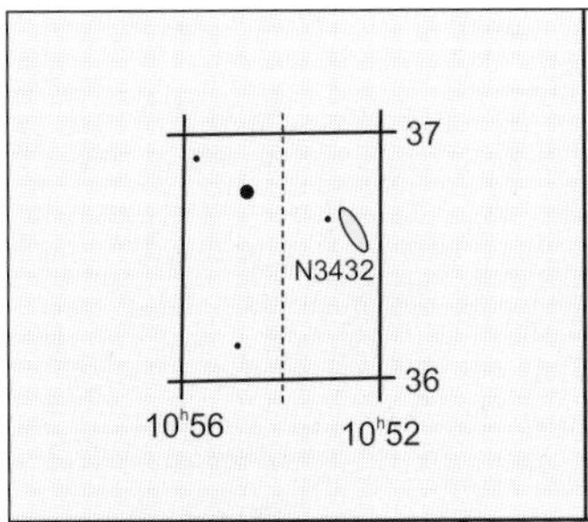

NGC 3432

RA:	$10^h\ 52^m\ 33.0^s$	Con:	Leo Minor
Dec:	36° 36' 48"	Type:	Spiral Galaxy
Size:	6.5' x 1.1'	Mag:	11.3

NGC 3432 is a spiral galaxy orientated edge-on to our line of sight. Object appears as a fuzzy elongated knot in the eyepiece.

Telescope Aperture:	4" f/5	4" f/9	6" f/7	6" f/9	8" f/6.3	8" f/10	10" f/6.3	10" f/10	12" f/6.3	12" f/10
FOV(35mm film):	2.7° x 4.1°	1.50° x 2.26°	1.29° x 1.93°	1.0° x 1.50°	1.07° x 1.61°	0.68° x 1.02°	0.86° x 1.29°	0.54° x 0.81°	0.72° x 1.07°	0.45° x 0.68°

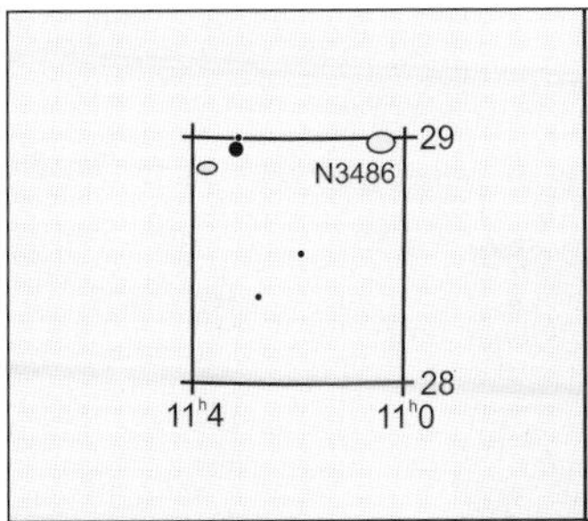

NGC 3486

RA:	$11^h\ 00^m\ 27.0^s$	Con:	Leo Minor
Dec:	28° 57' 47"	Type:	Spiral Galaxy
Size:	7.0' x 5.0'	Mag:	10.3

NGC 3486 is a round and non-distinct galaxy.

Telescope Aperture:	4" f/5	4" f/9	6" f/7	6" f/9	8" f/6.3	8" f/10	10" f/6.3	10" f/10	12" f/6.3	12" f/10
FOV(35mm film):	2.7° x 4.1°	1.50° x 2.26°	1.29° x 1.93°	1.0° x 1.50°	1.07° x 1.61°	0.68° x 1.02°	0.86° x 1.29°	0.54° x 0.81°	0.72° x 1.07°	0.45° x 0.68°

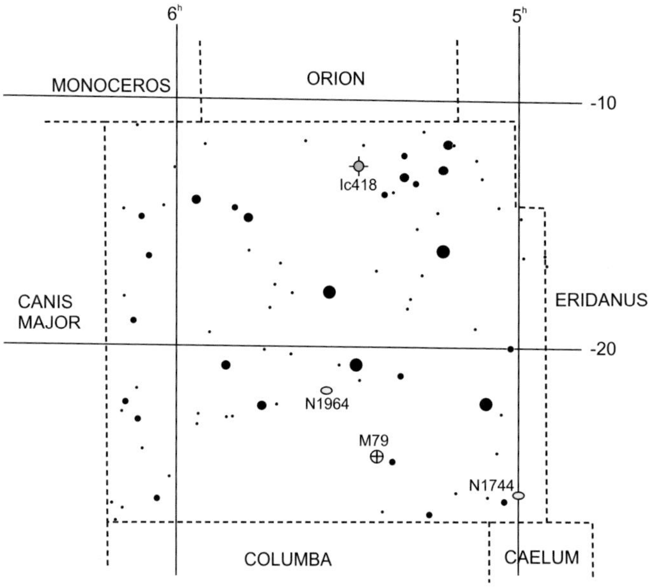

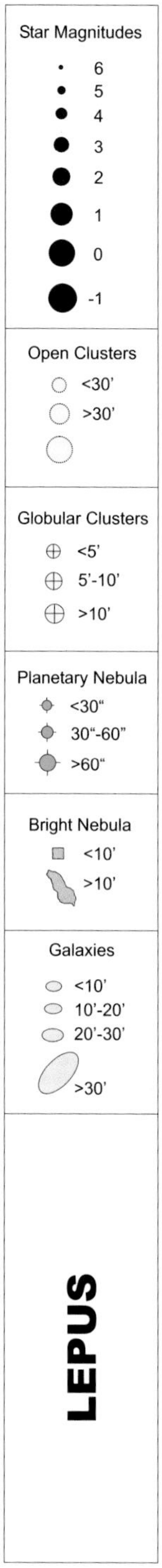

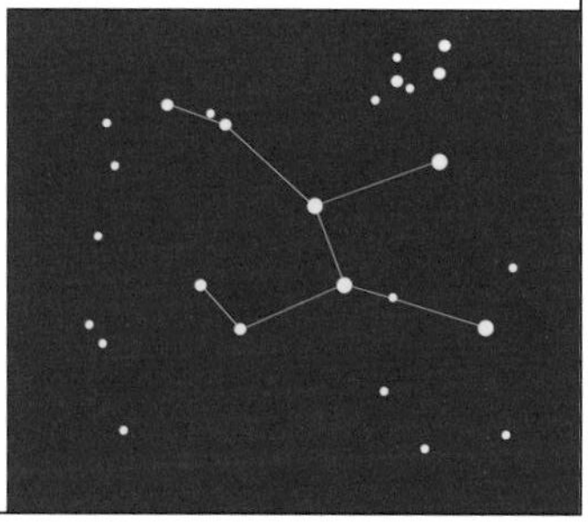

Constellation Facts:

Lepus; (LEE-pus)

Lepus, the Hare crosses low in the sky from southeast to the southwest. The constellation occupies the section of sky just south of Orion and west of Canis Major.
The constellation covers 290 square degrees.

Constellation is visible from 62° N to 900° S. Partially visible from 62° N to 80° N.

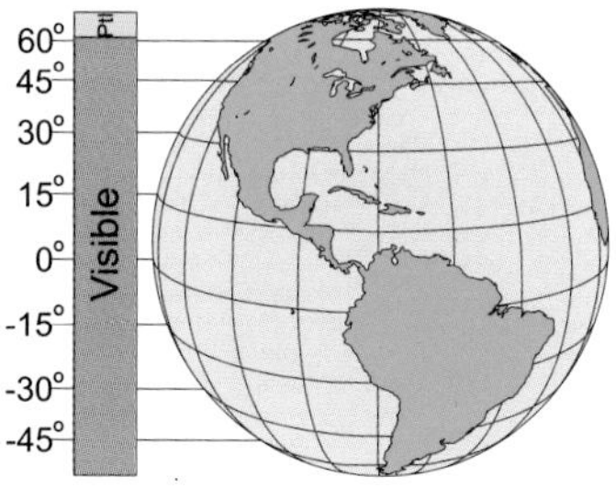

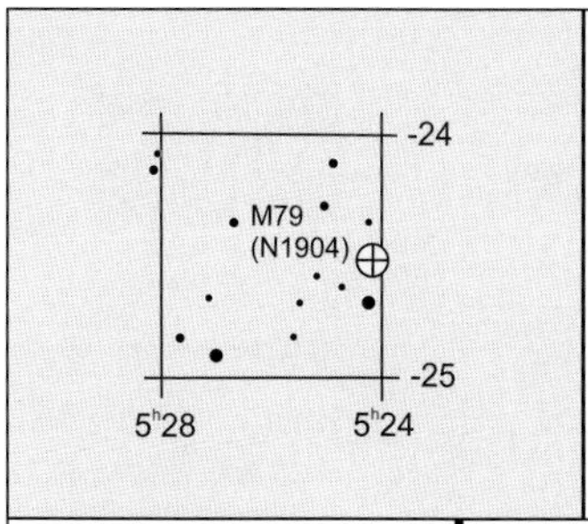

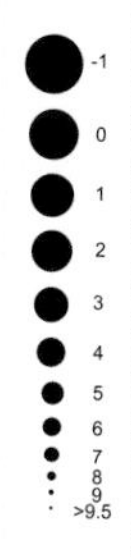

M79 (NGC 1904)

RA:	$05^h\ 24^m\ 31.7^s$	Con:	Lepus
Dec:	-24° 32' 51"	Type:	Globular Cluster
Size:	8.7'	Mag:	8.0

M79 (NGC 1904) is a highly resolved small globular cluster.

Telescope Aperture:	4" f/5	4" f/9	6" f/7	6" f/9	8" f/6.3	8" f/10	10" f/6.3	10" f/10	12" f/6.3	12" f/10
FOV(35mm film):	2.7° x 4.1°	1.50° x 2.26°	1.29° x 1.93°	1.0° x 1.50°	1.07° x 1.61°	0.68° x 1.02°	0.86° x 1.29°	0.54° x 0.81°	0.72° x 1.07°	0.45° x 0.68°

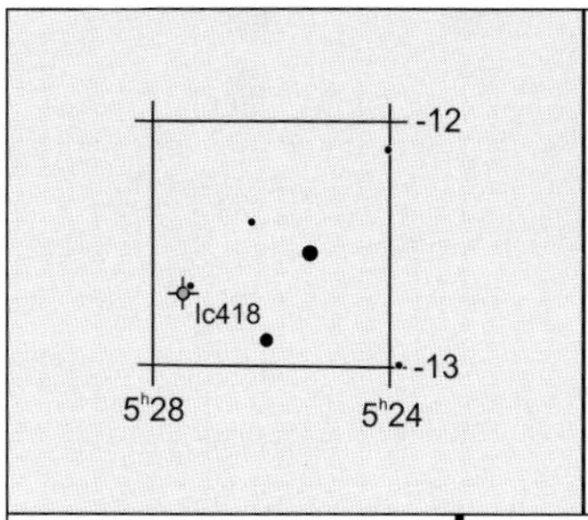

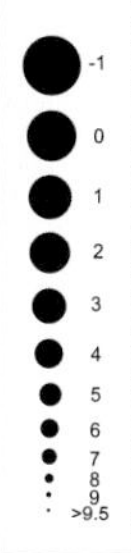

IC 418

RA:	$05^h\ 27^m\ 32.1^s$	Con:	Lepus
Dec:	-12° 41' 53"	Type:	Planetary Nebula
Size:	0.2'	Mag:	11.0

IC 418 is a small, difficult to resolve planetary nebula. Object has been named the Spirograph nebula, after Hubble images revealed fine filament structures at its center.

Telescope Aperture:	4" f/5	4" f/9	6" f/7	6" f/9	8" f/6.3	8" f/10	10" f/6.3	10" f/10	12" f/6.3	12" f/10
FOV(35mm film):	2.7° x 4.1°	1.50° x 2.26°	1.29° x 1.93°	1.0° x 1.50°	1.07° x 1.61°	0.68° x 1.02°	0.86° x 1.29°	0.54° x 0.81°	0.72° x 1.07°	0.45° x 0.68°

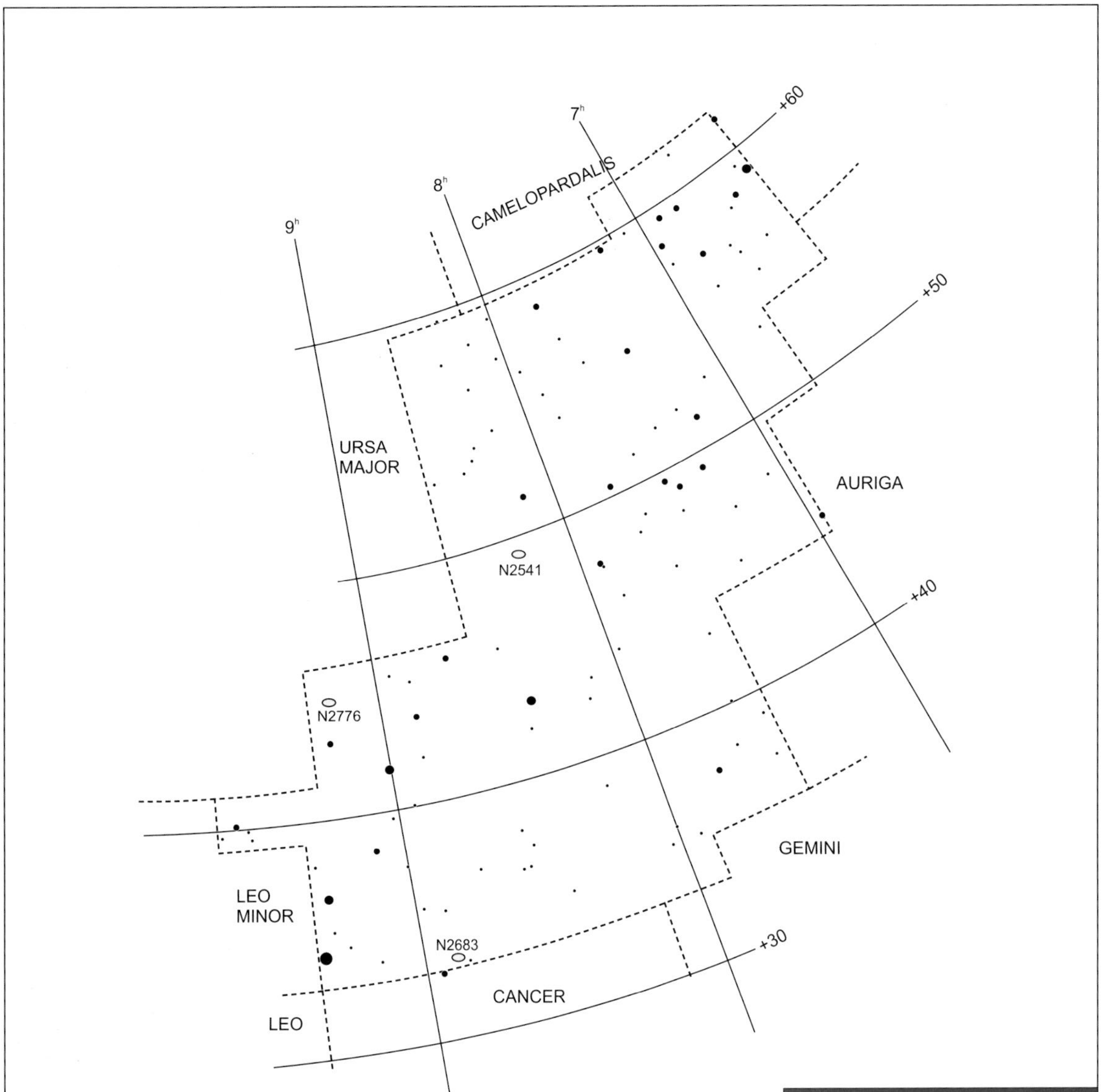

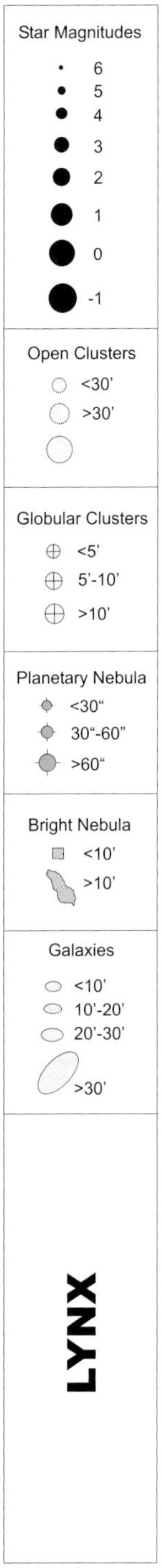

Constellation Facts:

Lynx; (links)

Lynx, the Lynx.
The stars of Lynx rise in the northeast cross the meridian and set in the northwest.
The constellation covers 545 square degrees.

Constellation is visible from 90° N to 28° S. Partially visible from 28° S to 90° S.

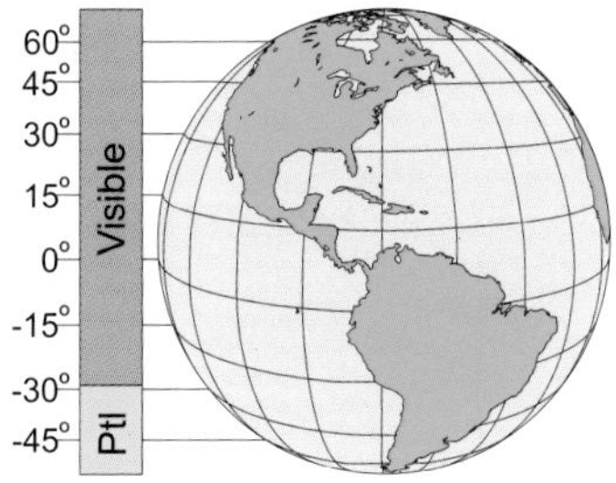

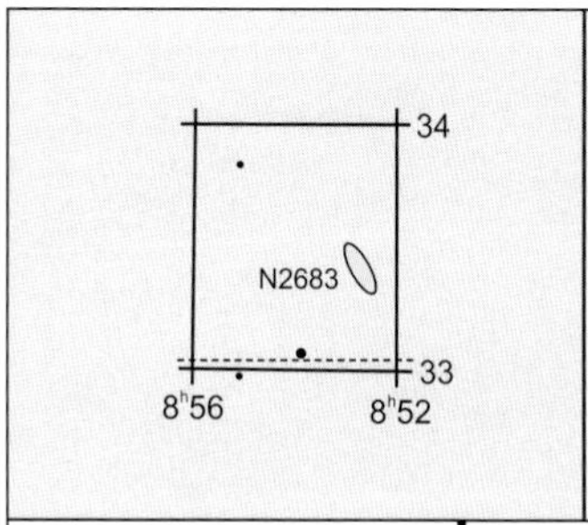

NGC 2683

RA:	$08^h\ 52^m\ 45.0^s$	Con:	Lynx
Dec:	$33^\circ\ 24'\ 48''$	Type:	Spiral Galaxy
Size:	8.0' x 1.8'	Mag:	9.7

NGC 2683 is the best appearing galaxy in this region of the sky. Object is a large Sb-type spiral galaxy, found just inside the Lynx border.

Telescope Aperture:	4" f/5	4" f/9	6" f/7	6" f/9	8" f/6.3	8" f/10	10" f/6.3	10" f/10	12" f/6.3	12" f/10
FOV(35mm film):	2.7° x 4.1°	1.50° x 2.26°	1.29° x 1.93°	1.0° x 1.50°	1.07° x 1.61°	0.68° x 1.02°	0.86° x 1.29°	0.54° x 0.81°	0.72° x 1.07°	0.45° x 0.68°

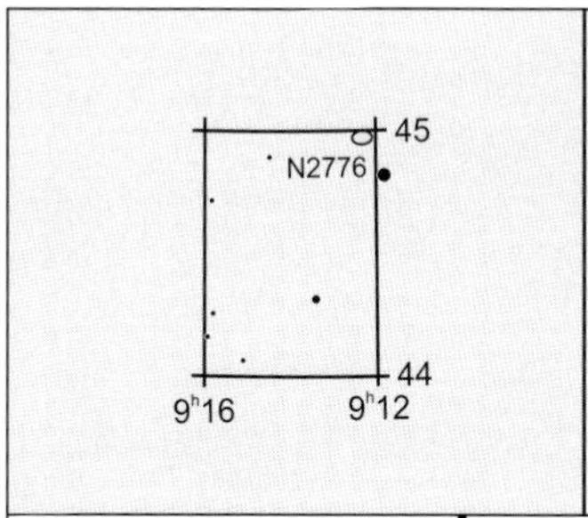

NGC 2776

RA:	$09^h\ 12^m\ 15.1^s$	Con:	Lynx
Dec:	$44^\circ\ 56'\ 50''$	Type:	Spiral Galaxy
Size:	1.8' x 1.5'	Mag:	11.6

NGC 2776 is a loosely wound Sc-type spiral galaxy. Object is oriented nearly face-on to our line of site.

Telescope Aperture:	4" f/5	4" f/9	6" f/7	6" f/9	8" f/6.3	8" f/10	10" f/6.3	10" f/10	12" f/6.3	12" f/10
FOV(35mm film):	2.7° x 4.1°	1.50° x 2.26°	1.29° x 1.93°	1.0° x 1.50°	1.07° x 1.61°	0.68° x 1.02°	0.86° x 1.29°	0.54° x 0.81°	0.72° x 1.07°	0.45° x 0.68°

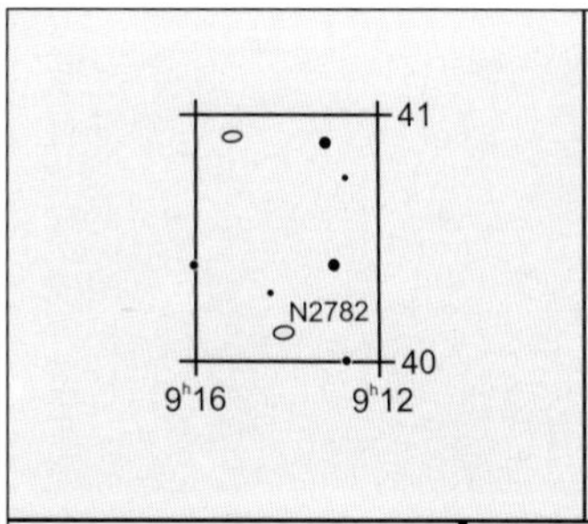

NGC 2782

RA:	$09^h\ 14^m\ 9.0^s$	Con:	Lynx
Dec:	$40^\circ\ 06'\ 50''$	Type:	Spiral Galaxy
Size:	3.7' x 2.2'	Mag:	11.5

NGC 2782 is an excellent example of an Sb-type spiral galaxy.

Telescope Aperture:	4" f/5	4" f/9	6" f/7	6" f/9	8" f/6.3	8" f/10	10" f/6.3	10" f/10	12" f/6.3	12" f/10
FOV(35mm film):	2.7° x 4.1°	1.50° x 2.26°	1.29° x 1.93°	1.0° x 1.50°	1.07° x 1.61°	0.68° x 1.02°	0.86° x 1.29°	0.54° x 0.81°	0.72° x 1.07°	0.45° x 0.68°

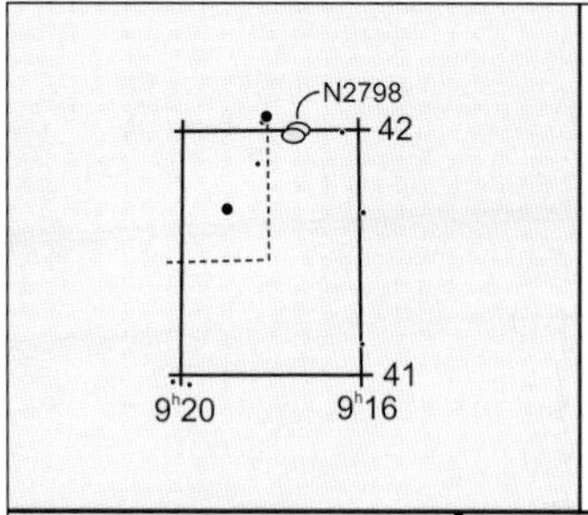

NGC 2798

RA:	$09^h\ 17^m\ 27.0^s$	Con:	Lynx
Dec:	$41^\circ\ 59'\ 50''$	Type:	Barred Spiral
Size:	2.5' x 0.8'	Mag:	12.3

NGC 2798 is a faint barred spiral galaxy.

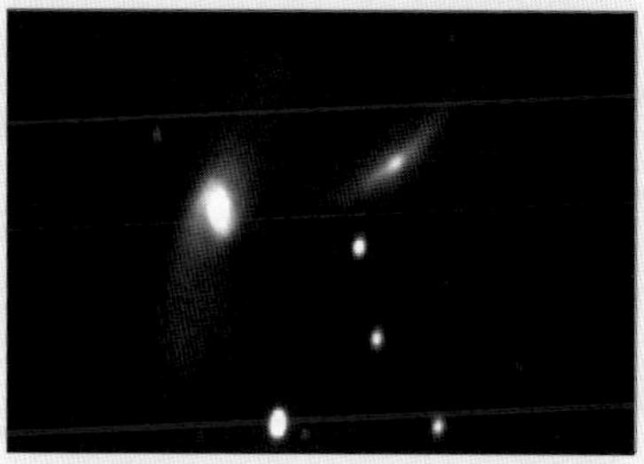

Telescope Aperture:	4" f/5	4" f/9	6" f/7	6" f/9	8" f/6.3	8" f/10	10" f/6.3	10" f/10	12" f/6.3	12" f/10
FOV(35mm film):	2.7° x 4.1°	1.50° x 2.26°	1.29° x 1.93°	1.0° x 1.50°	1.07° x 1.61°	0.68° x 1.02°	0.86° x 1.29°	0.54° x 0.81°	0.72° x 1.07°	0.45° x 0.68°

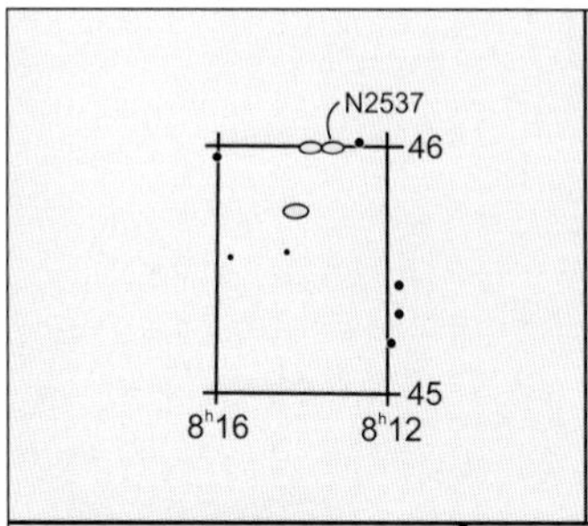

NGC 2537

RA:	$08^h\ 13^m\ 15.2^s$	Con:	Lynx
Dec:	$45^\circ\ 59'\ 51''$	Type:	Spiral Galaxy
Size:	0.5'	Mag:	11.7

NGC 2537 is located in the central region of the constellation.

Telescope Aperture:	4" f/5	4" f/9	6" f/7	6" f/9	8" f/6.3	8" f/10	10" f/6.3	10" f/10	12" f/6.3	12" f/10
FOV(35mm film):	2.7° x 4.1°	1.50° x 2.26°	1.29° x 1.93°	1.0° x 1.50°	1.07° x 1.61°	0.68° x 1.02°	0.86° x 1.29°	0.54° x 0.81°	0.72° x 1.07°	0.45° x 0.68°

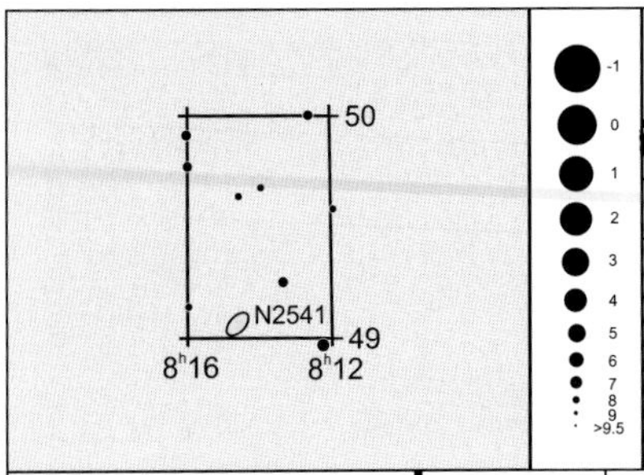

NGC 2541

RA:	$08^h\ 14^m\ 45.2^s$	Con:	Lynx
Dec:	49° 03' 51"	Type:	Spiral Galaxy
Size:	6.0' x 3.0'	Mag:	11.8

NGC 2541 is a large spiral galaxy.

Telescope Aperture:	4" f/5	4" f/9	6" f/7	6" f/9	8" f/6.3	8" f/10	10" f/6.3	10" f/10	12" f/6.3	12" f/10
FOV(35mm film):	2.7° x 4.1°	1.50° x 2.26°	1.29° x 1.93°	1.0° x 1.50°	1.07° x 1.61°	0.68° x 1.02°	0.86° x 1.29°	0.54° x 0.81°	0.72° x 1.07°	0.45° x 0.68°

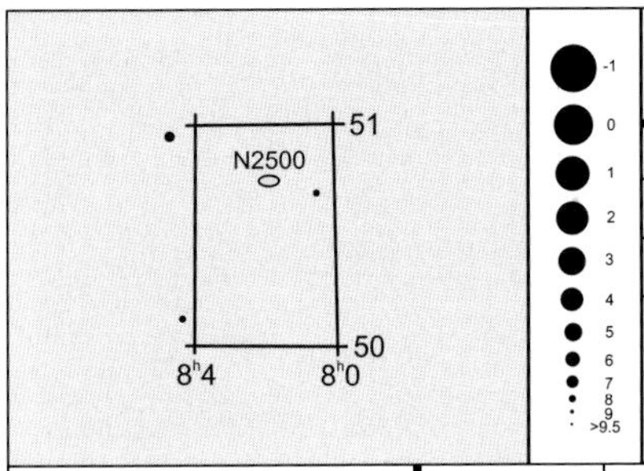

NGC 2500

RA:	$08^h\ 01^m\ 57.3^s$	Con:	Lynx
Dec:	50° 43' 51"	Type:	Barred Spiral
Size:	3.0' x 2.5'	Mag:	11.6

NGC 2500 is a barred spiral galaxy that exhibits a roundish shape and visible dust lanes.

Telescope Aperture:	4" f/5	4" f/9	6" f/7	6" f/9	8" f/6.3	8" f/10	10" f/6.3	10" f/10	12" f/6.3	12" f/10
FOV(35mm film):	2.7° x 4.1°	1.50° x 2.26°	1.29° x 1.93°	1.0° x 1.50°	1.07° x 1.61°	0.68° x 1.02°	0.86° x 1.29°	0.54° x 0.81°	0.72° x 1.07°	0.45° x 0.68°

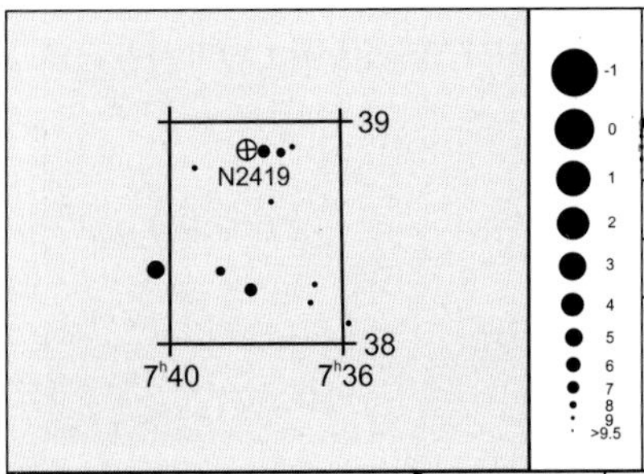

NGC 2419

RA:	$07^h\ 38^m\ 9.1^s$	Con:	Lynx
Dec:	38° 52' 52"	Type:	Globular Cluster
Size:	4.1'	Mag:	10.4

NGC 2419 is a small unresolved globular cluster. Object is located in the constellations southwestern edge.

Telescope Aperture:	4" f/5	4" f/9	6" f/7	6" f/9	8" f/6.3	8" f/10	10" f/6.3	10" f/10	12" f/6.3	12" f/10
FOV(35mm film):	2.7° x 4.1°	1.50° x 2.26°	1.29° x 1.93°	1.0° x 1.50°	1.07° x 1.61°	0.68° x 1.02°	0.86° x 1.29°	0.54° x 0.81°	0.72° x 1.07°	0.45° x 0.68°

Lyr

Crosses Prime Meridian: July thru August

SUMMER

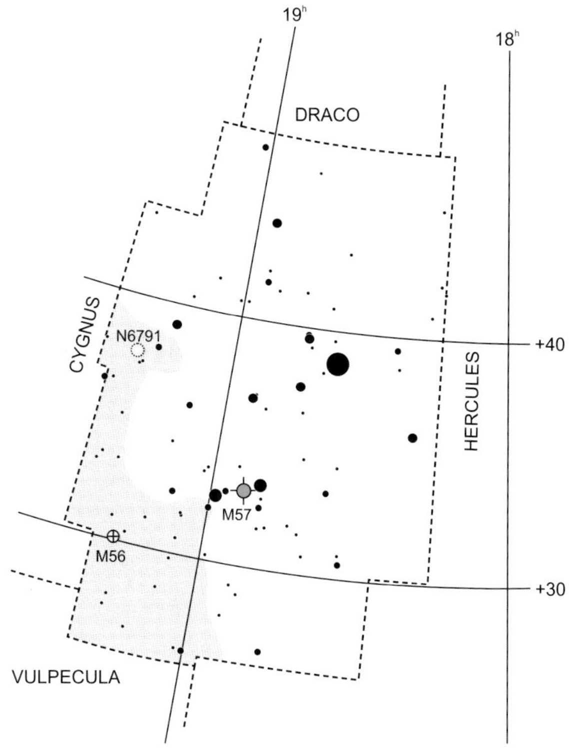

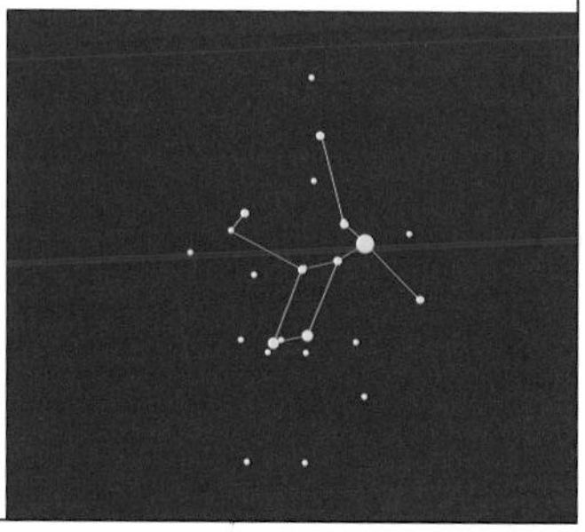

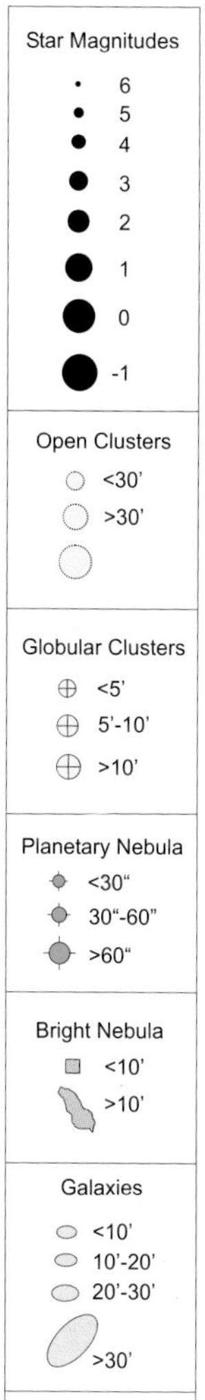

LYRA

Constellation Facts:

Lyra; (Lie-rah)

Lyra, the Lyre moves across the sky from northeast to northwest, crossing the meridian overhead. The constellation is small, but contains some remarkable objects.
The constellation covers 286 square degrees.

Constellation is visible from 90° N to 42° S. Partially visible from 42° S to 90° S.

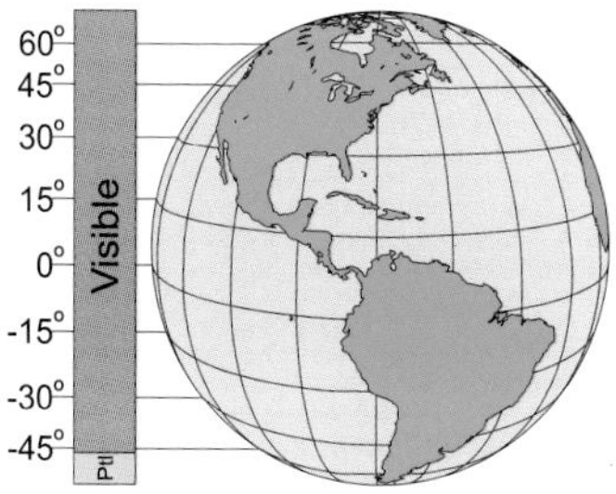

M57 (NGC 6720) "Ring Nebula"

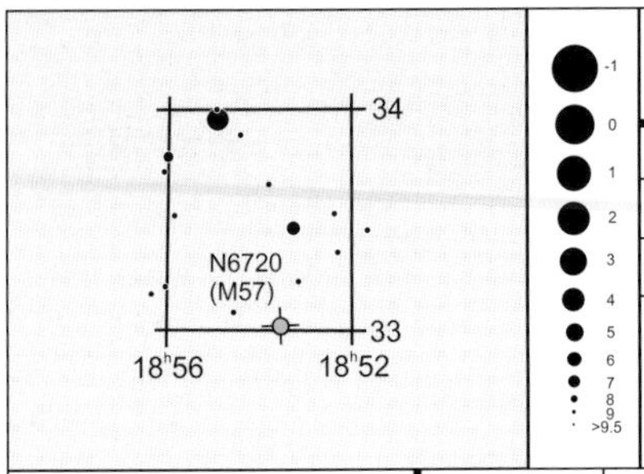

RA:	$18^h\ 53^m\ 40.3^s$	Con:	Lyra
Dec:	33° 02' 10"	Type:	Planetary Nebula
Size:	2.5'	Mag:	9.0

M57 (NGC 6720) is known as the Ring Nebula. Planetary has a visible central star.

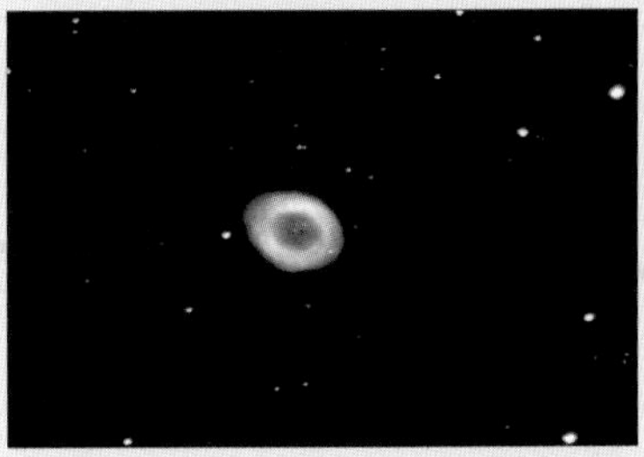

Telescope Aperture:	4" f/5	4" f/9	6" f/7	6" f/9	8" f/6.3	8" f/10	10" f/6.3	10" f/10	12" f/6.3	12" f/10
FOV(35mm film):	2.7°x 4.1°	1.50°x 2.26°	1.29° x 1.93°	1.0° x 1.50°	1.07° x 1.61°	0.68°x 1.02°	0.86° x 1.29°	0.54° x 0.81°	0.72° x 1.07°	0.45°x 0.68°

NGC 6765

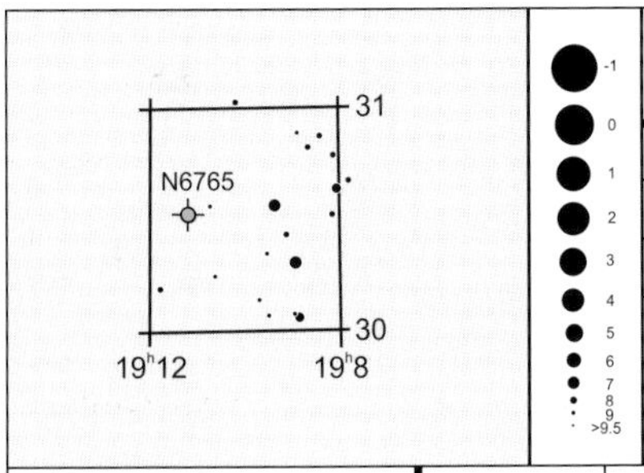

RA:	$19^h\ 11^m\ 10.3^s$	Con:	Lyra
Dec:	30° 33' 11"	Type:	Planetary Nebula
Size:	0.6'	Mag:	11.0

NGC 6765 is the second planetary nebula in the constellation Lyra.

Telescope Aperture:	4" f/5	4" f/9	6" f/7	6" f/9	8" f/6.3	8" f/10	10" f/6.3	10" f/10	12" f/6.3	12" f/10
FOV(35mm film):	2.7°x 4.1°	1.50°x 2.26°	1.29° x 1.93°	1.0° x 1.50°	1.07° x 1.61°	0.68°x 1.02°	0.86° x 1.29°	0.54° x 0.81°	0.72° x 1.07°	0.45°x 0.68°

NGC 6791

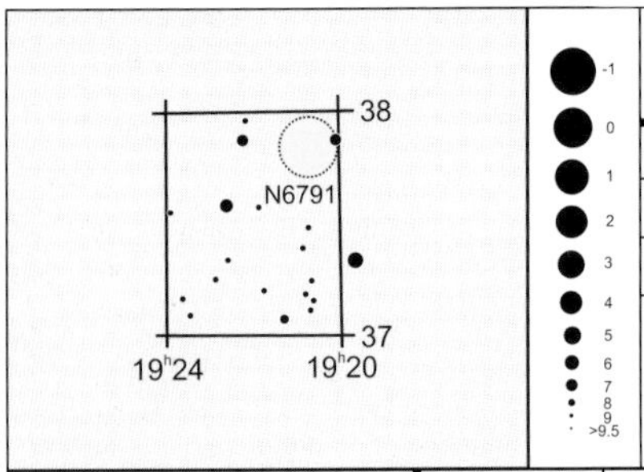

RA:	$19^h\ 20^m\ 46.2^s$	Con:	Lyra
Dec:	37° 51' 12"	Type:	Open Cluster
Size:	16.0'	Mag:	9.5

NGC 6791 is a rich open cluster found toward the eastern edge of the constellation. Object is comprised of some 300 stars.

Telescope Aperture:	4" f/5	4" f/9	6" f/7	6" f/9	8" f/6.3	8" f/10	10" f/6.3	10" f/10	12" f/6.3	12" f/10
FOV(35mm film):	2.7°x 4.1°	1.50°x 2.26°	1.29° x 1.93°	1.0° x 1.50°	1.07° x 1.61°	0.68°x 1.02°	0.86° x 1.29°	0.54° x 0.81°	0.72° x 1.07°	0.45°x 0.68°

M56 (NGC 6779)

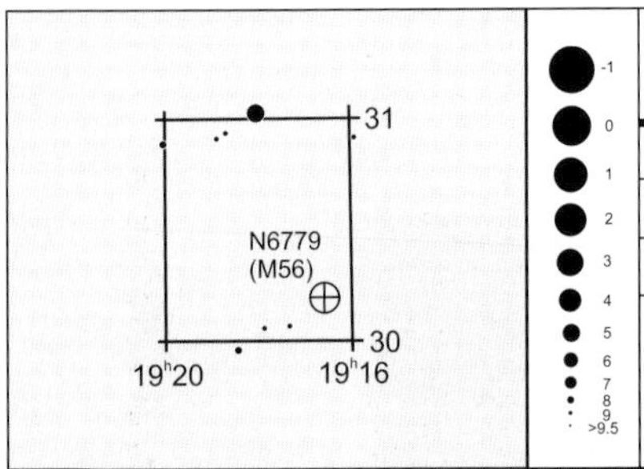

RA:	$19^h\ 16^m\ 40.4^s$	Con:	Lyra
Dec:	30° 11' 12"	Type:	Globular Cluster
Size:	7.1'	Mag:	8.3

M56 (NGC 6779) is a highly resolved small and bright globular cluster.

Telescope Aperture:	4" f/5	4" f/9	6" f/7	6" f/9	8" f/6.3	8" f/10	10" f/6.3	10" f/10	12" f/6.3	12" f/10
FOV(35mm film):	2.7°x 4.1°	1.50°x 2.26°	1.29° x 1.93°	1.0° x 1.50°	1.07° x 1.61°	0.68°x 1.02°	0.86° x 1.29°	0.54° x 0.81°	0.72° x 1.07°	0.45°x 0.68°

NGC 6710

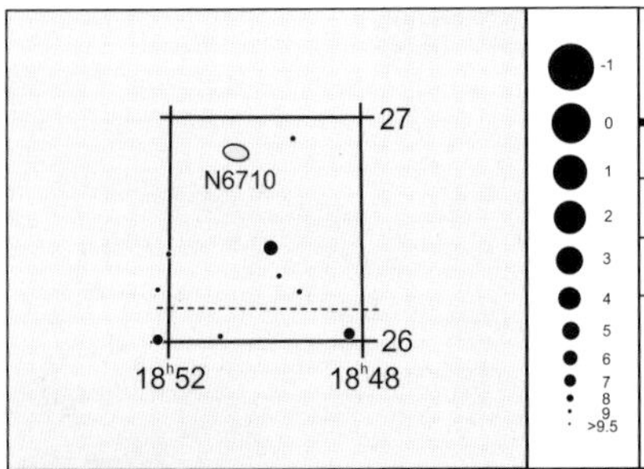

RA:	$18^h\ 50^m\ 40.3^s$	Con:	Lyra
Dec:	26° 50' 10"	Type:	Lenticular Galaxy
Size:	1.5' x 1.0'	Mag:	12.8

NGC 6710 is a lenticular galaxy.

Telescope Aperture:	4" f/5	4" f/9	6" f/7	6" f/9	8" f/6.3	8" f/10	10" f/6.3	10" f/10	12" f/6.3	12" f/10
FOV(35mm film):	2.7°x 4.1°	1.50°x 2.26°	1.29° x 1.93°	1.0° x 1.50°	1.07° x 1.61°	0.68°x 1.02°	0.86° x 1.29°	0.54° x 0.81°	0.72° x 1.07°	0.45°x 0.68°

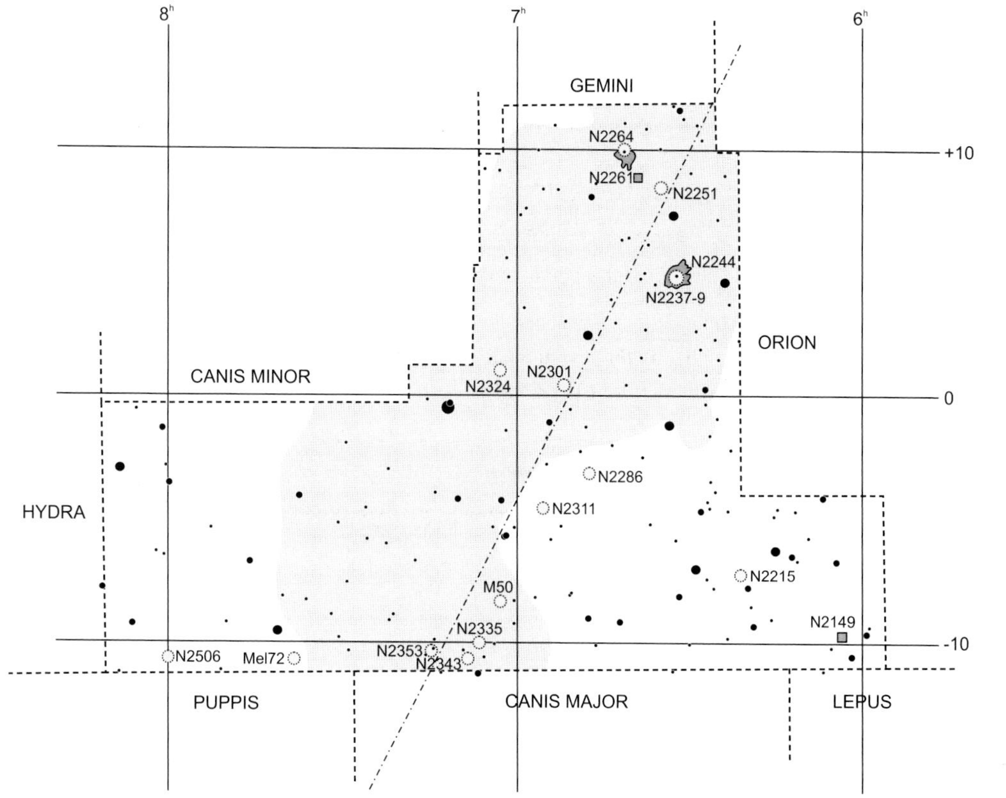

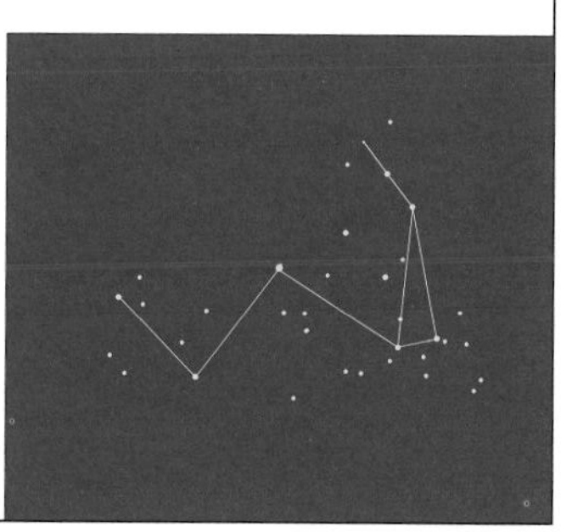

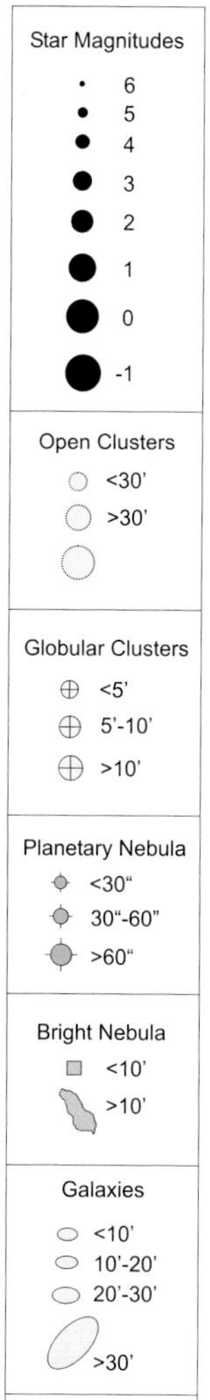

MONOCEROS

Constellation Facts:

Monoceros; (muh-NOS-er-us)

Monoceros, the Unicorn. The constellation rises at the eastern point of the horizon, crosses the meridian halfway between the horizon and the zenith, and sets directly to the west.
The constellation covers 482 square degrees.

Constellation is visible from 78° N to 78° S. Partially visible from 78° N to 90° N.

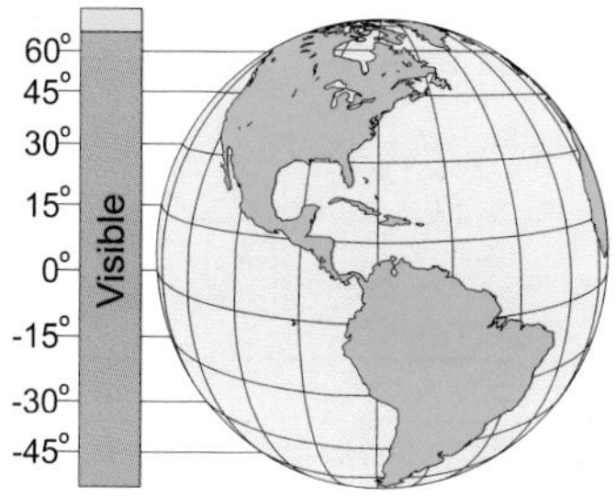

NGC 2244/46 "Rosette Nebula"

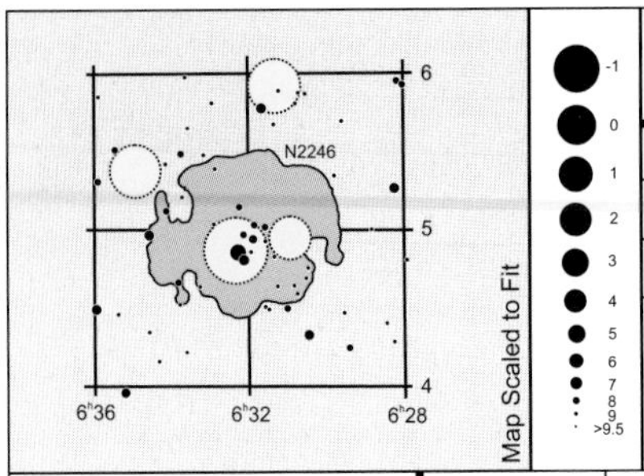

RA:	06h 32m 26.5s	Con:	Monoceros
Dec:	05° 06' 58"	Type:	Cluster & Nebula
Size:	80' x 60'	Mag:	4.8

NGC 2244/46 is a grouping of 100 stars surrounded by emission nebula of hydrogen gas.

Telescope Aperture:	4" f/5	4" f/9	6" f/7	6" f/9	8" f/6.3	8" f/10	10" f/6.3	10" f/10	12" f/6.3	12" f/10
FOV(35mm film):	2.7° x 4.1°	1.50° x 2.26°	1.29° x 1.93°	1.0° x 1.50°	1.07° x 1.61°	0.68° x 1.02°	0.86° x 1.29°	0.54° x 0.81°	0.72° x 1.07°	0.45° x 0.68°

NGC 2264 "Christmas Tree Cluster"

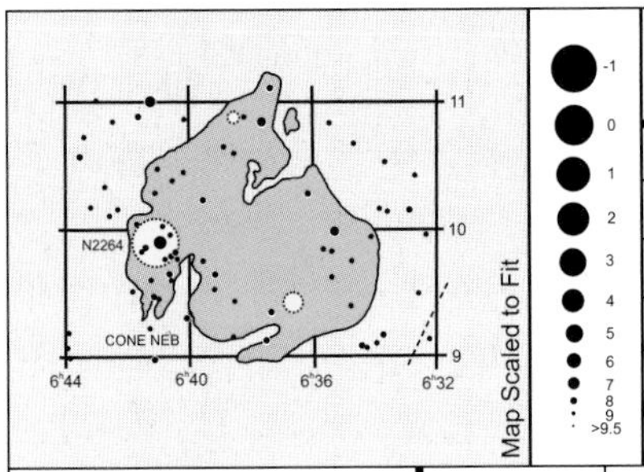

RA:	06h 41m 8.6s	Con:	Monoceros
Dec:	09° 52' 57"	Type:	Cluster & Nebula
Size:	60.0'	Mag:	3.9

NGC 2264 is located 5° north-northeast of the Rosette nebula. Object is a large complex of bright and dark nebulae. Cluster contains 40 stars embedded in a large complex of nebulae.

Telescope Aperture:	4" f/5	4" f/9	6" f/7	6" f/9	8" f/6.3	8" f/10	10" f/6.3	10" f/10	12" f/6.3	12" f/10
FOV(35mm film):	2.7° x 4.1°	1.50° x 2.26°	1.29° x 1.93°	1.0° x 1.50°	1.07° x 1.61°	0.68° x 1.02°	0.86° x 1.29°	0.54° x 0.81°	0.72° x 1.07°	0.45° x 0.68°

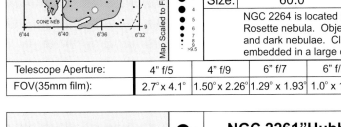

NGC 2261"Hubble's Variable Nebula"

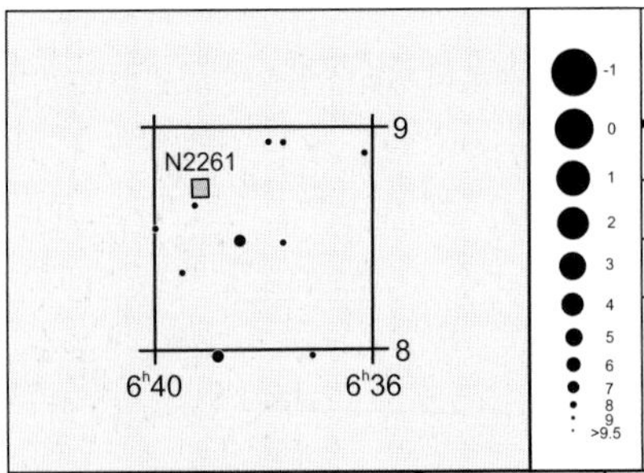

RA:	06h 39m 14.6s	Con:	Monoceros
Dec:	08° 43' 56"	Type:	Nebula
Size:	2.0'	Mag:	9.4

NGC 2261 is found 2° southwest of NGC 2264. Object is a triangular shaped nebula that changes as a result of dark material orbiting R Mon.

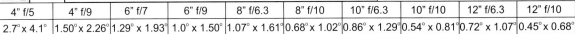

Telescope Aperture:	4" f/5	4" f/9	6" f/7	6" f/9	8" f/6.3	8" f/10	10" f/6.3	10" f/10	12" f/6.3	12" f/10
FOV(35mm film):	2.7° x 4.1°	1.50° x 2.26°	1.29° x 1.93°	1.0° x 1.50°	1.07° x 1.61°	0.68° x 1.02°	0.86° x 1.29°	0.54° x 0.81°	0.72° x 1.07°	0.45° x 0.68°

NGC 3745 "Cone Nebula"

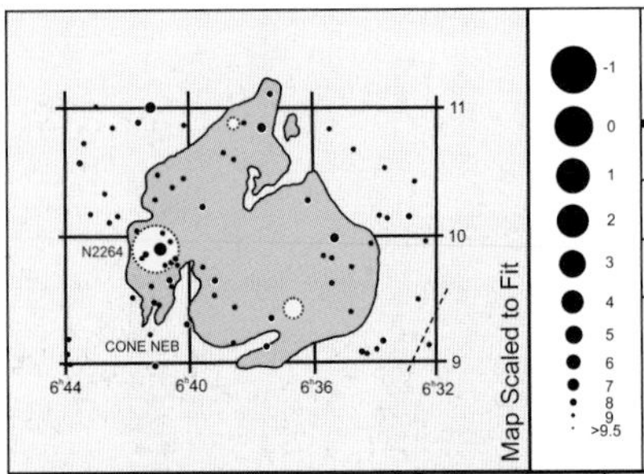

RA:	6h 37m 36s	Con:	Monoceros
Dec:	09° 52' 56"	Type:	Dark Nebula
Size:	5.0' x 3.0'	Mag:	

NGC 3745 is located 40' south of NGC 2264. Object is an opaque dark nebula.

Telescope Aperture:	4" f/5	4" f/9	6" f/7	6" f/9	8" f/6.3	8" f/10	10" f/6.3	10" f/10	12" f/6.3	12" f/10
FOV(35mm film):	2.7° x 4.1°	1.50° x 2.26°	1.29° x 1.93°	1.0° x 1.50°	1.07° x 1.61°	0.68° x 1.02°	0.86° x 1.29°	0.54° x 0.81°	0.72° x 1.07°	0.45° x 0.68°

M50 (NGC 2323)

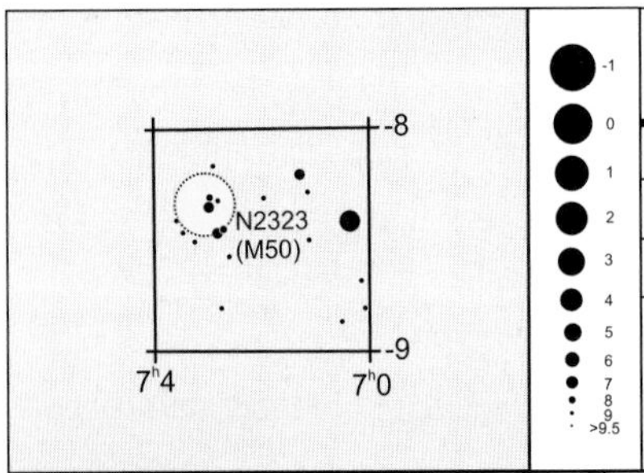

RA:	07h 03m 14.1s	Con:	Monoceros
Dec:	-08° 20' 04"	Type:	Open Cluster
Size:	16.0'	Mag:	5.9

M50 (NGC 2323) is a bright open cluster. Object is comprised of 80 stars.

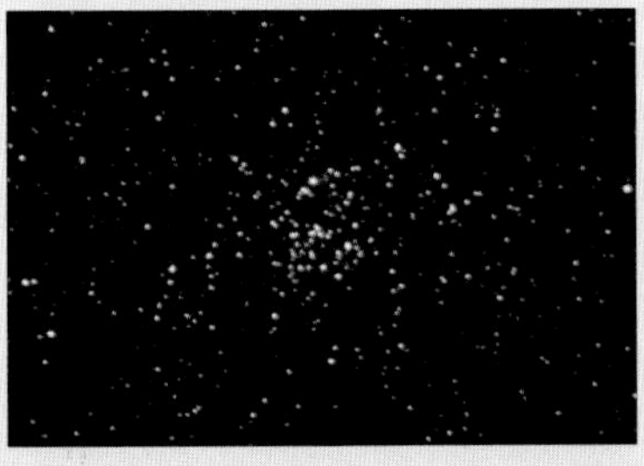

Telescope Aperture:	4" f/5	4" f/9	6" f/7	6" f/9	8" f/6.3	8" f/10	10" f/6.3	10" f/10	12" f/6.3	12" f/10
FOV(35mm film):	2.7° x 4.1°	1.50° x 2.26°	1.29° x 1.93°	1.0° x 1.50°	1.07° x 1.61°	0.68° x 1.02°	0.86° x 1.29°	0.54° x 0.81°	0.72° x 1.07°	0.45° x 0.68°

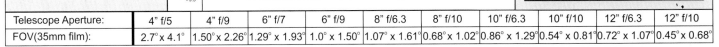

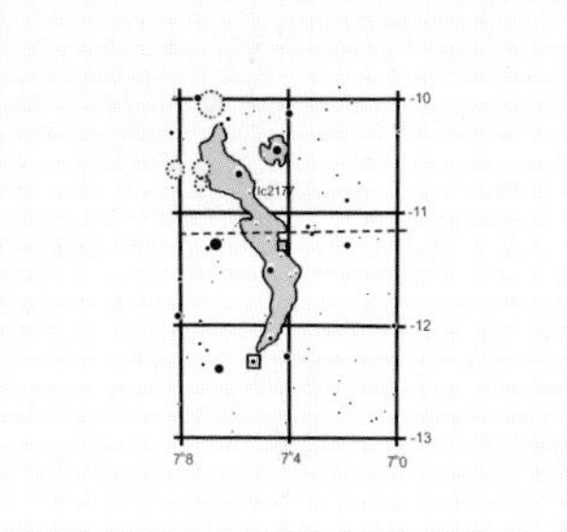

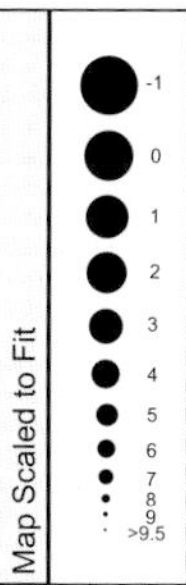

IC 2177 "Seagull Nebula"

RA:	07h 05m 8.1s	Con:	Monoceros
Dec:	-10° 42' 05"	Type:	Diffuse Nebula
Size:	120.0'	Mag:	

IC 2177 is known as the Seagull Nebula, and is a large diffuse complex of nebulae.

Telescope Aperture:	4" f/5	4" f/9	6" f/7	6" f/9	8" f/6.3	8" f/10	10" f/6.3	10" f/10	12" f/6.3	12" f/10
FOV(35mm film):	2.7° x 4.1°	1.50° x 2.26°	1.29° x 1.93°	1.0° x 1.50°	1.07° x 1.61°	0.68° x 1.02°	0.86° x 1.29°	0.54° x 0.81°	0.72° x 1.07°	0.45° x 0.68°

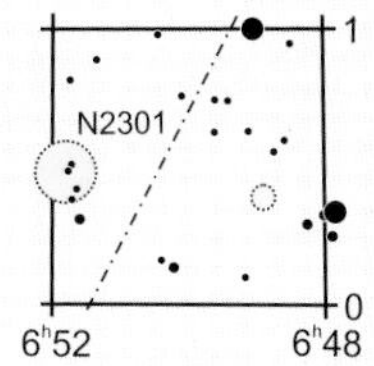

NGC 2301

RA:	06h 51m 50.4s	Con:	Monoceros
Dec:	00° 27' 56"	Type:	Open Cluster
Size:	12.0'	Mag:	6.0

NGC 2301 is a dense open cluster, comprised of a group of 80 stars.

Telescope Aperture:	4" f/5	4" f/9	6" f/7	6" f/9	8" f/6.3	8" f/10	10" f/6.3	10" f/10	12" f/6.3	12" f/10
FOV(35mm film):	2.7° x 4.1°	1.50° x 2.26°	1.29° x 1.93°	1.0° x 1.50°	1.07° x 1.61°	0.68° x 1.02°	0.86° x 1.29°	0.54° x 0.81°	0.72° x 1.07°	0.45° x 0.68°

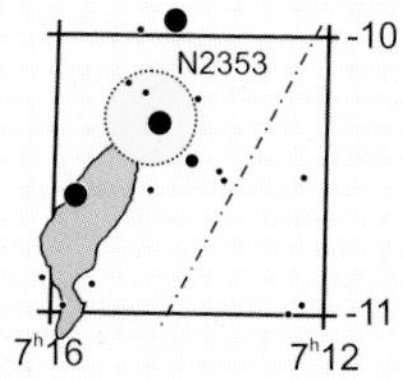

NGC 2353

RA:	07h 14m 38.0s	Con:	Monoceros
Dec:	-10° 18' 06"	Type:	Open Cluster
Size:	20.0'	Mag:	7.1

NGC 2353 is a rich open cluster comprised of 30 stars.

Telescope Aperture:	4" f/5	4" f/9	6" f/7	6" f/9	8" f/6.3	8" f/10	10" f/6.3	10" f/10	12" f/6.3	12" f/10
FOV(35mm film):	2.7° x 4.1°	1.50° x 2.26°	1.29° x 1.93°	1.0° x 1.50°	1.07° x 1.61°	0.68° x 1.02°	0.86° x 1.29°	0.54° x 0.81°	0.72° x 1.07°	0.45° x 0.68°

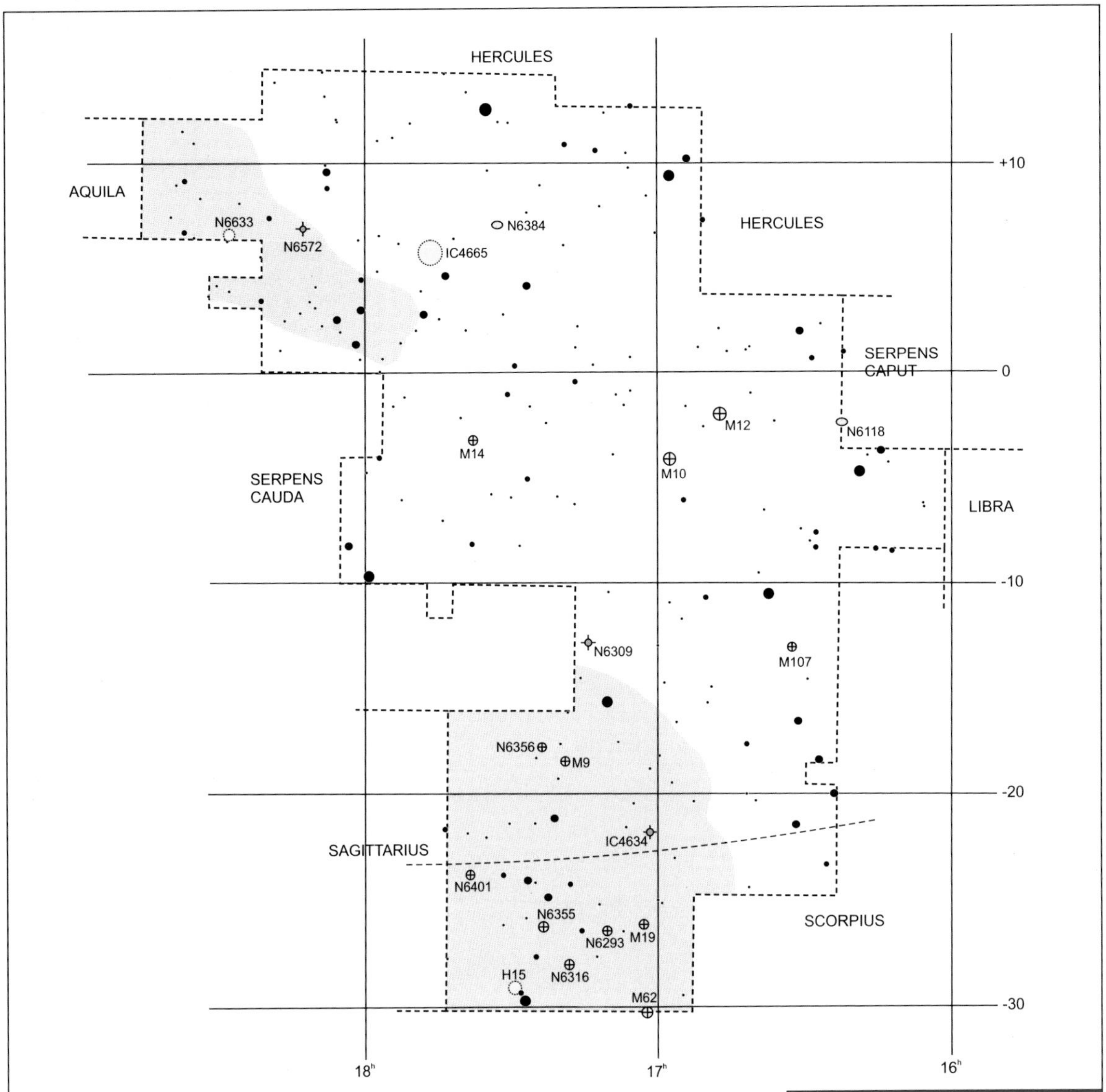

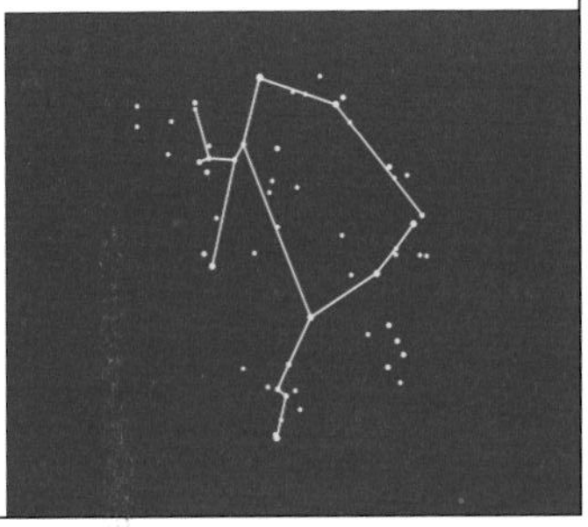

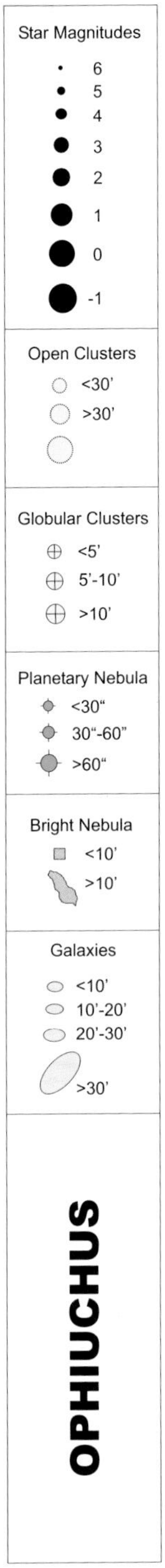

Constellation Facts:

Ophiuchus; (OFF-ih-YOU-kus)

Ophiuchus, the Serpent-Bearer.
The stars in the constellation are centered on the celestial equator. As a result, they rise in the east and set towards the west. Ophiuchus crosses the meridian about halfway between the horizon and the zenith.
The constellation covers 948 square degrees.

Constellation is visible from 59° N to 75° S. Partially visible from 59° N to 90° N.

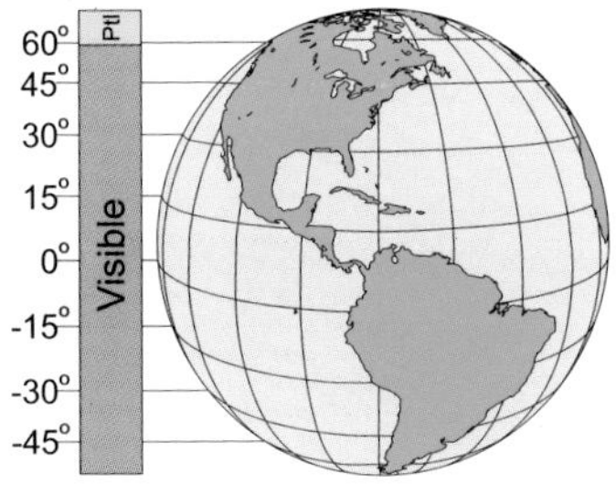

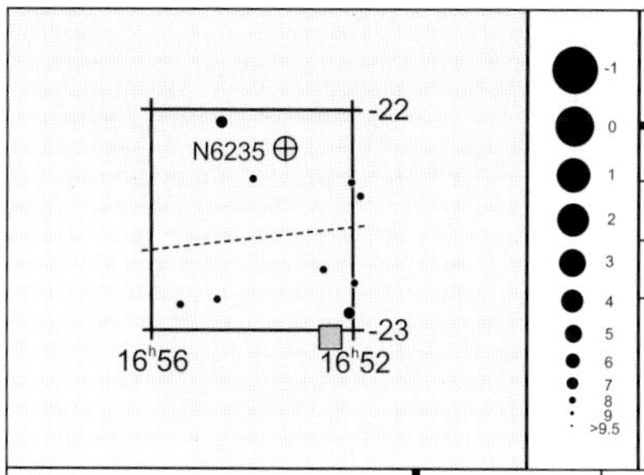

NGC 6235

RA:	$16^h\ 53^m\ 29.5^s$	Con:	Ophiuchus
Dec:	-22° 11' 08"	Type:	Globular Cluster
Size:	5.0'	Mag:	10.2

NGC 6235 is an unresolved and indistinct globular cluster.

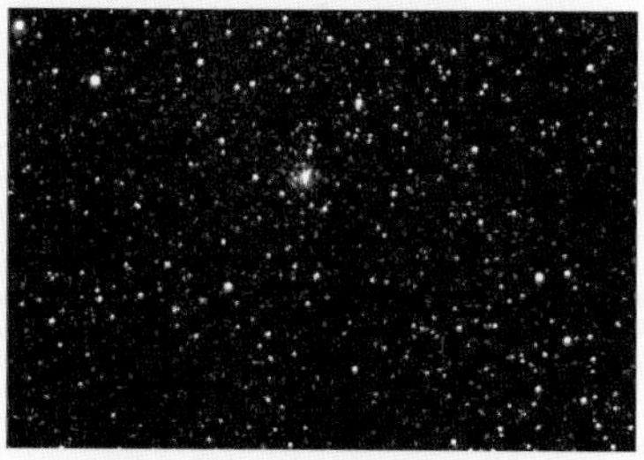

Telescope Aperture:	4" f/5	4" f/9	6" f/7	6" f/9	8" f/6.3	8" f/10	10" f/6.3	10" f/10	12" f/6.3	12" f/10
FOV(35mm film):	2.7°x 4.1°	1.50°x 2.26°	1.29° x 1.93°	1.0° x 1.50°	1.07° x 1.61°	0.68°x 1.02°	0.86° x 1.29°	0.54° x 0.81°	0.72° x 1.07°	0.45°x 0.68°

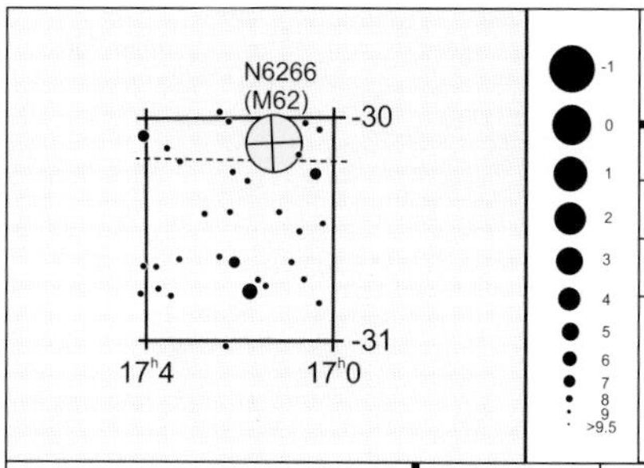

M62 (NGC 6266)

RA:	$17^h\ 01^m\ 17.8^s$	Con:	Ophiuchus
Dec:	-30° 07' 08"	Type:	Globular Cluster
Size:	13.5'	Mag:	6.6

M62 (NGC 6266) is a bright globular cluster. Object is found 7° southeast of Antares.

Telescope Aperture:	4" f/5	4" f/9	6" f/7	6" f/9	8" f/6.3	8" f/10	10" f/6.3	10" f/10	12" f/6.3	12" f/10
FOV(35mm film):	2.7°x 4.1°	1.50°x 2.26°	1.29° x 1.93°	1.0° x 1.50°	1.07° x 1.61°	0.68°x 1.02°	0.86° x 1.29°	0.54° x 0.81°	0.72° x 1.07°	0.45°x 0.68°

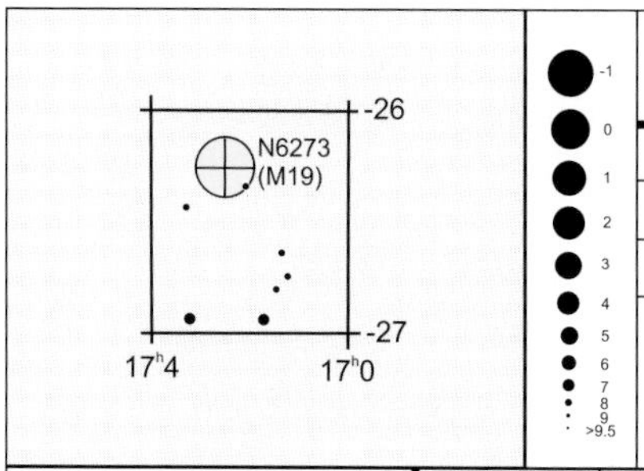

M19 (NGC 6273)

RA:	$17^h\ 02^m\ 41.7^s$	Con:	Ophiuchus
Dec:	-26° 16' 07"	Type:	Globular Cluster
Size:	13.5'	Mag:	7.2

M19 (NGC 6273) is located approximately 4.5° north of M62. Object is a bright globular cluster.

Telescope Aperture:	4" f/5	4" f/9	6" f/7	6" f/9	8" f/6.3	8" f/10	10" f/6.3	10" f/10	12" f/6.3	12" f/10
FOV(35mm film):	2.7°x 4.1°	1.50°x 2.26°	1.29° x 1.93°	1.0° x 1.50°	1.07° x 1.61°	0.68°x 1.02°	0.86° x 1.29°	0.54° x 0.81°	0.72° x 1.07°	0.45°x 0.68°

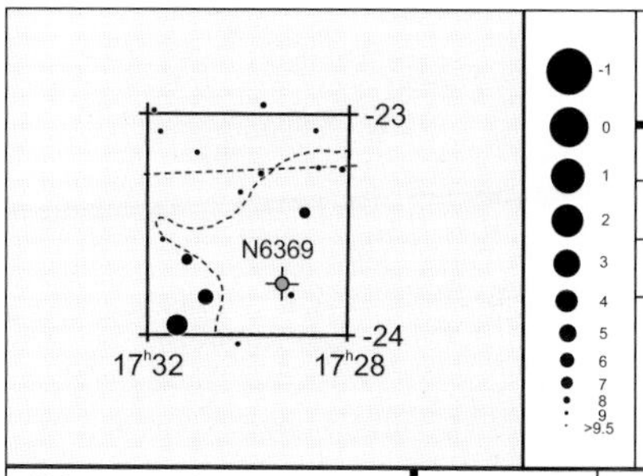

NGC 6369

RA:	$17^h\ 29^m\ 23.7^s$	Con:	Ophiuchus
Dec:	-23° 46' 03"	Type:	Planetary Nebula
Size:	1.1'	Mag:	13.0

NGC 6369 is a ring-shaped planetary nebula.

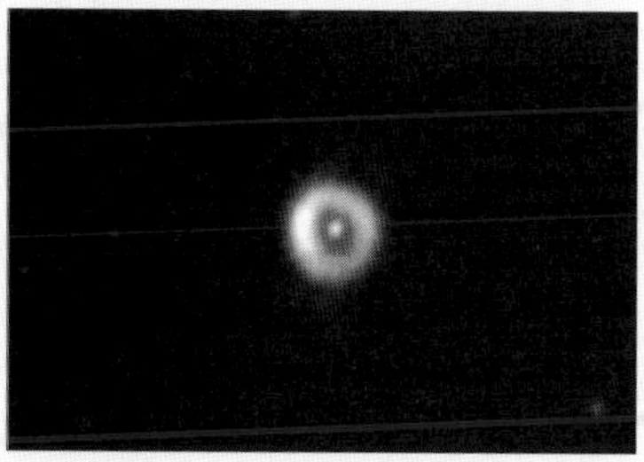

Telescope Aperture:	4" f/5	4" f/9	6" f/7	6" f/9	8" f/6.3	8" f/10	10" f/6.3	10" f/10	12" f/6.3	12" f/10
FOV(35mm film):	2.7°x 4.1°	1.50°x 2.26°	1.29° x 1.93°	1.0° x 1.50°	1.07° x 1.61°	0.68°x 1.02°	0.86° x 1.29°	0.54° x 0.81°	0.72° x 1.07°	0.45°x 0.68°

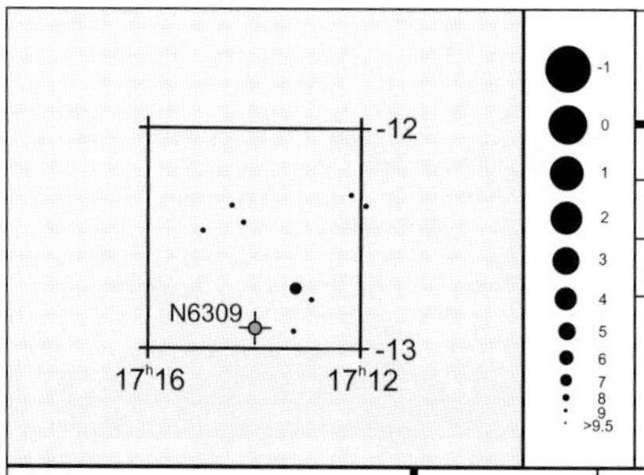

NGC 6309(Box Nebula)

RA:	$17^h\ 14^m\ 11.2^s$	Con:	Ophiuchus
Dec:	-12° 55' 03"	Type:	Planetary Nebula
Size:	1.1'	Mag:	11.0

NGC 6309 the "Box Nebula" is a small planetary nebula disk.

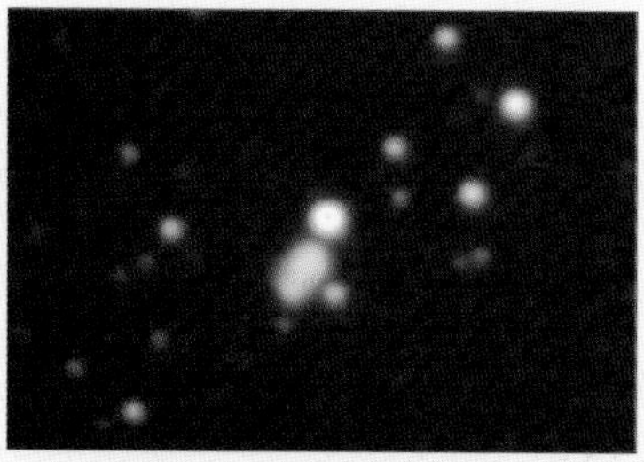

Telescope Aperture:	4" f/5	4" f/9	6" f/7	6" f/9	8" f/6.3	8" f/10	10" f/6.3	10" f/10	12" f/6.3	12" f/10
FOV(35mm film):	2.7°x 4.1°	1.50°x 2.26°	1.29° x 1.93°	1.0° x 1.50°	1.07° x 1.61°	0.68°x 1.02°	0.86° x 1.29°	0.54° x 0.81°	0.72° x 1.07°	0.45°x 0.68°

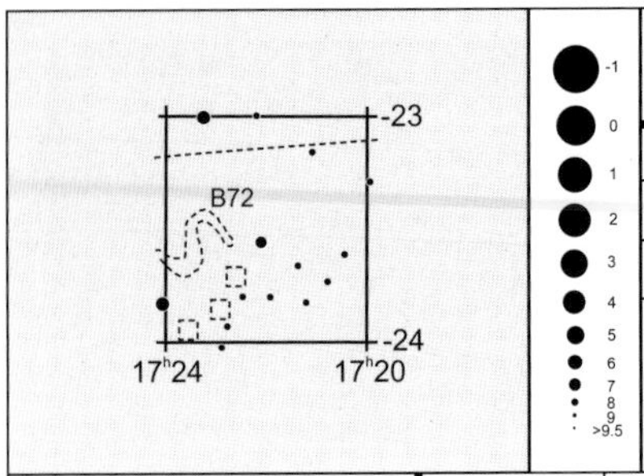

B-72 "Snake Nebula"

RA:	$17^h\ 23^m\ 30.0^s$	Con:	Ophiuchus
Dec:	-23° 37' 59"	Type:	Dark Nebula
Size:	30.0'	Mag:	

B-72 is a dark filamentary structure that resembles a snake.

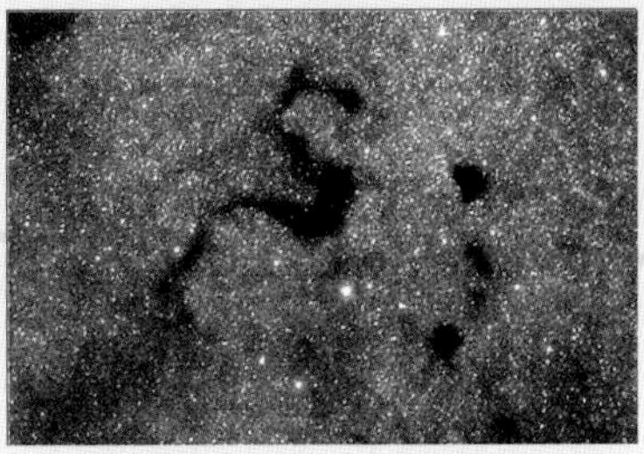

Telescope Aperture:	4" f/5	4" f/9	6" f/7	6" f/9	8" f/6.3	8" f/10	10" f/6.3	10" f/10	12" f/6.3	12" f/10
FOV(35mm film):	2.7° x 4.1°	1.50° x 2.26°	1.29° x 1.93°	1.0° x 1.50°	1.07° x 1.61°	0.68° x 1.02°	0.86° x 1.29°	0.54° x 0.81°	0.72° x 1.07°	0.45° x 0.68°

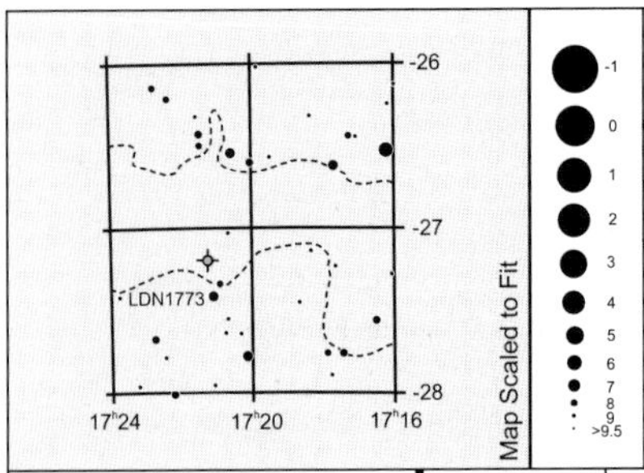

LDN1773 "Pipe Nebula"

RA:	$17^h\ 21^m\ 00^s$	Con:	Ophiuchus
Dec:	-27° 23' 00"	Type:	Dark Nebula
Size:	300'x60'	Mag:	

Telescope Aperture:	4" f/5	4" f/9	6" f/7	6" f/9	8" f/6.3	8" f/10	10" f/6.3	10" f/10	12" f/6.3	12" f/10
FOV(35mm film):	2.7° x 4.1°	1.50° x 2.26°	1.29° x 1.93°	1.0° x 1.50°	1.07° x 1.61°	0.68° x 1.02°	0.86° x 1.29°	0.54° x 0.81°	0.72° x 1.07°	0.45° x 0.68°

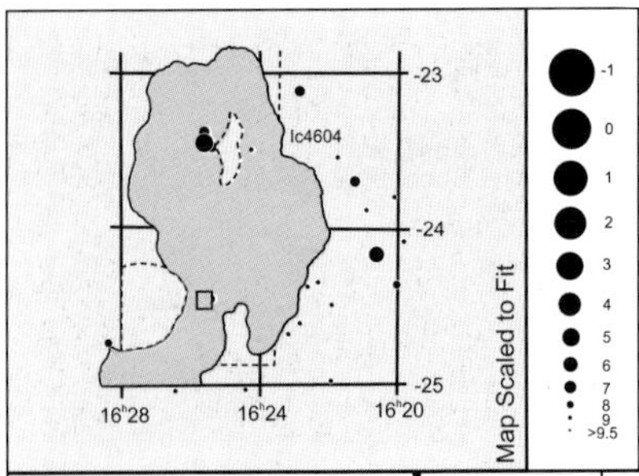

IC 4604

RA:	$16^h\ 25^m\ 41.3^s$	Con:	Ophiuchus
Dec:	-23° 26' 11"	Type:	Nebula
Size:	60.0'	Mag:	7.2

IC 4604 is a large, bright nebula.

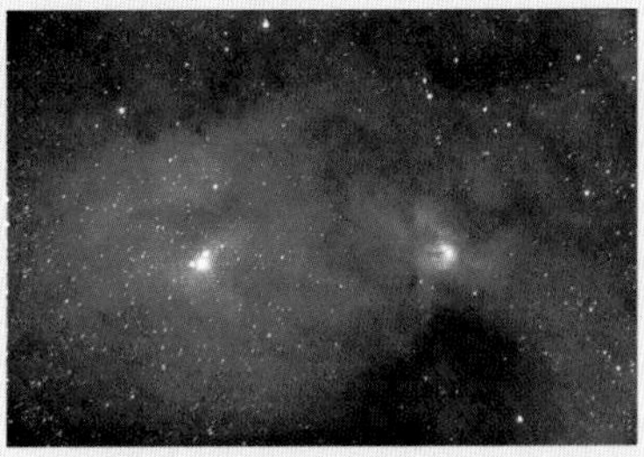

Telescope Aperture:	4" f/5	4" f/9	6" f/7	6" f/9	8" f/6.3	8" f/10	10" f/6.3	10" f/10	12" f/6.3	12" f/10
FOV(35mm film):	2.7° x 4.1°	1.50° x 2.26°	1.29° x 1.93°	1.0° x 1.50°	1.07° x 1.61°	0.68° x 1.02°	0.86° x 1.29°	0.54° x 0.81°	0.72° x 1.07°	0.45° x 0.68°

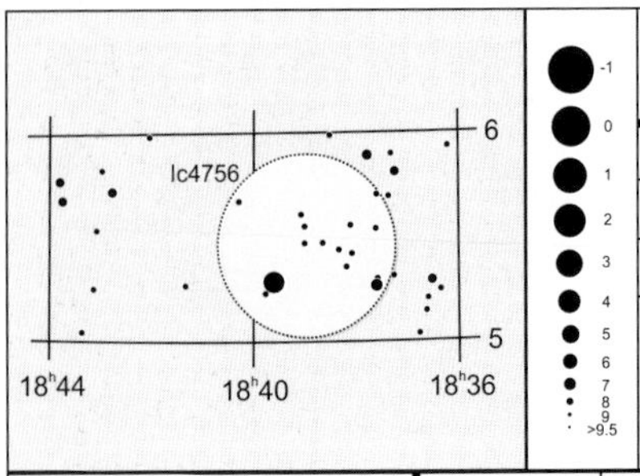

IC 4756

RA:	$18^h\ 39^m\ 4.9^s$	Con:	Ophiuchus
Dec:	05° 27' 09"	Type:	Open Cluster
Size:	52.0'	Mag:	5.0

IC 4756 is a large scattered open cluster. Object is comprised of some 80 stars.

Telescope Aperture:	4" f/5	4" f/9	6" f/7	6" f/9	8" f/6.3	8" f/10	10" f/6.3	10" f/10	12" f/6.3	12" f/10
FOV(35mm film):	2.7° x 4.1°	1.50° x 2.26°	1.29° x 1.93°	1.0° x 1.50°	1.07° x 1.61°	0.68° x 1.02°	0.86° x 1.29°	0.54° x 0.81°	0.72° x 1.07°	0.45° x 0.68°

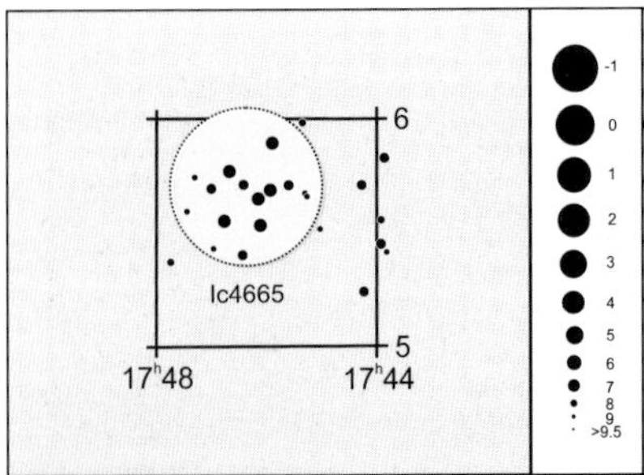

IC 4665

RA:	$17^h\ 46^m\ 22.8^s$	Con:	Ophiuchus
Dec:	05° 43' 04"	Type:	Open Cluster
Size:	41.0'	Mag:	4.0

IC 4665 is a sparse open cluster found northeast of Beta Oph. Object contains 20 stars.

Telescope Aperture:	4" f/5	4" f/9	6" f/7	6" f/9	8" f/6.3	8" f/10	10" f/6.3	10" f/10	12" f/6.3	12" f/10
FOV(35mm film):	2.7° x 4.1°	1.50° x 2.26°	1.29° x 1.93°	1.0° x 1.50°	1.07° x 1.61°	0.68° x 1.02°	0.86° x 1.29°	0.54° x 0.81°	0.72° x 1.07°	0.45° x 0.68°

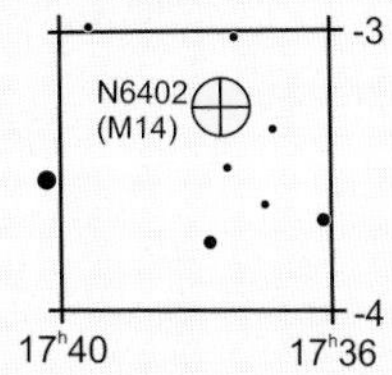

M14 (NGC 6402)

RA:	$17^h\ 37^m\ 40.9^s$	Con:	Ophiuchus
Dec:	-03° 14' 59"	Type:	Globular Cluster
Size:	11.7'	Mag:	7.6

M14 (NGC 6402) is a bright globular cluster located east of M10/12.

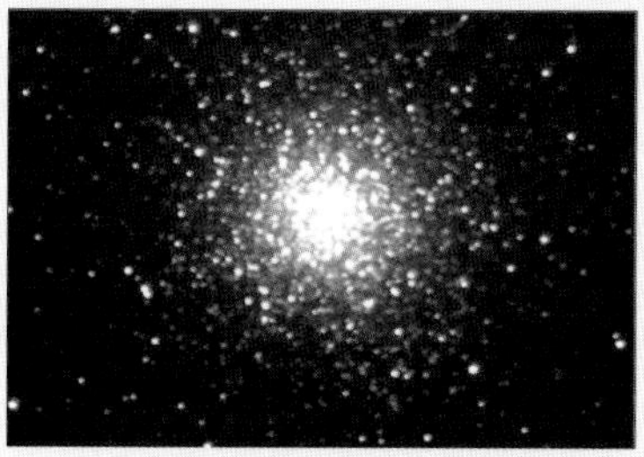

Telescope Aperture:	4" f/5	4" f/9	6" f/7	6" f/9	8" f/6.3	8" f/10	10" f/6.3	10" f/10	12" f/6.3	12" f/10
FOV(35mm film):	2.7° x 4.1°	1.50° x 2.26°	1.29° x 1.93°	1.0° x 1.50°	1.07° x 1.61°	0.68° x 1.02°	0.86° x 1.29°	0.54° x 0.81°	0.72° x 1.07°	0.45° x 0.68°

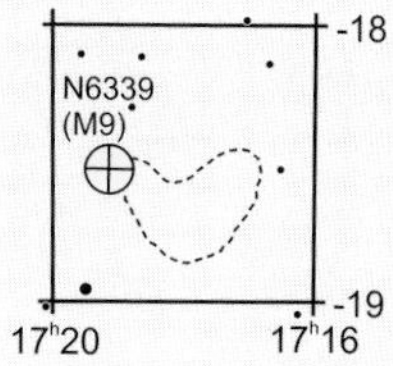

M9 (NGC 6333)

RA:	$17^h\ 19^m\ 17.4^s$	Con:	Ophiuchus
Dec:	-18° 31' 04"	Type:	Globular Cluster
Size:	9.3'	Mag:	7.9

M9 (NGC 6333) is a mottled, compact globular cluster. Object is located in the southern region of the constellation, atop a dust cloud called B-64.

Telescope Aperture:	4" f/5	4" f/9	6" f/7	6" f/9	8" f/6.3	8" f/10	10" f/6.3	10" f/10	12" f/6.3	12" f/10
FOV(35mm film):	2.7° x 4.1°	1.50° x 2.26°	1.29° x 1.93°	1.0° x 1.50°	1.07° x 1.61°	0.68° x 1.02°	0.86° x 1.29°	0.54° x 0.81°	0.72° x 1.07°	0.45° x 0.68°

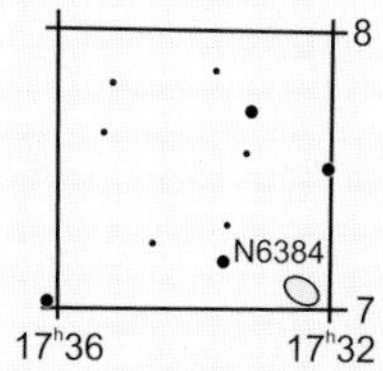

NGC 6384

RA:	$17^h\ 32^m\ 28.6^s$	Con:	Ophiuchus
Dec:	07° 04' 02"	Type:	Spiral Galaxy
Size:	4.6'x3.4'	Mag:	10.6

NGC 6384 is an Sb-type spiral galaxy. Object is located very close to IC 4665.

Telescope Aperture:	4" f/5	4" f/9	6" f/7	6" f/9	8" f/6.3	8" f/10	10" f/6.3	10" f/10	12" f/6.3	12" f/10
FOV(35mm film):	2.7° x 4.1°	1.50° x 2.26°	1.29° x 1.93°	1.0° x 1.50°	1.07° x 1.61°	0.68° x 1.02°	0.86° x 1.29°	0.54° x 0.81°	0.72° x 1.07°	0.45° x 0.68°

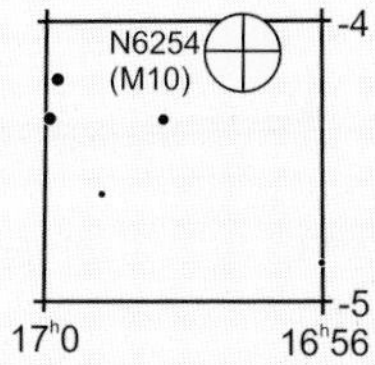

M10 (NGC 6254)

RA:	$16^h\ 57^m\ 10.9^s$	Con:	Ophiuchus
Dec:	-04° 06' 02"	Type:	Globular Cluster
Size:	15.1'	Mag:	6.6

M10 (NGC 6254) is a highly resolved globular cluster.

Telescope Aperture:	4" f/5	4" f/9	6" f/7	6" f/9	8" f/6.3	8" f/10	10" f/6.3	10" f/10	12" f/6.3	12" f/10
FOV(35mm film):	2.7° x 4.1°	1.50° x 2.26°	1.29° x 1.93°	1.0° x 1.50°	1.07° x 1.61°	0.68° x 1.02°	0.86° x 1.29°	0.54° x 0.81°	0.72° x 1.07°	0.45° x 0.68°

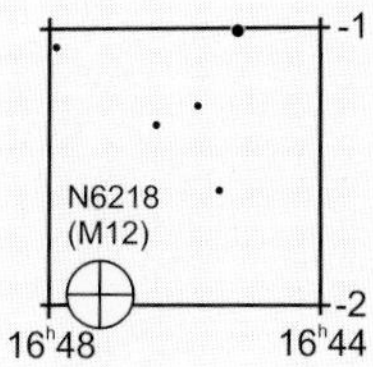

M12 (NGC 6218)

RA:	$16^h\ 47^m\ 16.7^s$	Con:	Ophiuchus
Dec:	-01° 57' 04"	Type:	Globular Cluster
Size:	14.5'	Mag:	6.6

M12 (NGC 6218) is a highly resolved globular cluster.

Telescope Aperture:	4" f/5	4" f/9	6" f/7	6" f/9	8" f/6.3	8" f/10	10" f/6.3	10" f/10	12" f/6.3	12" f/10
FOV(35mm film):	2.7° x 4.1°	1.50° x 2.26°	1.29° x 1.93°	1.0° x 1.50°	1.07° x 1.61°	0.68° x 1.02°	0.86° x 1.29°	0.54° x 0.81°	0.72° x 1.07°	0.45° x 0.68°

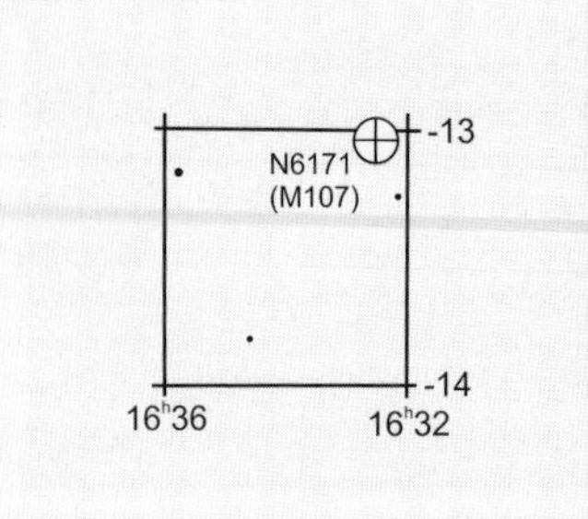

M107 (NGC 6171)

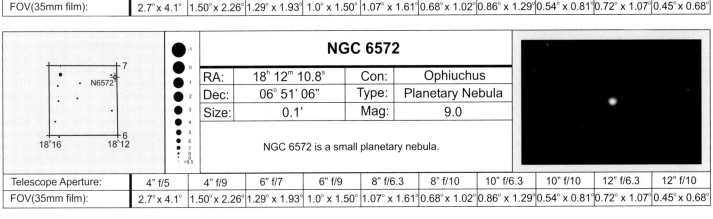

RA:	$16^h\ 32^m\ 35.1^s$	Con:	Ophiuchus
Dec:	-13° 03' 08"	Type:	Globular Cluster
Size:	10.0'	Mag:	8.1

M107 (NGC 6171) is a dense cluster found in southern region of the constellation.

Telescope Aperture:	4" f/5	4" f/9	6" f/7	6" f/9	8" f/6.3	8" f/10	10" f/6.3	10" f/10	12" f/6.3	12" f/10
FOV(35mm film):	2.7° x 4.1°	1.50° x 2.26°	1.29° x 1.93°	1.0° x 1.50°	1.07° x 1.61°	0.68° x 1.02°	0.86° x 1.29°	0.54° x 0.81°	0.72° x 1.07°	0.45° x 0.68°

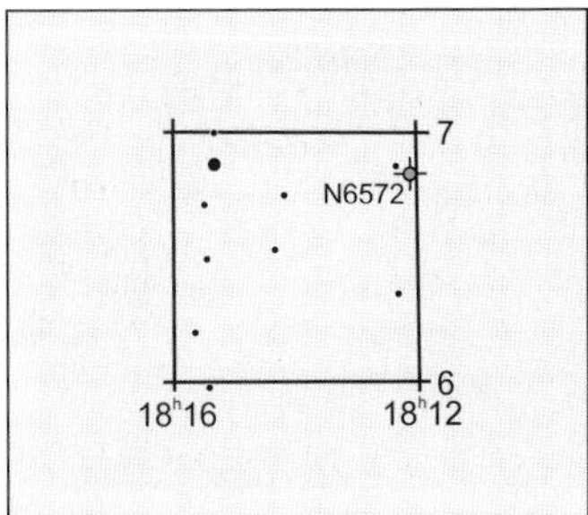

NGC 6572

RA:	$18^h\ 12^m\ 10.8^s$	Con:	Ophiuchus
Dec:	06° 51' 06"	Type:	Planetary Nebula
Size:	0.1'	Mag:	9.0

NGC 6572 is a small planetary nebula.

Telescope Aperture:	4" f/5	4" f/9	6" f/7	6" f/9	8" f/6.3	8" f/10	10" f/6.3	10" f/10	12" f/6.3	12" f/10
FOV(35mm film):	2.7° x 4.1°	1.50° x 2.26°	1.29° x 1.93°	1.0° x 1.50°	1.07° x 1.61°	0.68° x 1.02°	0.86° x 1.29°	0.54° x 0.81°	0.72° x 1.07°	0.45° x 0.68°

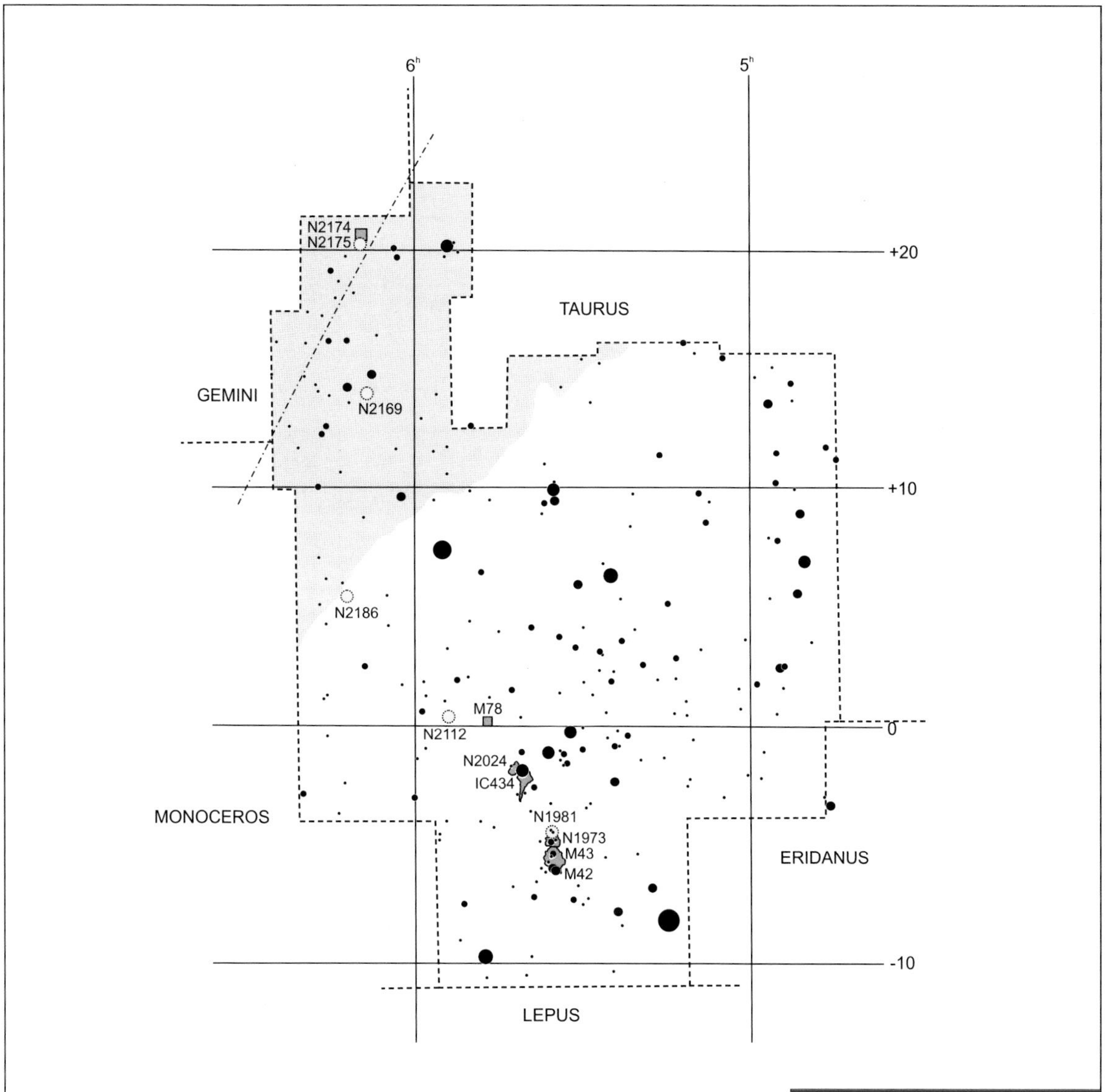

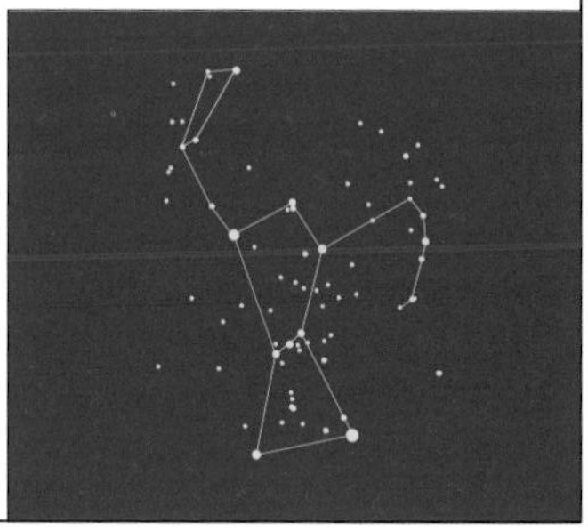

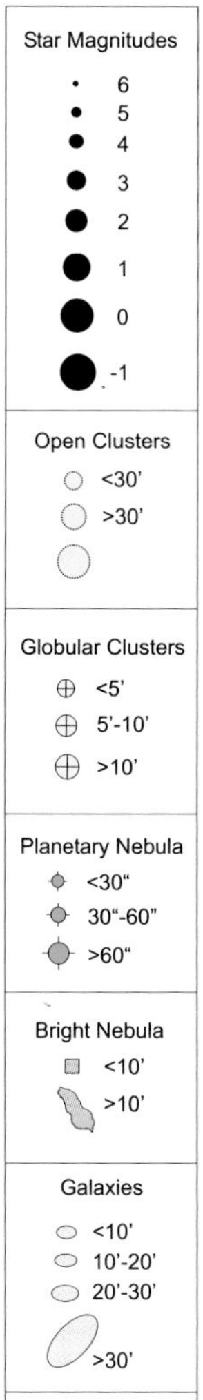

ORION

Constellation Facts:

Orion; (oh-RYE-un)

Orion, the Hunter.
Orion is an equatorial constellation that is bisected by the celestial equator. It rises in the eastern horizon, crosses the meridian halfway between the horizon and the zenith, then sets directly to the west.
The constellation covers 594 square degrees.

Constellation is visible from 79° N to 67° S. Partially visible from 79° N to 90° N.

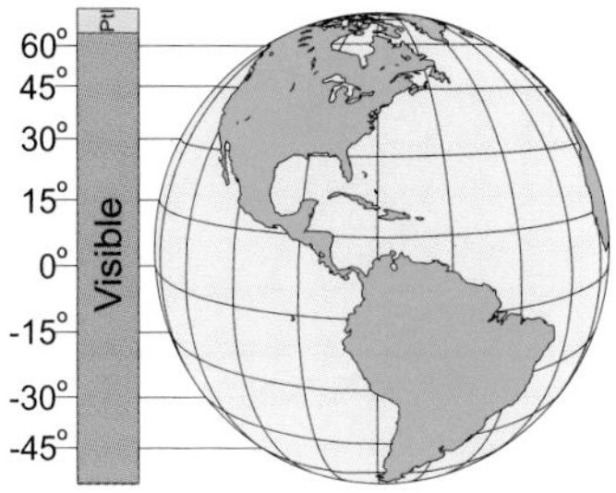

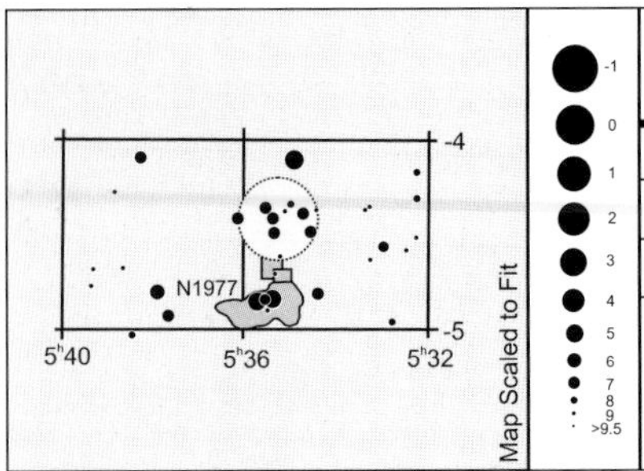

NGC 1977

RA:	05h 35m 32.4s	Con:	Orion
Dec:	-04° 51' 55"	Type:	Cluster & Nebula
Size:	20.0'	Mag:	4.5

NGC 1977 is an emission and reflection nebula associated with a loose grouping of about 12 stars.

Telescope Aperture:	4" f/5	4" f/9	6" f/7	6" f/9	8" f/6.3	8" f/10	10" f/6.3	10" f/10	12" f/6.3	12" f/10
FOV(35mm film):	2.7° x 4.1°	1.50° x 2.26°	1.29° x 1.93°	1.0° x 1.50°	1.07° x 1.61°	0.68° x 1.02°	0.86° x 1.29°	0.54° x 0.81°	0.72° x 1.07°	0.45° x 0.68°

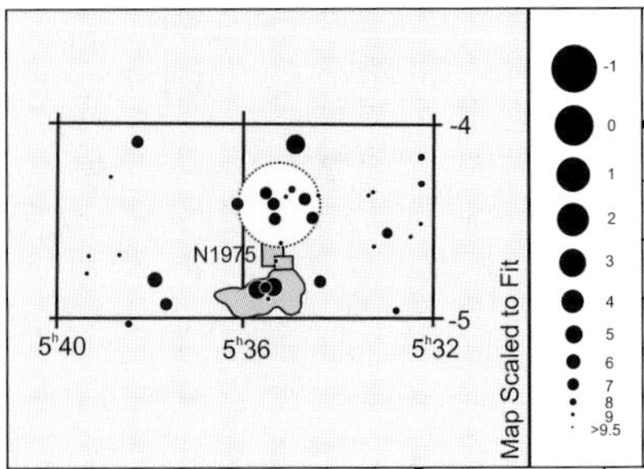

NGC 1975

RA:	05h 35m 26.4s	Con:	Orion
Dec:	-04° 40' 55"	Type:	E & R Nebula
Size:	10.0'	Mag:	9.0

NGC 1975 is a bright emission and reflection nebula that is part of the M42 nebular complex.

Telescope Aperture:	4" f/5	4" f/9	6" f/7	6" f/9	8" f/6.3	8" f/10	10" f/6.3	10" f/10	12" f/6.3	12" f/10
FOV(35mm film):	2.7° x 4.1°	1.50° x 2.26°	1.29° x 1.93°	1.0° x 1.50°	1.07° x 1.61°	0.68° x 1.02°	0.86° x 1.29°	0.54° x 0.81°	0.72° x 1.07°	0.45° x 0.68°

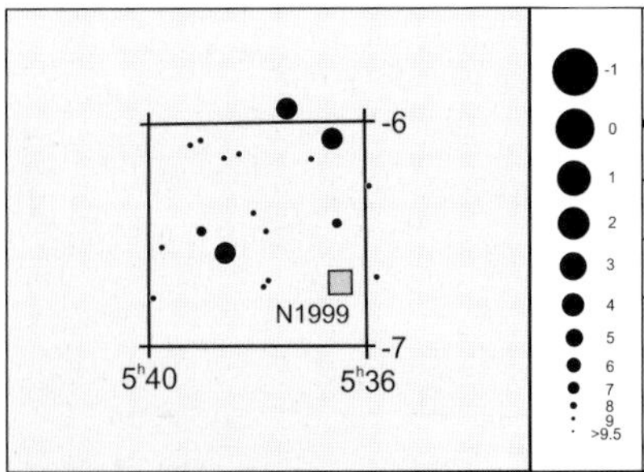

NGC 1999

RA:	05h 36m 32.3s	Con:	Orion
Dec:	-06° 41' 55"	Type:	Emission Nebula
Size:	16.0'	Mag:	9.4

NGC 1999 is a roundish, bright emission nebula. Object displays visible dust lanes.

Telescope Aperture:	4" f/5	4" f/9	6" f/7	6" f/9	8" f/6.3	8" f/10	10" f/6.3	10" f/10	12" f/6.3	12" f/10
FOV(35mm film):	2.7° x 4.1°	1.50° x 2.26°	1.29° x 1.93°	1.0° x 1.50°	1.07° x 1.61°	0.68° x 1.02°	0.86° x 1.29°	0.54° x 0.81°	0.72° x 1.07°	0.45° x 0.68°

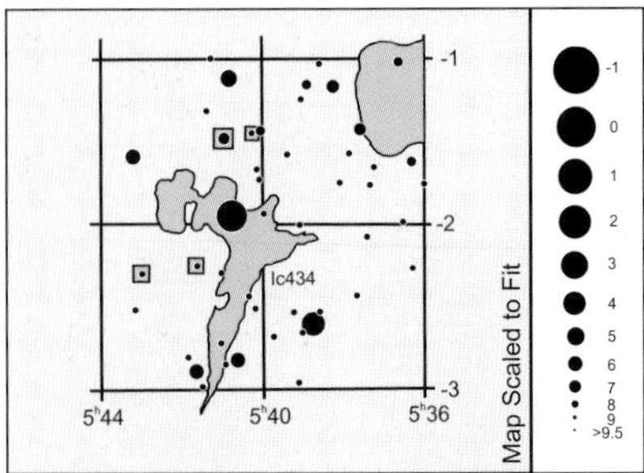

IC 434

RA:	05h 41m 2.4s	Con:	Orion
Dec:	-02° 23' 56"	Type:	Emission Nebula
Size:	60.0'	Mag:	7.3

IC 434 is the emission nebula that contains the dark "Horsehead Nebula".

Telescope Aperture:	4" f/5	4" f/9	6" f/7	6" f/9	8" f/6.3	8" f/10	10" f/6.3	10" f/10	12" f/6.3	12" f/10
FOV(35mm film):	2.7° x 4.1°	1.50° x 2.26°	1.29° x 1.93°	1.0° x 1.50°	1.07° x 1.61°	0.68° x 1.02°	0.86° x 1.29°	0.54° x 0.81°	0.72° x 1.07°	0.45° x 0.68°

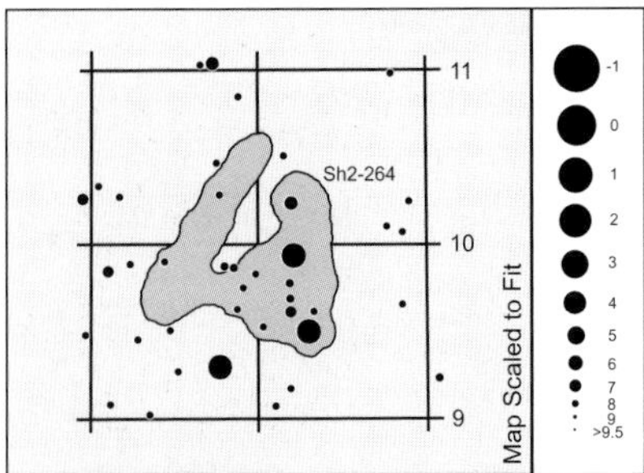

Sh2-264

RA:	05h 35m 17.0s	Con:	Orion
Dec:	09° 56' 38"	Type:	Emission Nebula
Size:	300'	Mag:	

Sh2-264 is a large, faint emission nebula.

Telescope Aperture:	4" f/5	4" f/9	6" f/7	6" f/9	8" f/6.3	8" f/10	10" f/6.3	10" f/10	12" f/6.3	12" f/10
FOV(35mm film):	2.7° x 4.1°	1.50° x 2.26°	1.29° x 1.93°	1.0° x 1.50°	1.07° x 1.61°	0.68° x 1.02°	0.86° x 1.29°	0.54° x 0.81°	0.72° x 1.07°	0.45° x 0.68°

NGC 2023

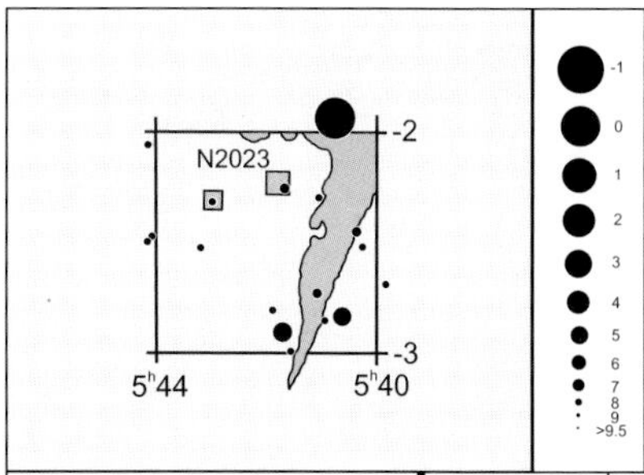

RA:	$05^h\ 41^m\ 38.4^s$	Con:	Orion
Dec:	-02° 13' 56"	Type:	E & R Nebula
Size:	10.0'	Mag:	

NGC 2023 is a bright emission and reflection nebula.

Telescope Aperture:	4" f/5	4" f/9	6" f/7	6" f/9	8" f/6.3	8" f/10	10" f/6.3	10" f/10	12" f/6.3	12" f/10
FOV(35mm film):	2.7° x 4.1°	1.50° x 2.26°	1.29° x 1.93°	1.0° x 1.50°	1.07° x 1.61°	0.68° x 1.02°	0.86° x 1.29°	0.54° x 0.81°	0.72° x 1.07°	0.45° x 0.68°

NGC 1990

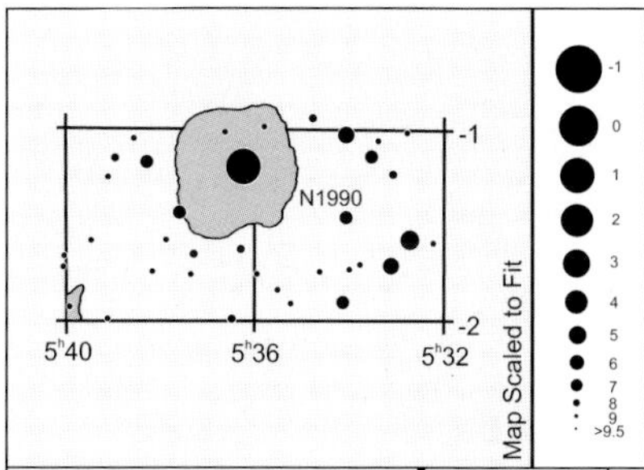

RA:	$05^h\ 36^m\ 14.5^s$	Con:	Orion
Dec:	-01° 11' 55"	Type:	E & R Nebula
Size:	50.0'	Mag:	

NGC 1990 is an emission and reflection nebula. Object has low surface brightness.

Telescope Aperture:	4" f/5	4" f/9	6" f/7	6" f/9	8" f/6.3	8" f/10	10" f/6.3	10" f/10	12" f/6.3	12" f/10
FOV(35mm film):	2.7° x 4.1°	1.50° x 2.26°	1.29° x 1.93°	1.0° x 1.50°	1.07° x 1.61°	0.68° x 1.02°	0.86° x 1.29°	0.54° x 0.81°	0.72° x 1.07°	0.45° x 0.68°

NGC 2169

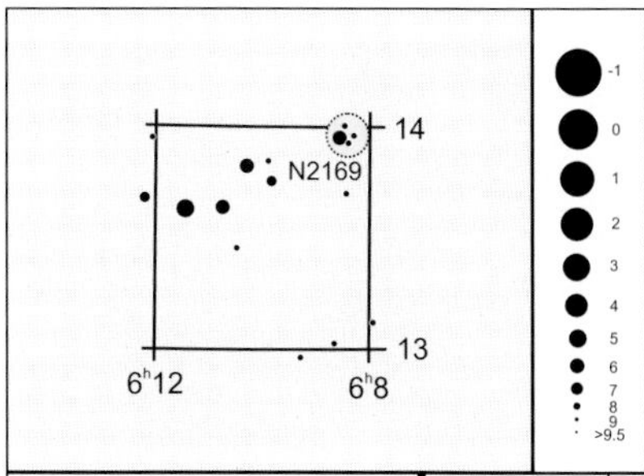

RA:	$06^h\ 08^m\ 26.8^s$	Con:	Orion
Dec:	13° 56' 59"	Type:	Open Cluster
Size:	7.0'	Mag:	5.9

NGC 2169 is a bright, open cluster with no central concentration. Object is moderately rich in stars.

Telescope Aperture:	4" f/5	4" f/9	6" f/7	6" f/9	8" f/6.3	8" f/10	10" f/6.3	10" f/10	12" f/6.3	12" f/10
FOV(35mm film):	2.7° x 4.1°	1.50° x 2.26°	1.29° x 1.93°	1.0° x 1.50°	1.07° x 1.61°	0.68° x 1.02°	0.86° x 1.29°	0.54° x 0.81°	0.72° x 1.07°	0.45° x 0.68°

NGC 2141

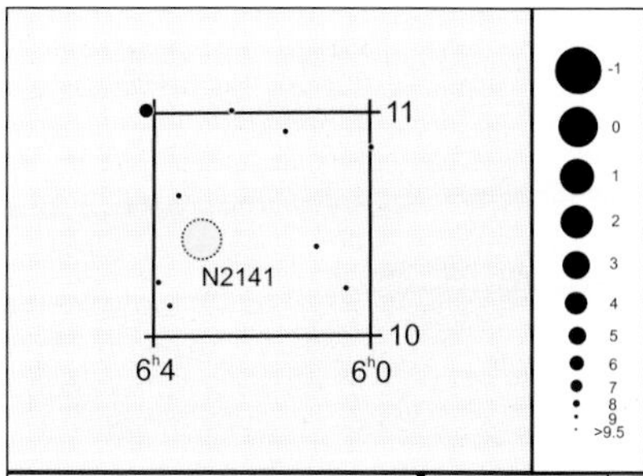

RA:	$06^h\ 03^m\ 8.7^s$	Con:	Orion
Dec:	10° 26' 00"	Type:	Open Cluster
Size:	10.0'	Mag:	9.4

NGC 2141 is a rich open cluster.

Telescope Aperture:	4" f/5	4" f/9	6" f/7	6" f/9	8" f/6.3	8" f/10	10" f/6.3	10" f/10	12" f/6.3	12" f/10
FOV(35mm film):	2.7° x 4.1°	1.50° x 2.26°	1.29° x 1.93°	1.0° x 1.50°	1.07° x 1.61°	0.68° x 1.02°	0.86° x 1.29°	0.54° x 0.81°	0.72° x 1.07°	0.45° x 0.68°

M42 (NGC 1976)

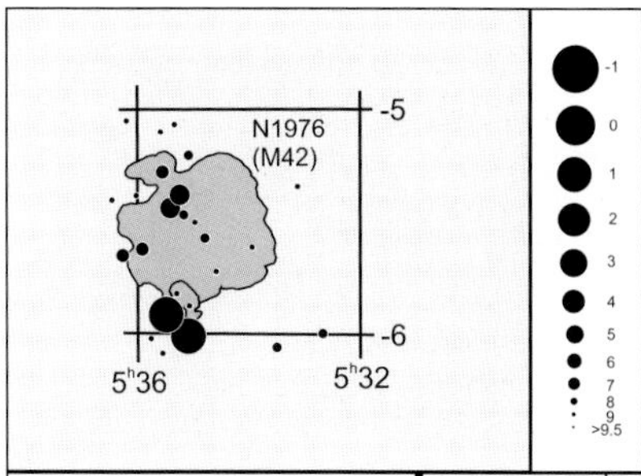

RA:	$05^h\ 35^m\ 26.4^s$	Con:	Orion
Dec:	-05° 26' 55"	Type:	Emission Nebula
Size:	66.0'	Mag:	4.0

M42 (NGC 1976) "the Great Orion Nebula" is a glowing cloud of ionized gas, energized by hot young stars in its central region.

Telescope Aperture:	4" f/5	4" f/9	6" f/7	6" f/9	8" f/6.3	8" f/10	10" f/6.3	10" f/10	12" f/6.3	12" f/10
FOV(35mm film):	2.7° x 4.1°	1.50° x 2.26°	1.29° x 1.93°	1.0° x 1.50°	1.07° x 1.61°	0.68° x 1.02°	0.86° x 1.29°	0.54° x 0.81°	0.72° x 1.07°	0.45° x 0.68°

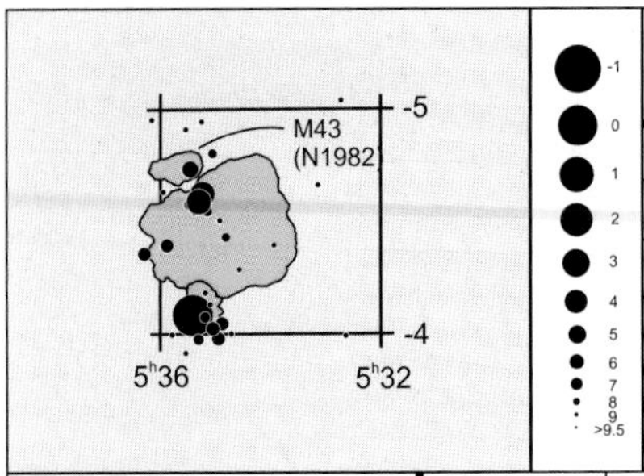

M43 (NGC 1982)

RA:	$05^h\ 35^m\ 38.4^s$	Con:	Orion
Dec:	-05° 15' 55"	Type:	E & R Nebula
Size:	20.0'	Mag:	9.0

M43 (NGC 1982) is actually part of the M42 "Orion Nebula" complex. It appears detached because of a dark dust lane crossing the nebula obscuring their connection.

Telescope Aperture:	4" f/5	4" f/9	6" f/7	6" f/9	8" f/6.3	8" f/10	10" f/6.3	10" f/10	12" f/6.3	12" f/10
FOV(35mm film):	2.7° x 4.1°	1.50° x 2.26°	1.29° x 1.93°	1.0° x 1.50°	1.07° x 1.61°	0.68° x 1.02°	0.86° x 1.29°	0.54° x 0.81°	0.72° x 1.07°	0.45° x 0.68°

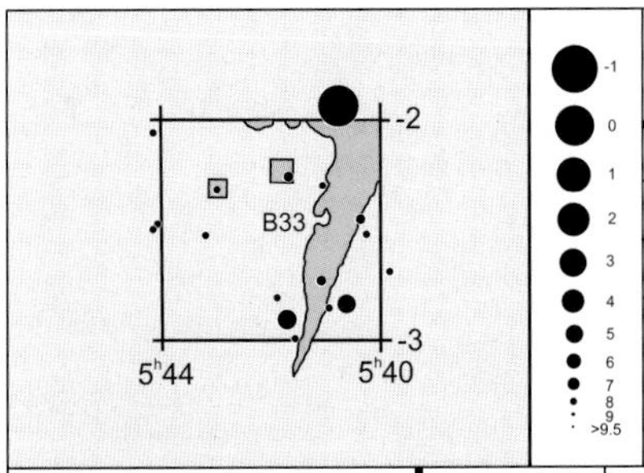

B-33 (Horsehead Nebula)

RA:	$05^h\ 36^m\ 14.5^s$	Con:	Orion
Dec:	-01° 11' 55"	Type:	Dark Nebula
Size:	5.0'	Mag:	

B-33 is a dark nebular mass silhouetted against the bright IC 434 nebula structure.

Telescope Aperture:	4" f/5	4" f/9	6" f/7	6" f/9	8" f/6.3	8" f/10	10" f/6.3	10" f/10	12" f/6.3	12" f/10
FOV(35mm film):	2.7° x 4.1°	1.50° x 2.26°	1.29° x 1.93°	1.0° x 1.50°	1.07° x 1.61°	0.68° x 1.02°	0.86° x 1.29°	0.54° x 0.81°	0.72° x 1.07°	0.45° x 0.68°

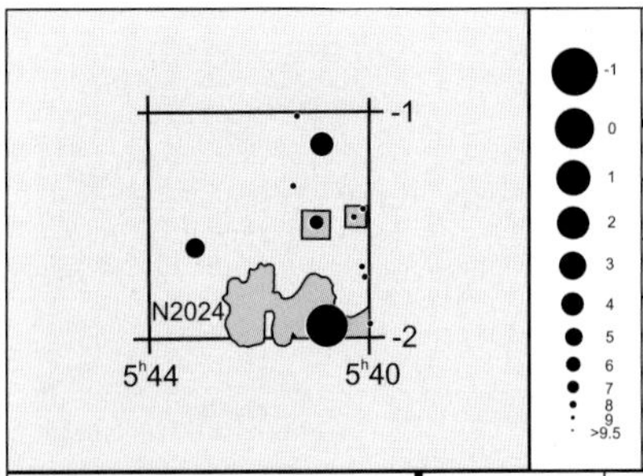

NGC 2024

RA:	$05^h\ 41^m\ 56.5^s$	Con:	Orion
Dec:	-01° 50' 56"	Type:	Emission Nebula
Size:	30.0'	Mag:	

NGC 2024 is a large emission nebula bisected by a broad, dark nebular band.

Telescope Aperture:	4" f/5	4" f/9	6" f/7	6" f/9	8" f/6.3	8" f/10	10" f/6.3	10" f/10	12" f/6.3	12" f/10
FOV(35mm film):	2.7° x 4.1°	1.50° x 2.26°	1.29° x 1.93°	1.0° x 1.50°	1.07° x 1.61°	0.68° x 1.02°	0.86° x 1.29°	0.54° x 0.81°	0.72° x 1.07°	0.45° x 0.68°

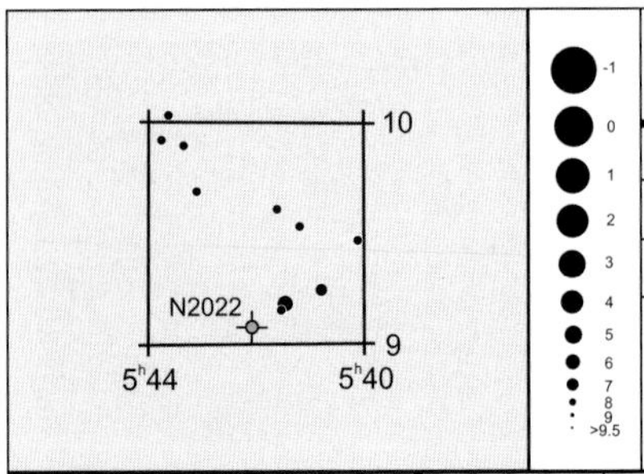

NGC 2022

RA:	$05^h\ 42^m\ 8.7^s$	Con:	Orion
Dec:	09° 05' 02"	Type:	Planetary Nebula
Size:	0.3'	Mag:	12.0

NGC 2022 is a planetary ring nebula.

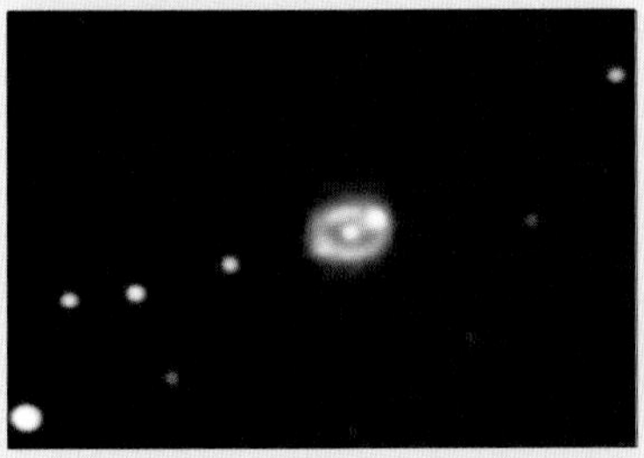

Telescope Aperture:	4" f/5	4" f/9	6" f/7	6" f/9	8" f/6.3	8" f/10	10" f/6.3	10" f/10	12" f/6.3	12" f/10
FOV(35mm film):	2.7° x 4.1°	1.50° x 2.26°	1.29° x 1.93°	1.0° x 1.50°	1.07° x 1.61°	0.68° x 1.02°	0.86° x 1.29°	0.54° x 0.81°	0.72° x 1.07°	0.45° x 0.68°

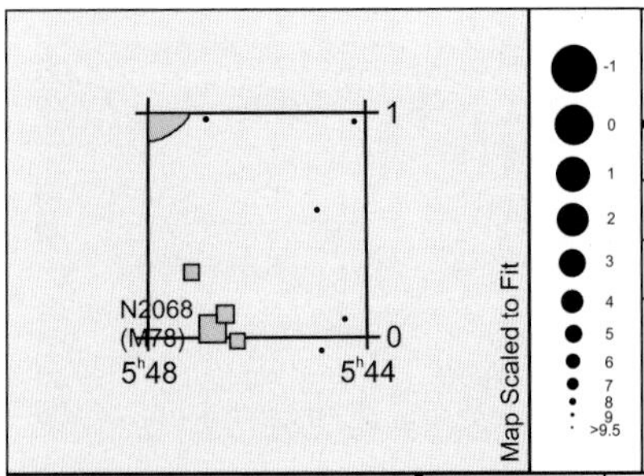

M78 (NGC 2068)

RA:	$05^h\ 49^m\ 44.5^s$	Con:	Orion
Dec:	00° 03' 03"	Type:	Reflection Nebula
Size:	8.0'	Mag:	8.0

M78 (NCG 2068) is found northeast of Orion's Belt.

Telescope Aperture:	4" f/5	4" f/9	6" f/7	6" f/9	8" f/6.3	8" f/10	10" f/6.3	10" f/10	12" f/6.3	12" f/10
FOV(35mm film):	2.7° x 4.1°	1.50° x 2.26°	1.29° x 1.93°	1.0° x 1.50°	1.07° x 1.61°	0.68° x 1.02°	0.86° x 1.29°	0.54° x 0.81°	0.72° x 1.07°	0.45° x 0.68°

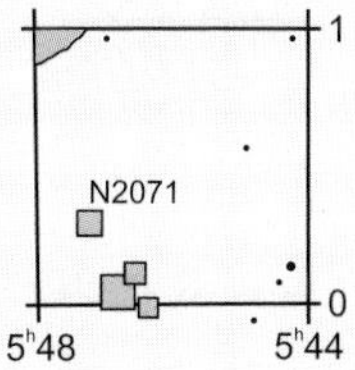

NGC 2071

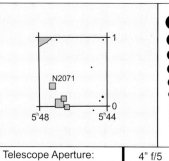

RA:	$05^h\ 47^m\ 14.5^s$	Con:	Orion
Dec:	00° 18' 03"	Type:	Reflection Nebula
Size:	4.0'	Mag:	

NGC 2071 is located north of M78 and is a faint reflection nebula.

Telescope Aperture:	4" f/5	4" f/9	6" f/7	6" f/9	8" f/6.3	8" f/10	10" f/6.3	10" f/10	12" f/6.3	12" f/10
FOV(35mm film):	2.7° x 4.1°	1.50° x 2.26°	1.29° x 1.93°	1.0° x 1.50°	1.07° x 1.61°	0.68° x 1.02°	0.86° x 1.29°	0.54° x 0.81°	0.72° x 1.07°	0.45° x 0.68°

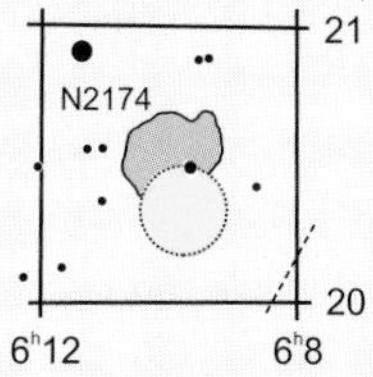

NGC 2174/75 (Monkey Head Nebula)

RA:	$06^h\ 09^m\ 45.0^s$	Con:	Orion
Dec:	20° 29' 59"	Type:	Emission Nebula
Size:	40.0'	Mag:	

NGC 2174 is found in the northern regions of the constellation. Object is a faint emission nebula surrounding the open cluster NGC 2175.

Telescope Aperture:	4" f/5	4" f/9	6" f/7	6" f/9	8" f/6.3	8" f/10	10" f/6.3	10" f/10	12" f/6.3	12" f/10
FOV(35mm film):	2.7° x 4.1°	1.50° x 2.26°	1.29° x 1.93°	1.0° x 1.50°	1.07° x 1.61°	0.68° x 1.02°	0.86° x 1.29°	0.54° x 0.81°	0.72° x 1.07°	0.45° x 0.68°

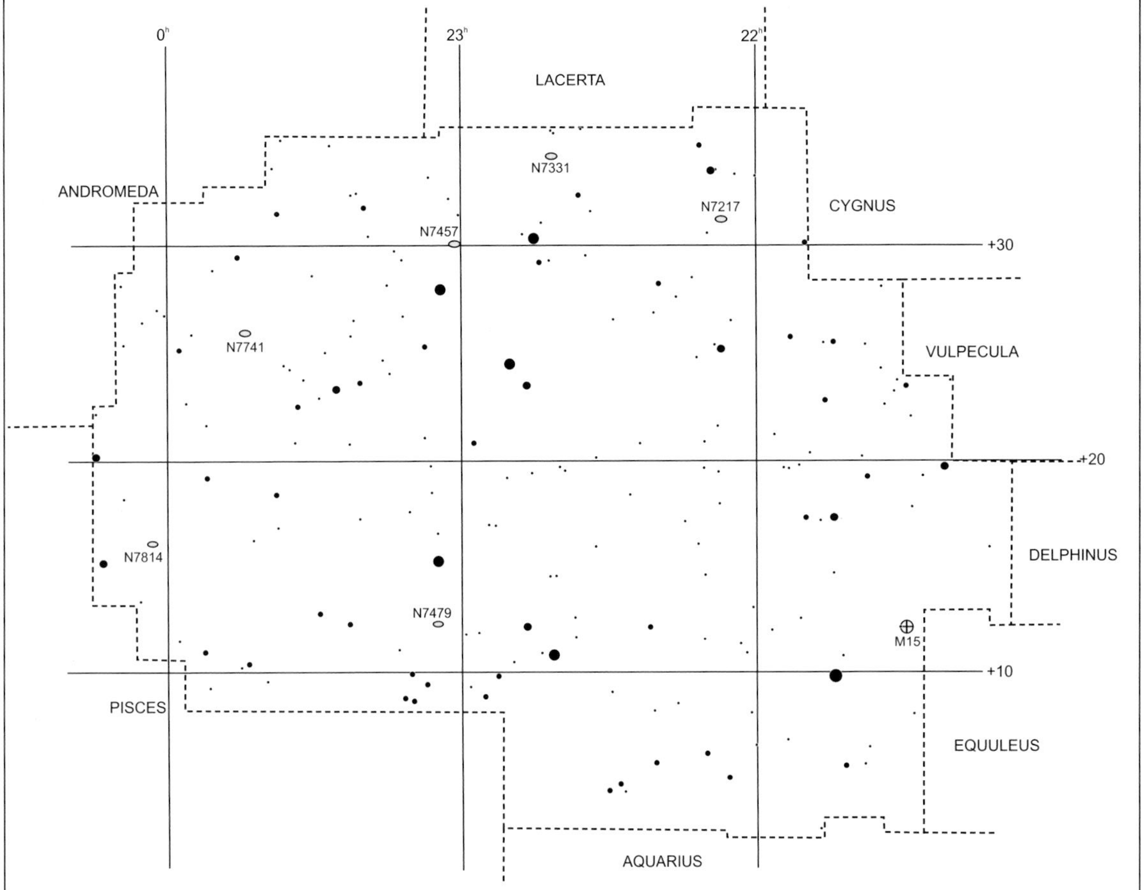

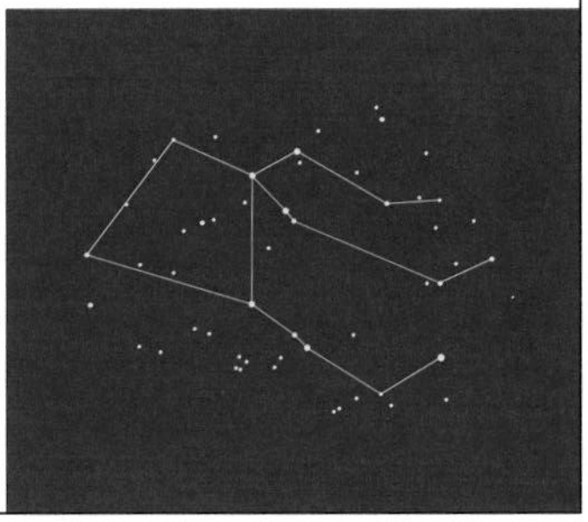

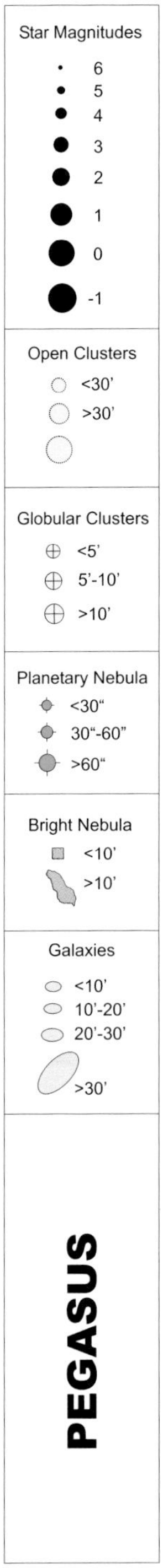

Constellation Facts:

Pegasus; (PEG-uh-suss)

Pegasus, the Winged Horse.
Pegasus rises in the northeastern sky, passes overhead near the zenith, and sets towards the northwest.
The constellation covers 1121 square degrees.

Constellation is visible from 90° N to 53° S. Partially visible from 53° S to 90° S.

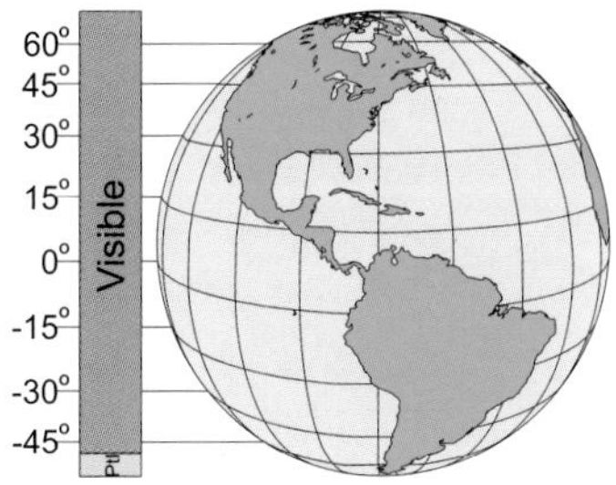

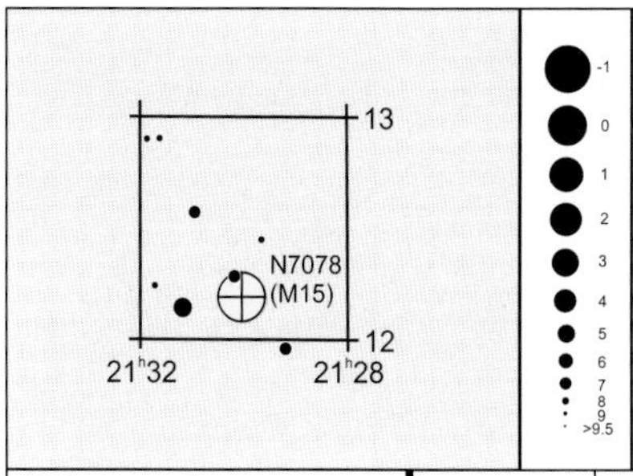

M15 (NGC 7078)

RA:	$21^h\ 30^m\ 4.7^s$	Con:	Pegasus
Dec:	12° 10' 21"	Type:	Globular Cluster
Size:	12.3'	Mag:	6.4

M15 (NGC 7078) is a highly resolved globular cluster. M15 is not only an excellent example of a globular cluster, but is the only known cluster to contain a planetary nebula.

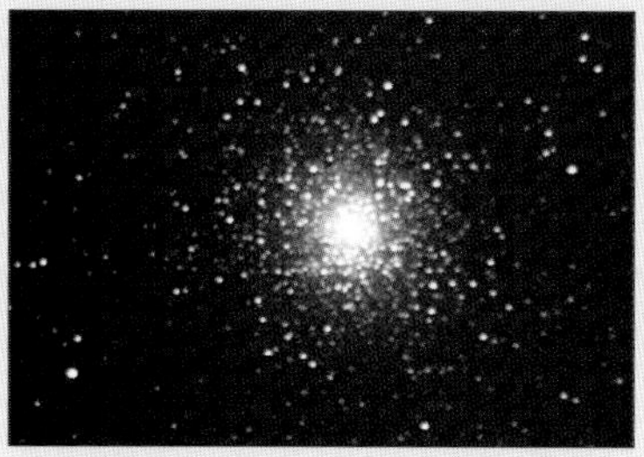

Telescope Aperture:	4" f/5	4" f/9	6" f/7	6" f/9	8" f/6.3	8" f/10	10" f/6.3	10" f/10	12" f/6.3	12" f/10
FOV(35mm film):	2.7° x 4.1°	1.50° x 2.26°	1.29° x 1.93°	1.0° x 1.50°	1.07° x 1.61°	0.68° x 1.02°	0.86° x 1.29°	0.54° x 0.81°	0.72° x 1.07°	0.45° x 0.68°

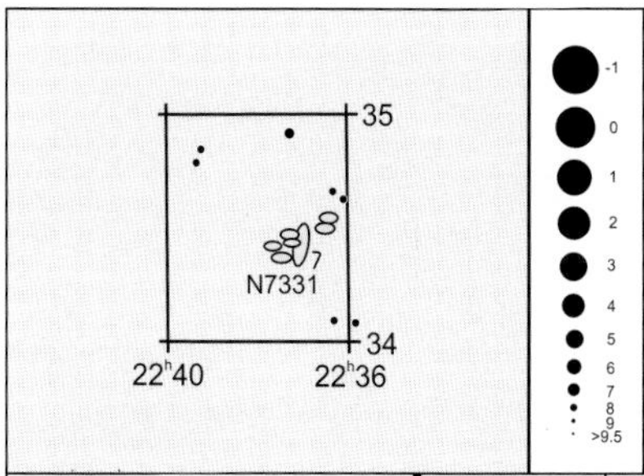

NGC 7331

RA:	$22^h\ 37^m\ 10.4^s$	Con:	Pegasus
Dec:	34° 25' 16"	Type:	Spiral Galaxy
Size:	10.0' x 3.0'	Mag:	9.5

NGC 7331 is an Sb-type spiral galaxy that in the eyepiece appears elongated with a dusty, bright core.

Telescope Aperture:	4" f/5	4" f/9	6" f/7	6" f/9	8" f/6.3	8" f/10	10" f/6.3	10" f/10	12" f/6.3	12" f/10
FOV(35mm film):	2.7° x 4.1°	1.50° x 2.26°	1.29° x 1.93°	1.0° x 1.50°	1.07° x 1.61°	0.68° x 1.02°	0.86° x 1.29°	0.54° x 0.81°	0.72° x 1.07°	0.45° x 0.68°

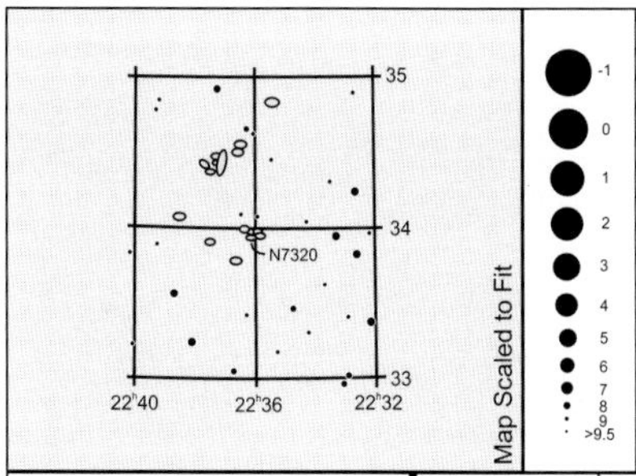

NGC 7320

RA:	$22^h\ 36^m\ 10.4^s$	Con:	Pegasus
Dec:	33° 57' 17"	Type:	Spiral Galaxy
Size:	0.4'	Mag:	12.7

NGC 7320 appears as an elongated galaxy in the eyepiece. Object is the brightest member of Stephan's Quintet.

Telescope Aperture:	4" f/5	4" f/9	6" f/7	6" f/9	8" f/6.3	8" f/10	10" f/6.3	10" f/10	12" f/6.3	12" f/10
FOV(35mm film):	2.7° x 4.1°	1.50° x 2.26°	1.29° x 1.93°	1.0° x 1.50°	1.07° x 1.61°	0.68° x 1.02°	0.86° x 1.29°	0.54° x 0.81°	0.72° x 1.07°	0.45° x 0.68°

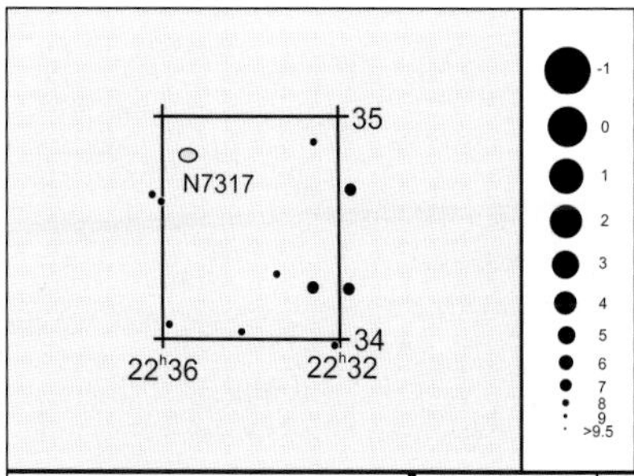

NGC 7317

RA:	$22^h\ 25^m\ 58.4^s$	Con:	Pegasus
Dec:	33° 57' 17"	Type:	Elliptical Galaxy
Size:	0.4'	Mag:	13.6

NGC 7317 is the second galaxy of Stephan's Quintet. Object is an elliptical galaxy and appears round in the eyepiece.

Telescope Aperture:	4" f/5	4" f/9	6" f/7	6" f/9	8" f/6.3	8" f/10	10" f/6.3	10" f/10	12" f/6.3	12" f/10
FOV(35mm film):	2.7° x 4.1°	1.50° x 2.26°	1.29° x 1.93°	1.0° x 1.50°	1.07° x 1.61°	0.68° x 1.02°	0.86° x 1.29°	0.54° x 0.81°	0.72° x 1.07°	0.45° x 0.68°

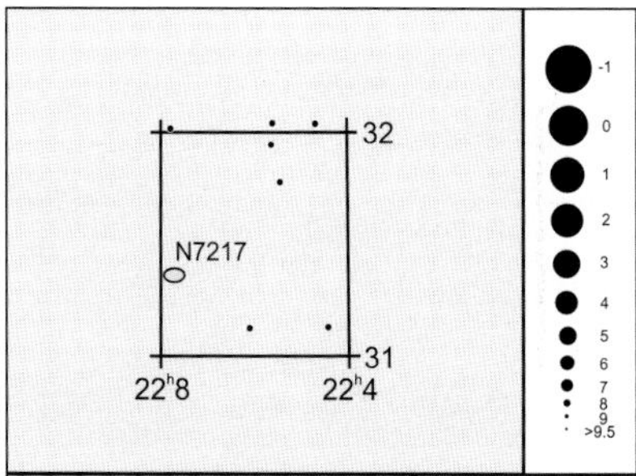

NGC 7217

RA:	$22^h\ 07^m\ 58.5^s$	Con:	Pegasus
Dec:	31° 22' 18"	Type:	Spiral Galaxy
Size:	3.2' x 2.6'	Mag:	10.2

NGC 7217 appears as a round galaxy with a bright core, but is in-fact an Sb-type spiral galaxy, appearing nearly face-on.

Telescope Aperture:	4" f/5	4" f/9	6" f/7	6" f/9	8" f/6.3	8" f/10	10" f/6.3	10" f/10	12" f/6.3	12" f/10
FOV(35mm film):	2.7° x 4.1°	1.50° x 2.26°	1.29° x 1.93°	1.0° x 1.50°	1.07° x 1.61°	0.68° x 1.02°	0.86° x 1.29°	0.54° x 0.81°	0.72° x 1.07°	0.45° x 0.68°

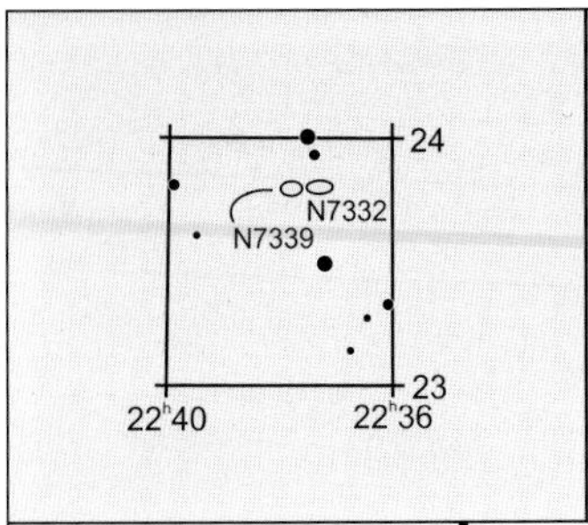

NGC 7332/39

RA:	22h 37m 28.5s	Con:	Pegasus
Dec:	23° 48' 20"	Type:	Galaxies
Size:	3.5' x 1.0'/2.8' x 0.6'	Mag:	11.0/12.1

NGC 7332 is an elliptical galaxy. Found 10° south of Stephan's Quintet.
NGC 7339 is an edge-on barred spiral galaxy.

Telescope Aperture:	4" f/5	4" f/9	6" f/7	6" f/9	8" f/6.3	8" f/10	10" f/6.3	10" f/10	12" f/6.3	12" f/10
FOV(35mm film):	2.7° x 4.1°	1.50° x 2.26°	1.29° x 1.93°	1.0° x 1.50°	1.07° x 1.61°	0.68° x 1.02°	0.86° x 1.29°	0.54° x 0.81°	0.72° x 1.07°	0.45° x 0.68°

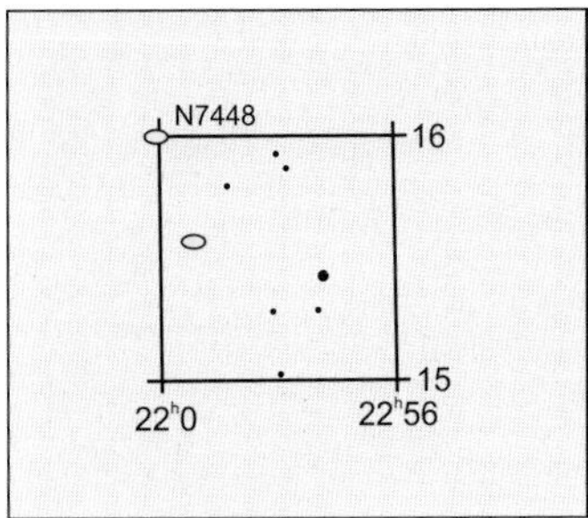

NGC 7448

RA:	23h 00m 10.4s	Con:	Pegasus
Dec:	15° 59' 22"	Type:	Spiral Galaxy
Size:	2.3' x 0.8'	Mag:	11.7

NGC 7448 is an Sc-type spiral galaxy, and is the brightest galaxy in its local group.

Telescope Aperture:	4" f/5	4" f/9	6" f/7	6" f/9	8" f/6.3	8" f/10	10" f/6.3	10" f/10	12" f/6.3	12" f/10
FOV(35mm film):	2.7° x 4.1°	1.50° x 2.26°	1.29° x 1.93°	1.0° x 1.50°	1.07° x 1.61°	0.68° x 1.02°	0.86° x 1.29°	0.54° x 0.81°	0.72° x 1.07°	0.45° x 0.68°

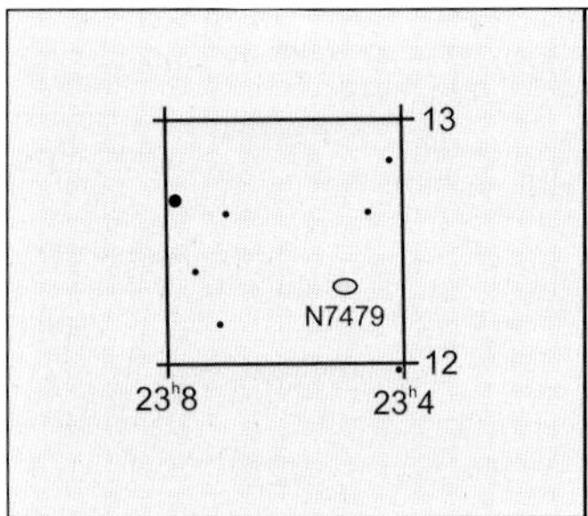

NGC 7479

RA:	23h 04m 58.4s	Con:	Pegasus
Dec:	12° 19' 23"	Type:	Spiral Galaxy
Size:	4.0' x 3.3'	Mag:	11.0

NGC 7479 is a large and impressive barred spiral galaxy.

Telescope Aperture:	4" f/5	4" f/9	6" f/7	6" f/9	8" f/6.3	8" f/10	10" f/6.3	10" f/10	12" f/6.3	12" f/10
FOV(35mm film):	2.7° x 4.1°	1.50° x 2.26°	1.29° x 1.93°	1.0° x 1.50°	1.07° x 1.61°	0.68° x 1.02°	0.86° x 1.29°	0.54° x 0.81°	0.72° x 1.07°	0.45° x 0.68°

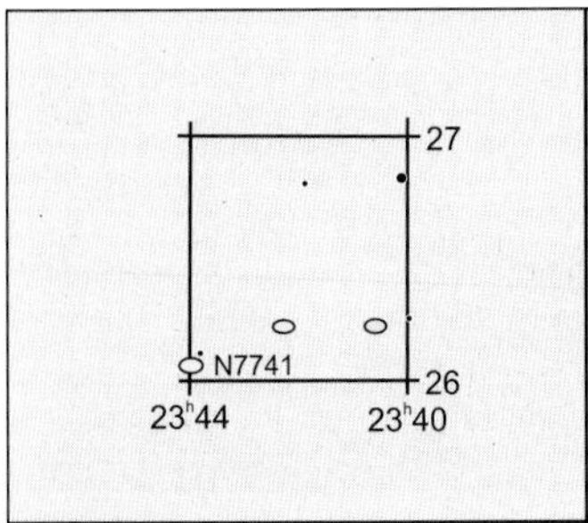

NGC 7741

RA:	23h 43m 58.3s	Con:	Pegasus
Dec:	26° 05' 19"	Type:	Spiral Galaxy
Size:	3.8' x 2.9'	Mag:	11.4

NGC 7741 is a barred spiral.

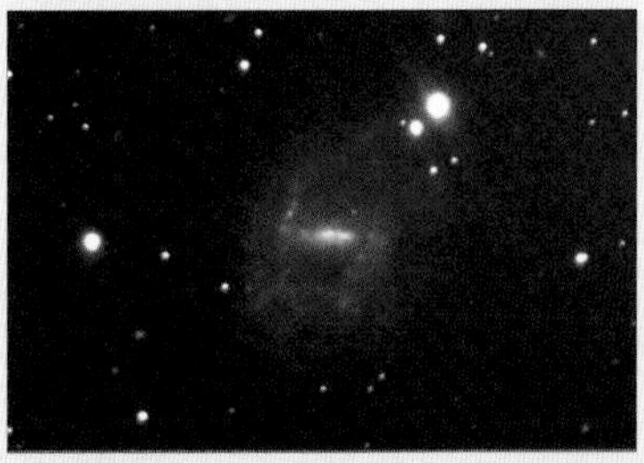

Telescope Aperture:	4" f/5	4" f/9	6" f/7	6" f/9	8" f/6.3	8" f/10	10" f/6.3	10" f/10	12" f/6.3	12" f/10
FOV(35mm film):	2.7° x 4.1°	1.50° x 2.26°	1.29° x 1.93°	1.0° x 1.50°	1.07° x 1.61°	0.68° x 1.02°	0.86° x 1.29°	0.54° x 0.81°	0.72° x 1.07°	0.45° x 0.68°

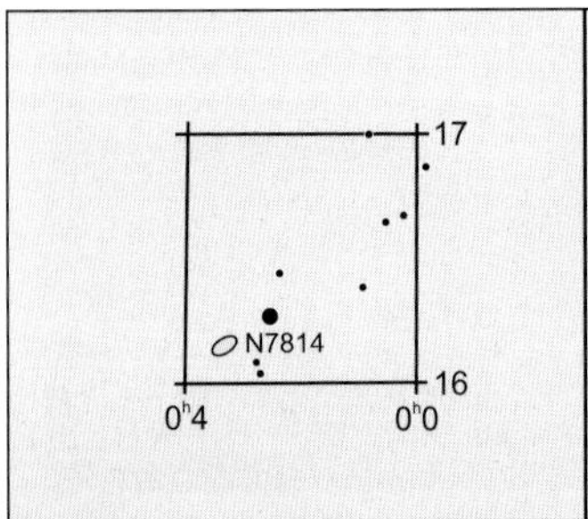

NGC 7814

RA:	00h 03m 22.2s	Con:	Pegasus
Dec:	16° 09' 21"	Type:	Spiral Galaxy
Size:	5.0' x 2.5'	Mag:	10.5

NGC 7814 is a bright edge-on Sb-type spiral galaxy.

Telescope Aperture:	4" f/5	4" f/9	6" f/7	6" f/9	8" f/6.3	8" f/10	10" f/6.3	10" f/10	12" f/6.3	12" f/10
FOV(35mm film):	2.7° x 4.1°	1.50° x 2.26°	1.29° x 1.93°	1.0° x 1.50°	1.07° x 1.61°	0.68° x 1.02°	0.86° x 1.29°	0.54° x 0.81°	0.72° x 1.07°	0.45° x 0.68°

Per **Crosses Prime Meridian:** November thru December

AUTUMN

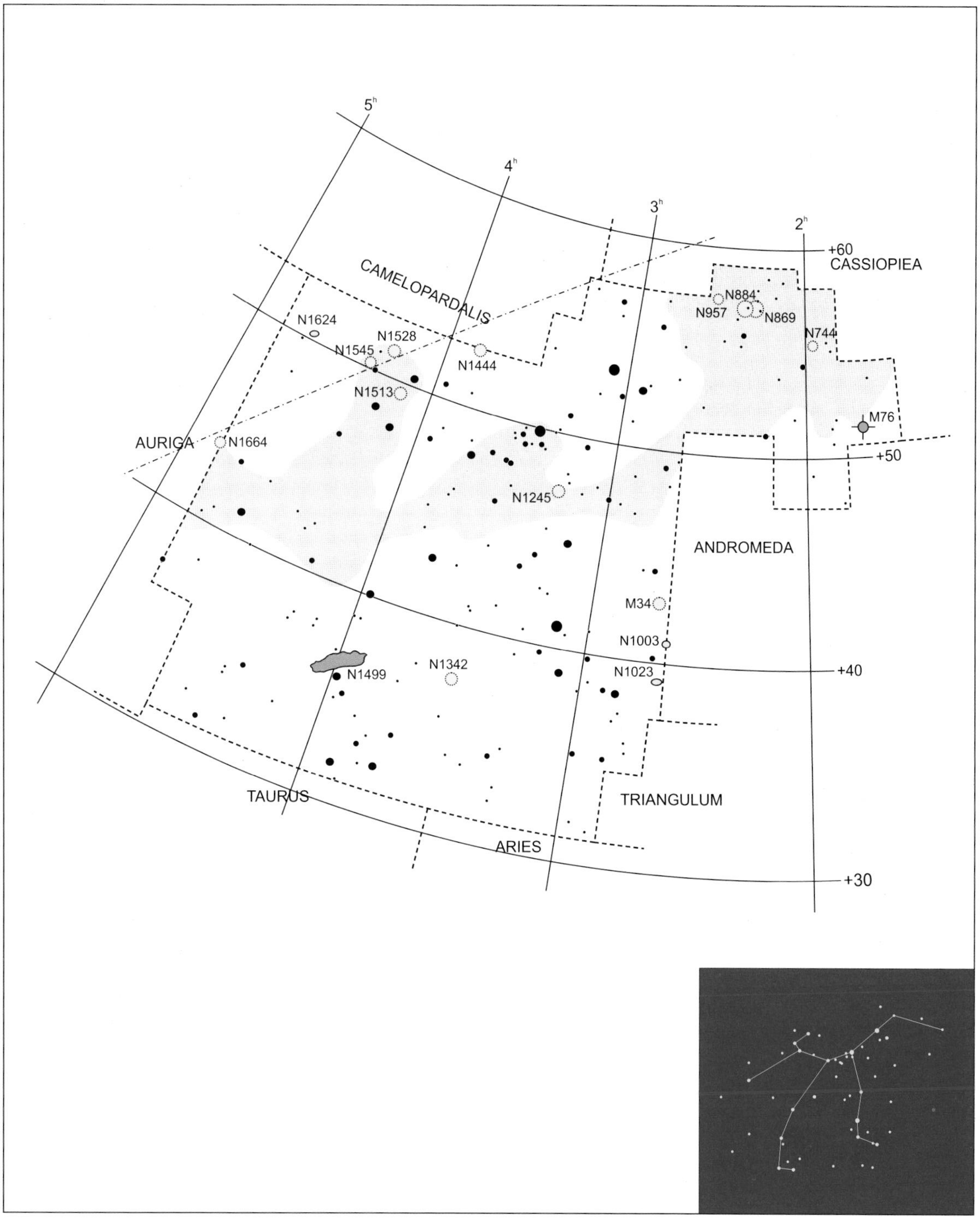

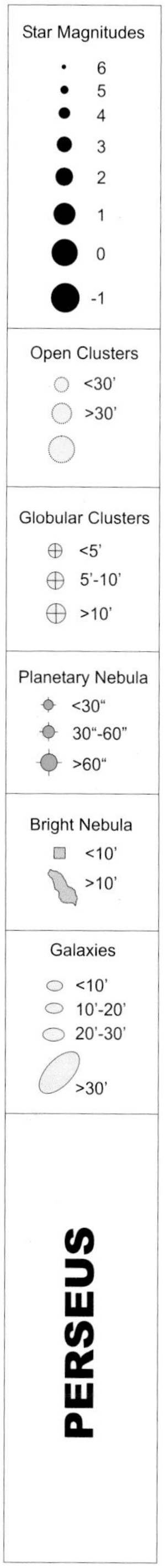

Constellation Facts:

Perseus; (PURR-see-us)

Perseus.
The constellation moves from the northeast to the northwest. The constellation is at its zenith as it crosses the meridian when viewed from mid-northern latitudes.
The constellation covers 615 square degrees.

Constellation is visible from 90° N to 31° S. Partially visible from 31° S to 90° S.

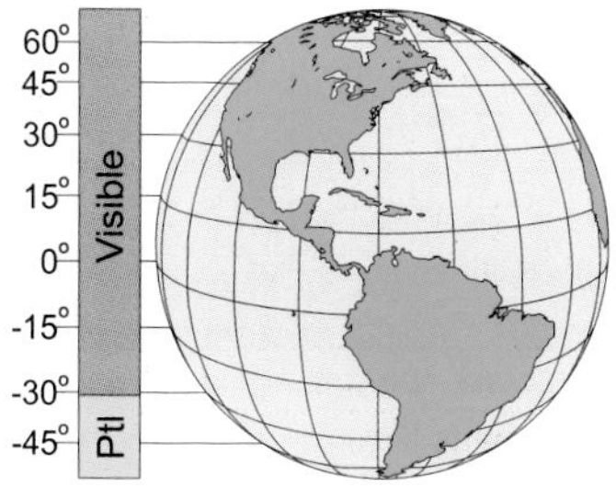

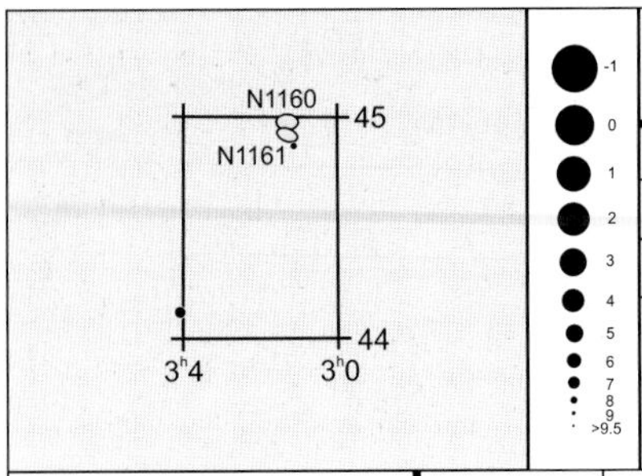

NGC 1160

RA:	03^h 01^m 16.1^s	Con:	Perseus
Dec:	44° 58' 06"	Type:	Spiral Galaxy
Size:	1.3' x 0.6'	Mag:	13.0

NGC 1160 appears through the eyepiece as an elongated galaxy with a close companion.

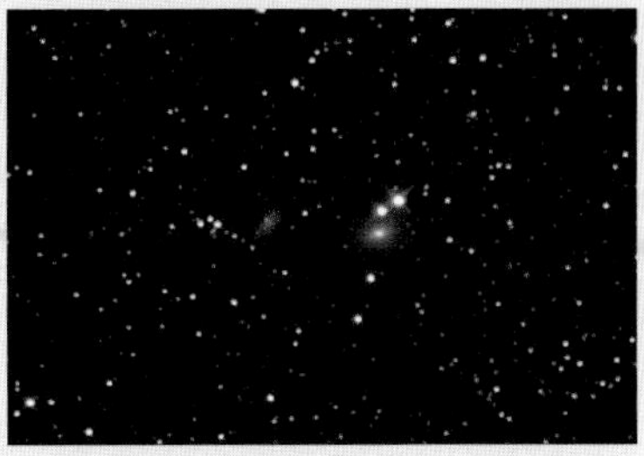

Telescope Aperture:	4" f/5	4" f/9	6" f/7	6" f/9	8" f/6.3	8" f/10	10" f/6.3	10" f/10	12" f/6.3	12" f/10
FOV(35mm film):	2.7°x 4.1°	1.50°x 2.26°	1.29° x 1.93°	1.0° x 1.50°	1.07° x 1.61°	0.68°x 1.02°	0.86° x 1.29°	0.54° x 0.81°	0.72° x 1.07°	0.45°x 0.68°

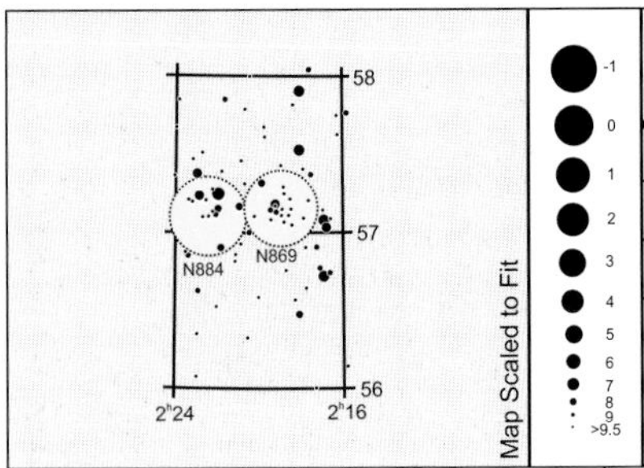

NGC 869/884 (Double Cluster)

RA:	02^h 19^m 4.7^s	Con:	Perseus
Dec:	57° 09' 05"	Type:	Open Cluster
Size:	30.0'	Mag:	4.0

NGC 869/884 the Double Cluster are two dense open clusters arranged side-by-side.

Telescope Aperture:	4" f/5	4" f/9	6" f/7	6" f/9	8" f/6.3	8" f/10	10" f/6.3	10" f/10	12" f/6.3	12" f/10
FOV(35mm film):	2.7°x 4.1°	1.50°x 2.26°	1.29° x 1.93°	1.0° x 1.50°	1.07° x 1.61°	0.68°x 1.02°	0.86° x 1.29°	0.54° x 0.81°	0.72° x 1.07°	0.45°x 0.68°

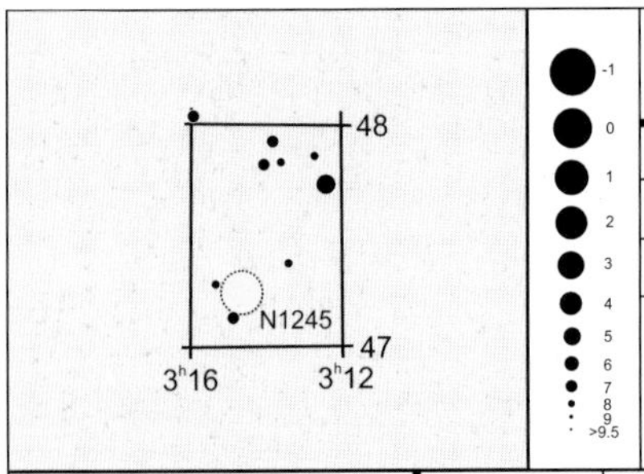

NGC 1245

RA:	03^h 14^m 46.2^s	Con:	Perseus
Dec:	47° 15' 04"	Type:	Open Cluster
Size:	10.0'	Mag:	8.4

NGC 1245 is a dense open cluster, containing 40 stars.

Telescope Aperture:	4" f/5	4" f/9	6" f/7	6" f/9	8" f/6.3	8" f/10	10" f/6.3	10" f/10	12" f/6.3	12" f/10
FOV(35mm film):	2.7°x 4.1°	1.50°x 2.26°	1.29° x 1.93°	1.0° x 1.50°	1.07° x 1.61°	0.68°x 1.02°	0.86° x 1.29°	0.54° x 0.81°	0.72° x 1.07°	0.45°x 0.68°

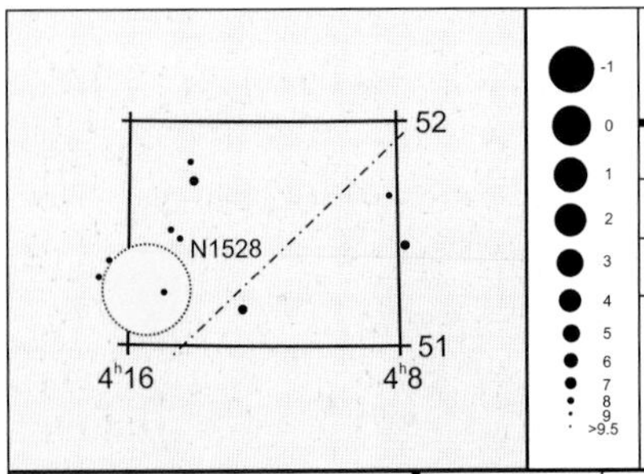

NGC 1528

RA:	04^h 15^m 28.1^s	Con:	Perseus
Dec:	51° 14' 00"	Type:	Open Cluster
Size:	24.0'	Mag:	6.4

NGC 1528 is a dense open cluster. Object is comprised of 80 stars, and is an photographic opportunity.

Telescope Aperture:	4" f/5	4" f/9	6" f/7	6" f/9	8" f/6.3	8" f/10	10" f/6.3	10" f/10	12" f/6.3	12" f/10
FOV(35mm film):	2.7°x 4.1°	1.50°x 2.26°	1.29° x 1.93°	1.0° x 1.50°	1.07° x 1.61°	0.68°x 1.02°	0.86° x 1.29°	0.54° x 0.81°	0.72° x 1.07°	0.45°x 0.68°

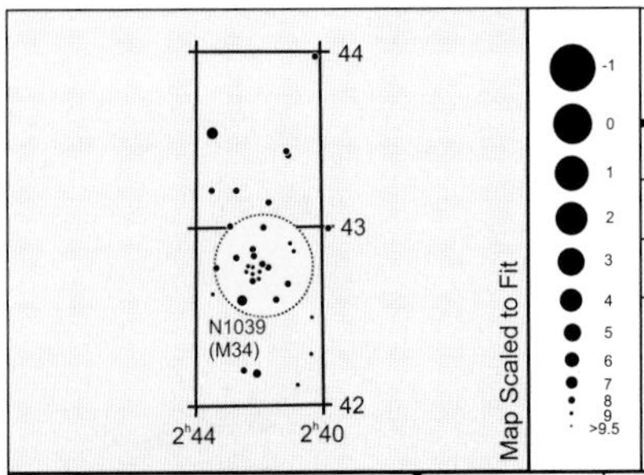

M34 (NGC 1039)

RA:	02^h 42^m 4.1^s	Con:	Perseus
Dec:	42° 47' 07"	Type:	Open Cluster
Size:	35.0'	Mag:	5.2

M34 (NGC 1039) is a rich open cluster, containing 80 stars.

Telescope Aperture:	4" f/5	4" f/9	6" f/7	6" f/9	8" f/6.3	8" f/10	10" f/6.3	10" f/10	12" f/6.3	12" f/10
FOV(35mm film):	2.7°x 4.1°	1.50°x 2.26°	1.29° x 1.93°	1.0° x 1.50°	1.07° x 1.61°	0.68°x 1.02°	0.86° x 1.29°	0.54° x 0.81°	0.72° x 1.07°	0.45°x 0.68°

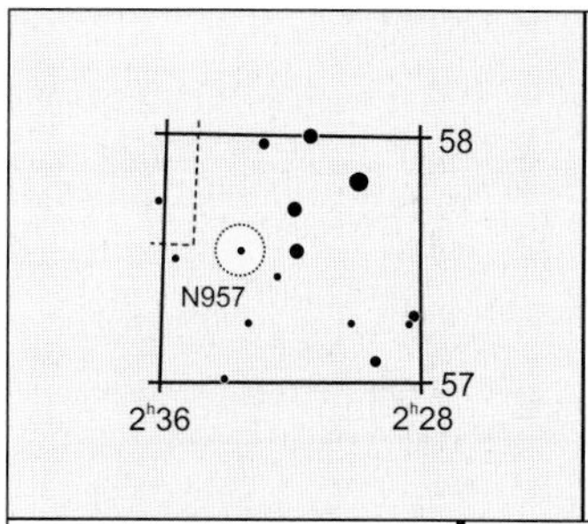

NGC 957

RA:	$02^h 33^m 40.6^s$	Con:	Perseus
Dec:	57° 32' 04"	Type:	Open Cluster
Size:	11.0'	Mag:	7.6

NGC 957 is a dense open cluster located 2° northwest of the Double Cluster.

Telescope Aperture:	4" f/5	4" f/9	6" f/7	6" f/9	8" f/6.3	8" f/10	10" f/6.3	10" f/10	12" f/6.3	12" f/10
FOV(35mm film):	2.7° x 4.1°	1.50° x 2.26°	1.29° x 1.93°	1.0° x 1.50°	1.07° x 1.61°	0.68° x 1.02°	0.86° x 1.29°	0.54° x 0.81°	0.72° x 1.07°	0.45° x 0.68°

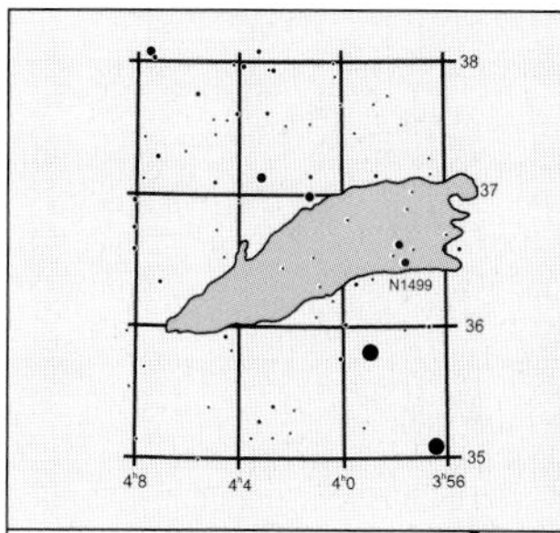

Map Scaled to Fit

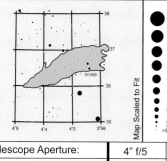

NGC 1499 (California Nebula)

RA:	$04^h 00^m 45.7^s$	Con:	Perseus
Dec:	36° 37' 04"	Type:	Emission Nebula
Size:	145.0'	Mag:	

NGC 1499 the California Nebula is one of the largest emission nebulae in the northern sky.

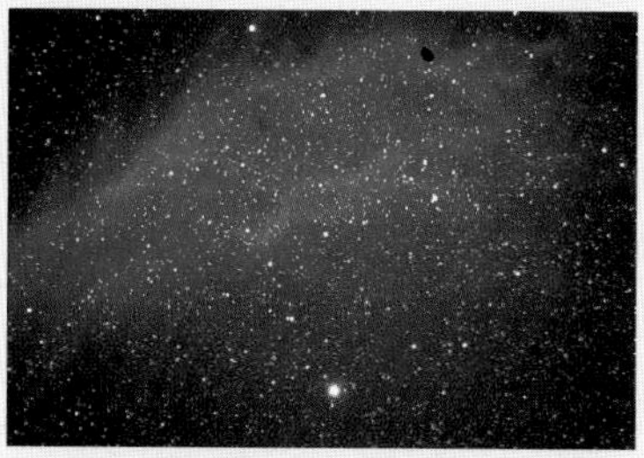

Telescope Aperture:	4" f/5	4" f/9	6" f/7	6" f/9	8" f/6.3	8" f/10	10" f/6.3	10" f/10	12" f/6.3	12" f/10
FOV(35mm film):	2.7° x 4.1°	1.50° x 2.26°	1.29° x 1.93°	1.0° x 1.50°	1.07° x 1.61°	0.68° x 1.02°	0.86° x 1.29°	0.54° x 0.81°	0.72° x 1.07°	0.45° x 0.68°

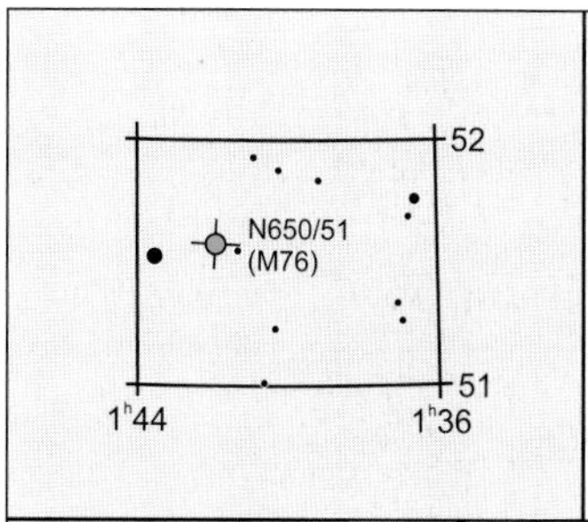

M76 (NGC 650/1) "Little Dumbell"

RA:	$01^h 42^m 22.5^s$	Con:	Perseus
Dec:	51° 34' 08"	Type:	Planetary Nebula
Size:	4.8'	Mag:	12.0

M76 (NGC 650/1) is known as the Little Dumbell. Object is an irregular planetary nebula.

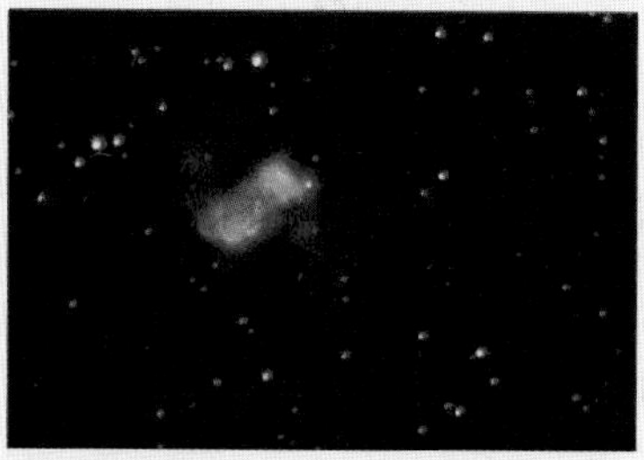

Telescope Aperture:	4" f/5	4" f/9	6" f/7	6" f/9	8" f/6.3	8" f/10	10" f/6.3	10" f/10	12" f/6.3	12" f/10
FOV(35mm film):	2.7° x 4.1°	1.50° x 2.26°	1.29° x 1.93°	1.0° x 1.50°	1.07° x 1.61°	0.68° x 1.02°	0.86° x 1.29°	0.54° x 0.81°	0.72° x 1.07°	0.45° x 0.68°

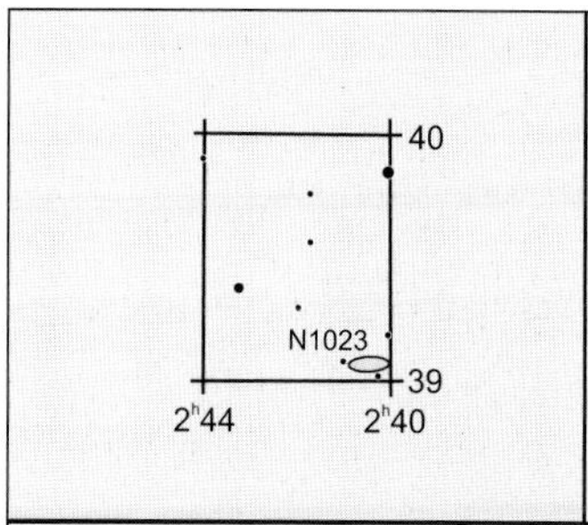

NGC 1023

RA:	$02^h 40^m 28.0^s$	Con:	Perseus
Dec:	39° 04' 08"	Type:	Galaxy
Size:	3.5' x 1.5'	Mag:	9.5

NGC 1023 appears as a very elongated object with a bright core.

Telescope Aperture:	4" f/5	4" f/9	6" f/7	6" f/9	8" f/6.3	8" f/10	10" f/6.3	10" f/10	12" f/6.3	12" f/10
FOV(35mm film):	2.7° x 4.1°	1.50° x 2.26°	1.29° x 1.93°	1.0° x 1.50°	1.07° x 1.61°	0.68° x 1.02°	0.86° x 1.29°	0.54° x 0.81°	0.72° x 1.07°	0.45° x 0.68°

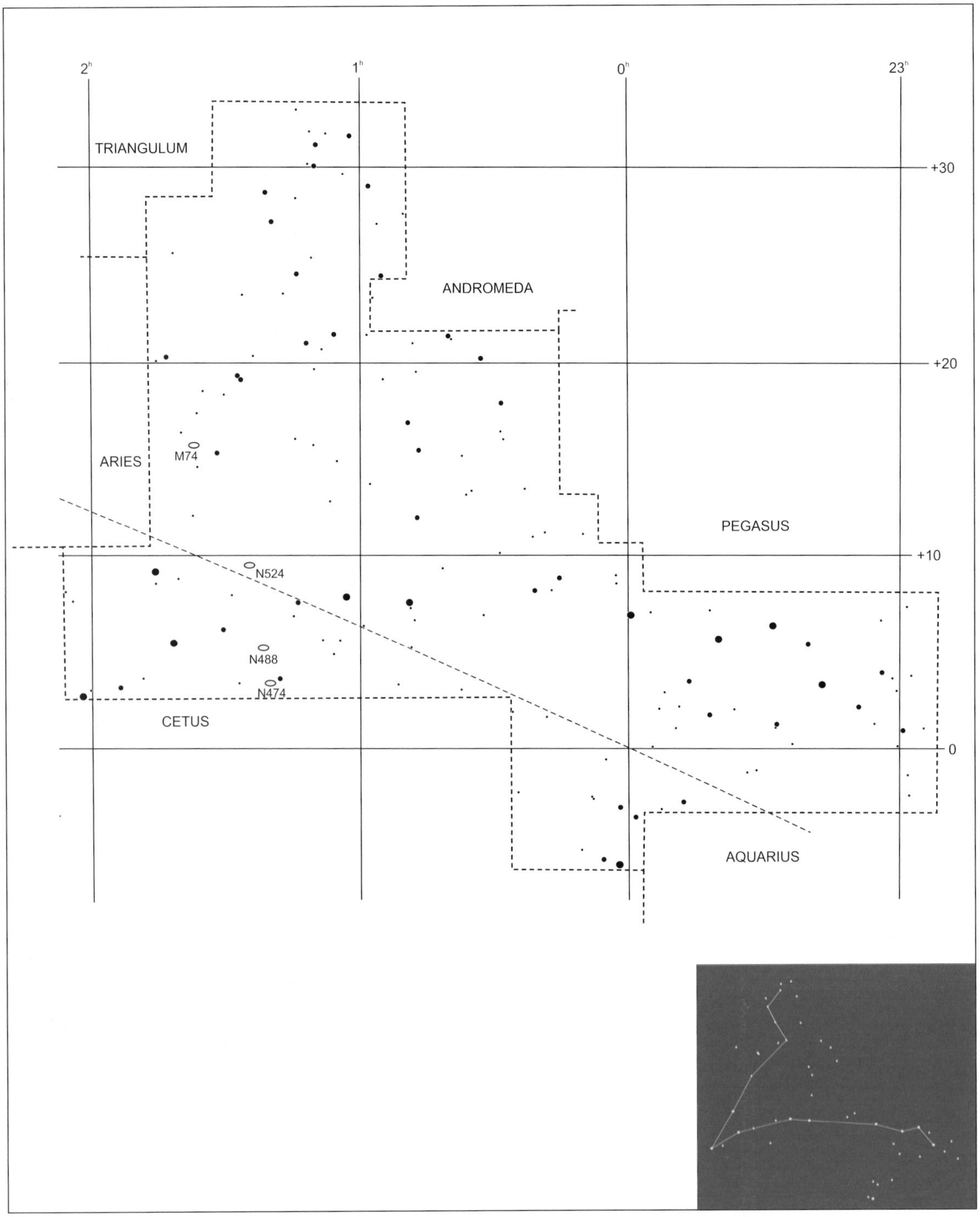

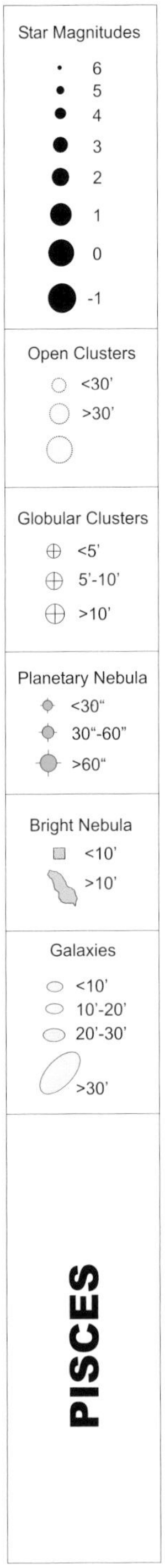

Constellation Facts:

Pisces; (PIE-sees)

Pisces, the Fish.
Constellation moves across the sky from east to west, crossing the meridian halfway between the horizon, and the zenith.
The constellation covers 889 square degrees.

Constellation is visible from 83° N to 56° S. Partially visible from 56° S to 90° S.

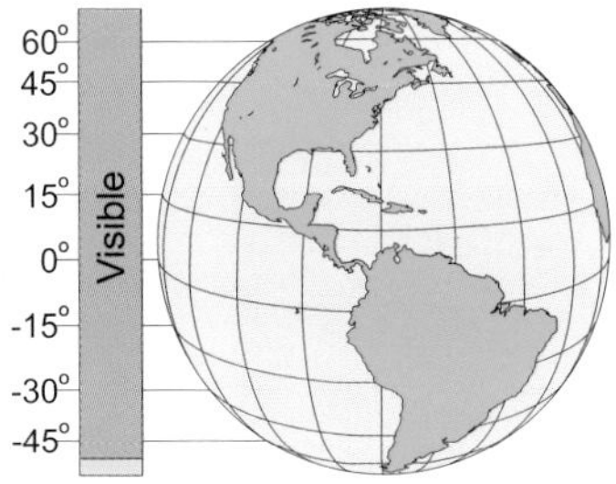

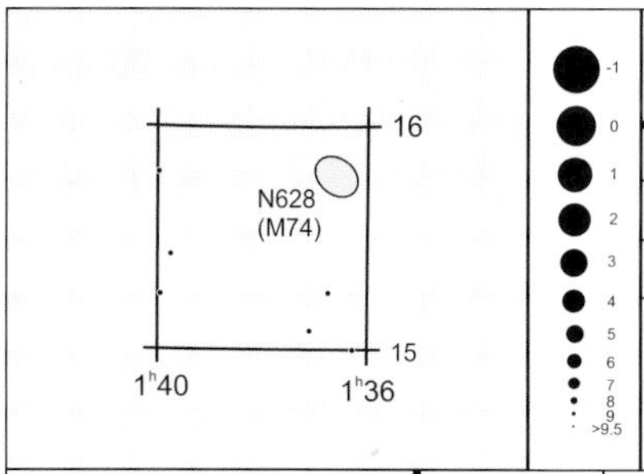

M74 (NGC 628)

RA:	$01^h 36^m 45.8^s$	Con:	Pisces
Dec:	15° 47' 19"	Type:	Spiral Galaxy
Size:	11.0' x 9.0'	Mag:	9.2

M74 (NGC 628) is a large and impressive face-on spiral galaxy.

Telescope Aperture:	4" f/5	4" f/9	6" f/7	6" f/9	8" f/6.3	8" f/10	10" f/6.3	10" f/10	12" f/6.3	12" f/10
FOV(35mm film):	2.7° x 4.1°	1.50° x 2.26°	1.29° x 1.93°	1.0° x 1.50°	1.07° x 1.61°	0.68° x 1.02°	0.86° x 1.29°	0.54° x 0.81°	0.72° x 1.07°	0.45° x 0.68°

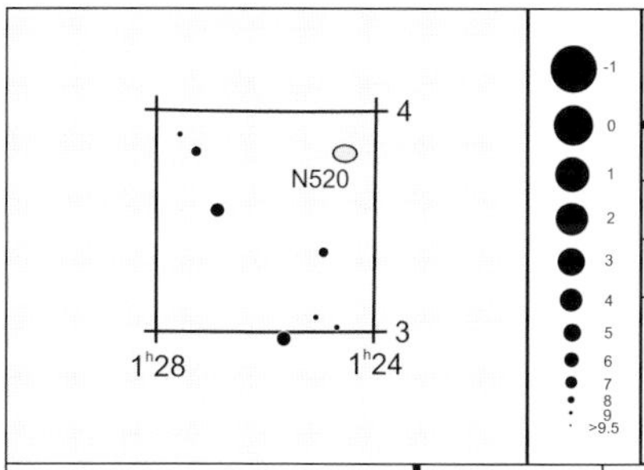

NGC 520

RA:	$01^h 24^m 39.8^s$	Con:	Pisces
Dec:	03° 48' 24"	Type:	Peculiar Galaxy
Size:	4.0' x 2.0'	Mag:	11.2

NGC 520 appears as an elongated galaxy with a bright core. Object is found 8° south of NGC 514.

Telescope Aperture:	4" f/5	4" f/9	6" f/7	6" f/9	8" f/6.3	8" f/10	10" f/6.3	10" f/10	12" f/6.3	12" f/10
FOV(35mm film):	2.7° x 4.1°	1.50° x 2.26°	1.29° x 1.93°	1.0° x 1.50°	1.07° x 1.61°	0.68° x 1.02°	0.86° x 1.29°	0.54° x 0.81°	0.72° x 1.07°	0.45° x 0.68°

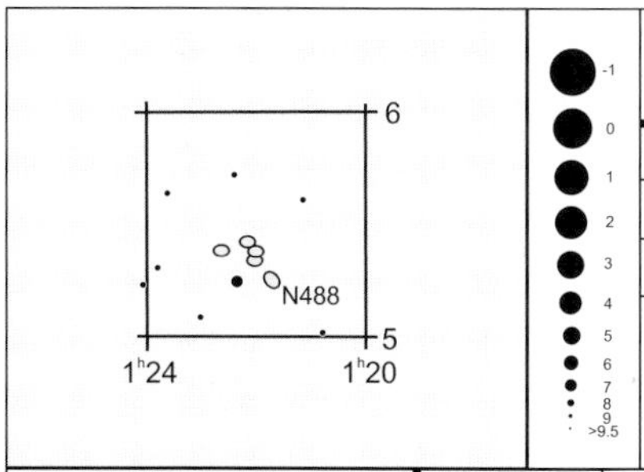

NGC 488

RA:	$01^h 21^m 51.8^s$	Con:	Pisces
Dec:	05° 15' 24"	Type:	Spiral Galaxy
Size:	5.0' x 4.0'	Mag:	10.3

NGC 488 appears as an elongated galaxy with a bright core. However, the object is an Sb-type spiral galaxy that is face-on to our line of sight.

Telescope Aperture:	4" f/5	4" f/9	6" f/7	6" f/9	8" f/6.3	8" f/10	10" f/6.3	10" f/10	12" f/6.3	12" f/10
FOV(35mm film):	2.7° x 4.1°	1.50° x 2.26°	1.29° x 1.93°	1.0° x 1.50°	1.07° x 1.61°	0.68° x 1.02°	0.86° x 1.29°	0.54° x 0.81°	0.72° x 1.07°	0.45° x 0.68°

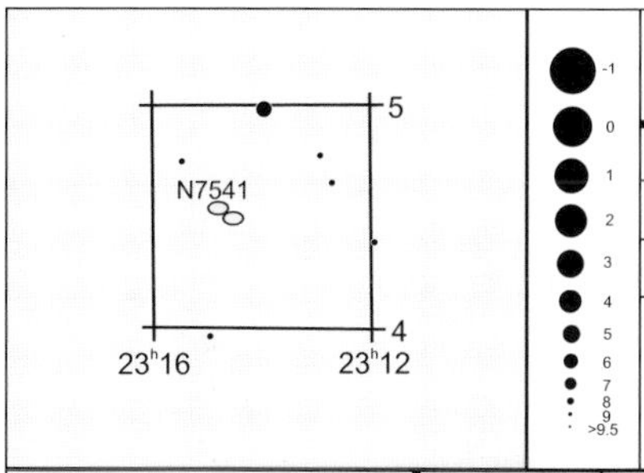

NGC 7541

RA:	$23^h 14^m 46.4^s$	Con:	Pisces
Dec:	04° 32' 26"	Type:	Spiral Galaxy
Size:	3.2' x 1.0'	Mag:	11.7

NGC 7541 is an Sc-type spiral galaxy that appears very elongated in the eyepiece.

Telescope Aperture:	4" f/5	4" f/9	6" f/7	6" f/9	8" f/6.3	8" f/10	10" f/6.3	10" f/10	12" f/6.3	12" f/10
FOV(35mm film):	2.7° x 4.1°	1.50° x 2.26°	1.29° x 1.93°	1.0° x 1.50°	1.07° x 1.61°	0.68° x 1.02°	0.86° x 1.29°	0.54° x 0.81°	0.72° x 1.07°	0.45° x 0.68°

Sge

Crosses Prime Meridian:
August thru September

SUMMER

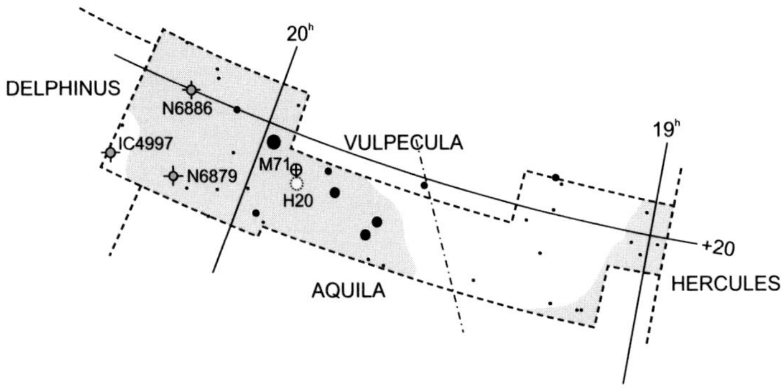

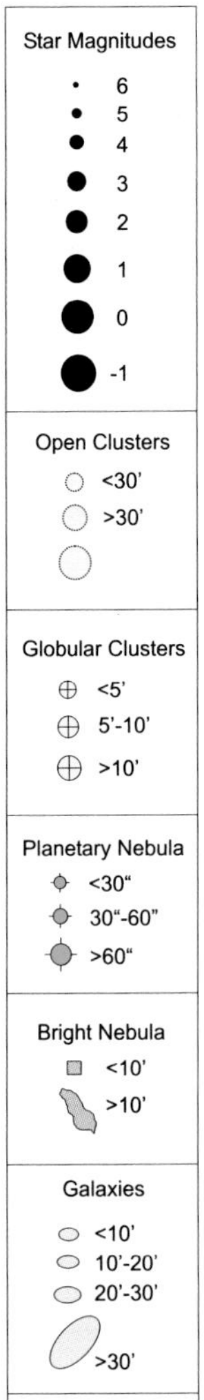

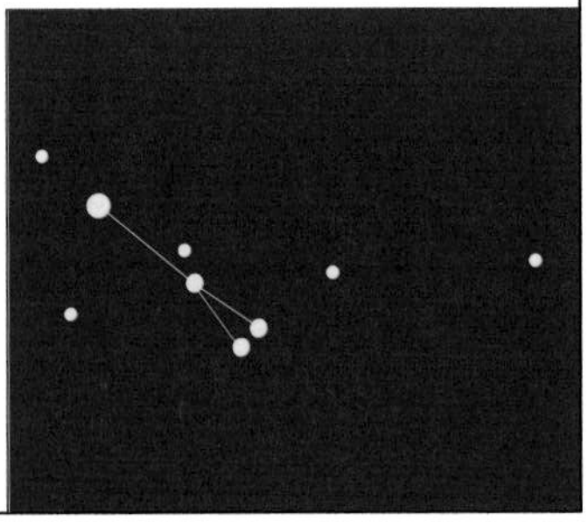

SAGITTA

Constellation Facts:

Sagitta; (suh-JIT-uh)

Sagitta, the Arrow.
Sagitta rises in the northeast, passes overhead, and sets in the northwest.
The constellation covers 80 square degrees.

Constellation is visible from 90° N to 69° S. Partially visible from 69° S to 90° S.

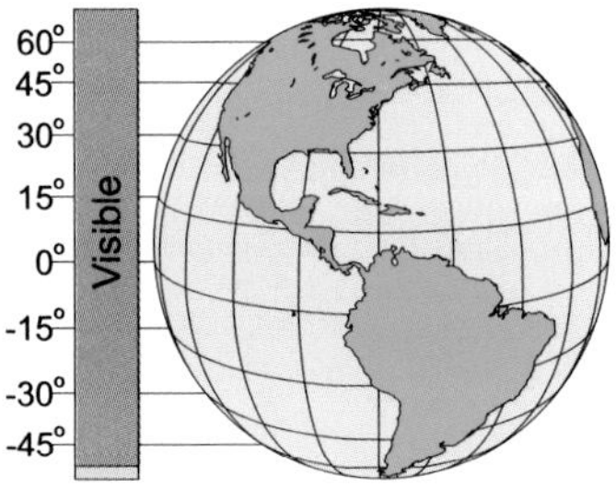

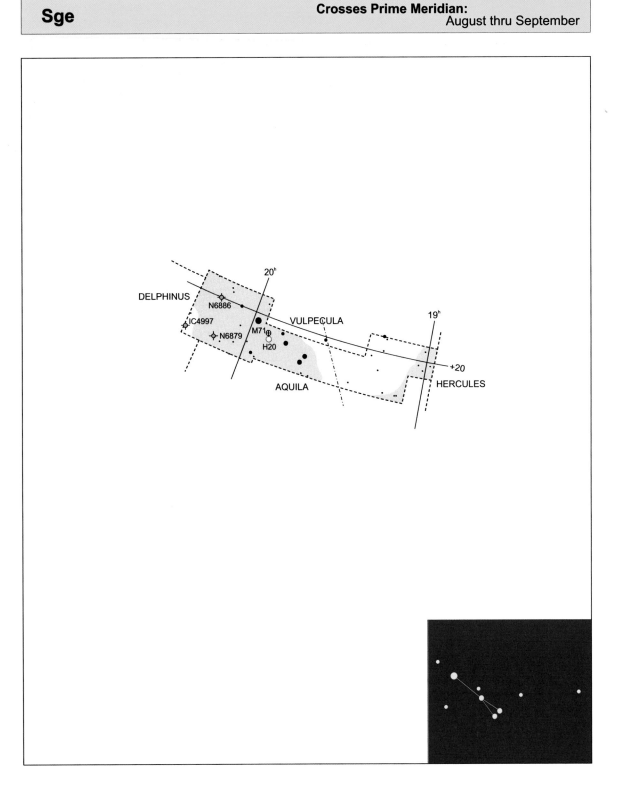

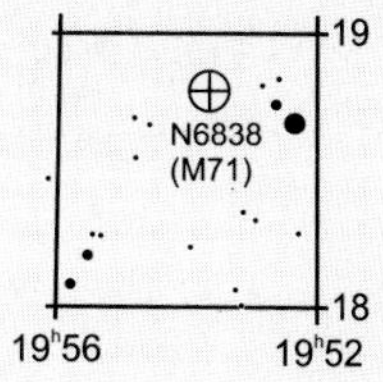

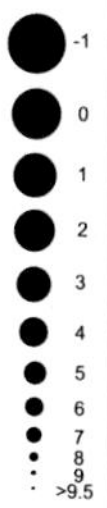

M71 (NGC 6838)

RA:	19^h 53^m 52.6^s	Con:	Sagitta
Dec:	18° 47' 15"	Type:	Globular Cluster
Size:	7.2'	Mag:	8.3

M71 (NGC 6838) is a highly resolved globular cluster.

Telescope Aperture:	4" f/5	4" f/9	6" f/7	6" f/9	8" f/6.3	8" f/10	10" f/6.3	10" f/10	12" f/6.3	12" f/10
FOV(35mm film):	2.7° x 4.1°	1.50° x 2.26°	1.29° x 1.93°	1.0° x 1.50°	1.07° x 1.61°	0.68° x 1.02°	0.86° x 1.29°	0.54° x 0.81°	0.72° x 1.07°	0.45° x 0.68°

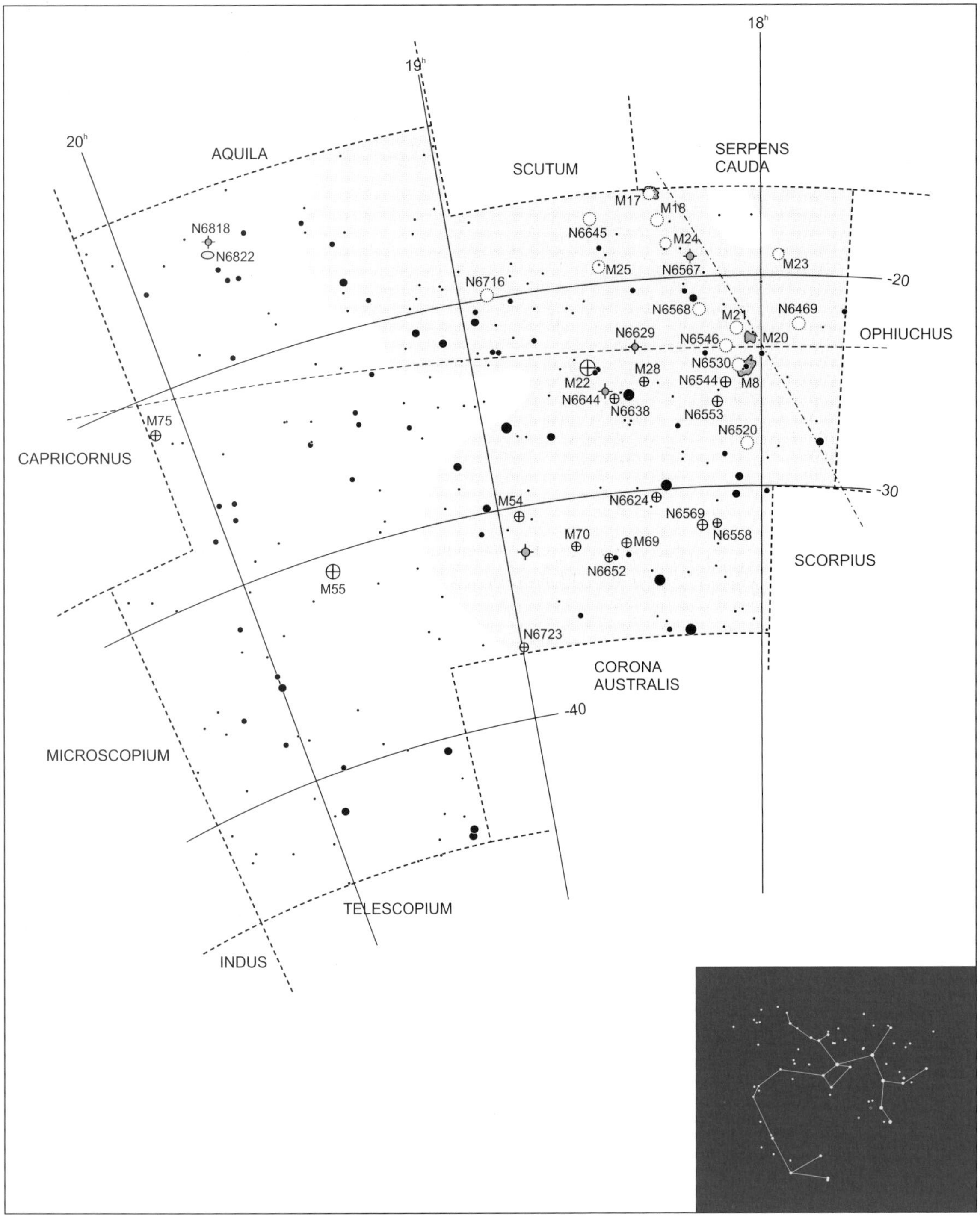

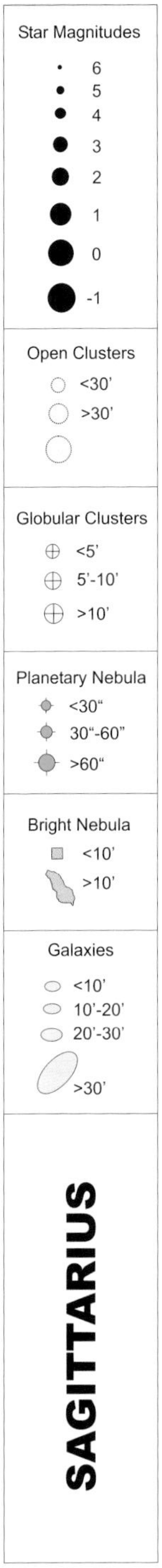

Constellation Facts:

Sagittarius; (saj-ih-TAY-rih-us)

Sagittarius, the Archer.
The constellation because of its southern declination appears to move southeast to southwest, never reaching far above the horizon.
The constellation covers 867 square degrees.

Constellation is visible from 44° N to 90° S. Partially visible from 44° N to 80° N.

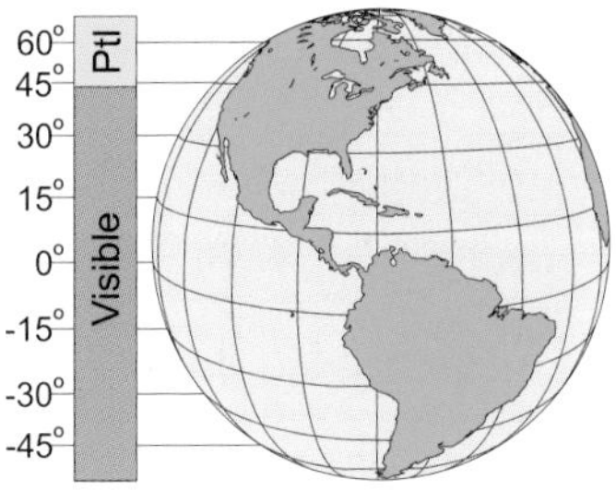

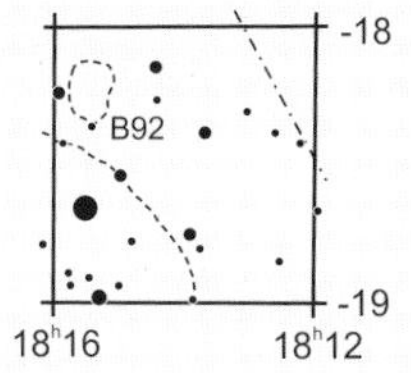

B-92

RA:	$18^h\ 15^m\ 5.0^s$	Con:	Sagittarius
Dec:	-18° 14' 00"	Type:	Dark Nebula
Size:	15' x 9'	Mag:	

Dark nebula elongated north and south. Sharply defined on its eastern edge.

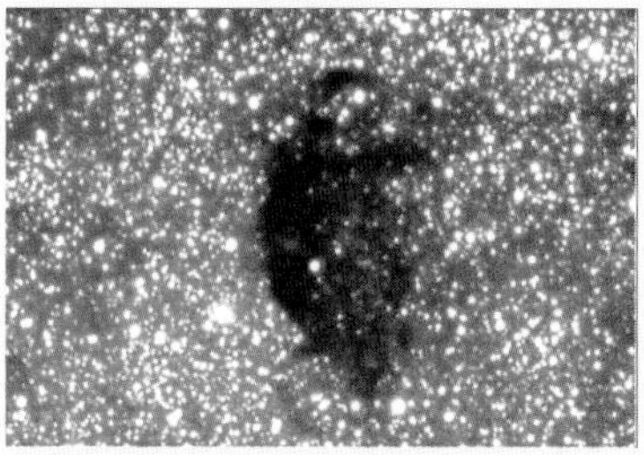

Telescope Aperture:	4" f/5	4" f/9	6" f/7	6" f/9	8" f/6.3	8" f/10	10" f/6.3	10" f/10	12" f/6.3	12" f/10
FOV(35mm film):	2.7° x 4.1°	1.50° x 2.26°	1.29° x 1.93°	1.0° x 1.50°	1.07° x 1.61°	0.68° x 1.02°	0.86° x 1.29°	0.54° x 0.81°	0.72° x 1.07°	0.45° x 0.68°

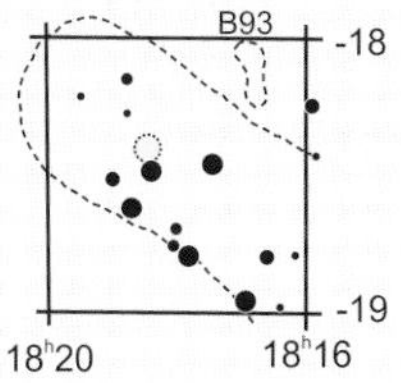

B-93

RA:	$18^h\ 16^m\ 9.0^s$	Con:	Sagittarius
Dec:	-18° 04' 00"	Type:	Dark Nebula
Size:		Mag:	

Sharply defined dark head, with a diffused tail that travels south.

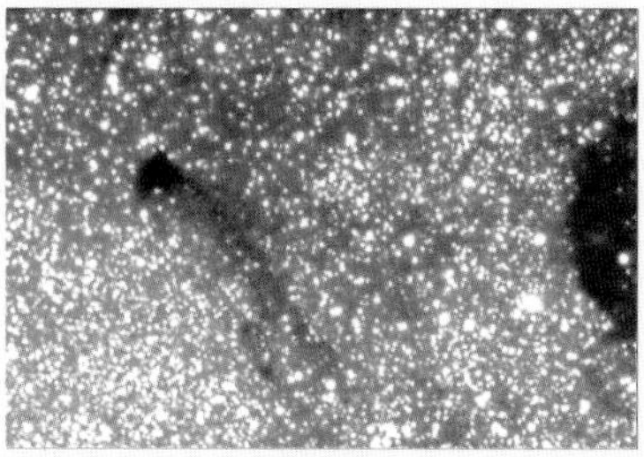

Telescope Aperture:	4" f/5	4" f/9	6" f/7	6" f/9	8" f/6.3	8" f/10	10" f/6.3	10" f/10	12" f/6.3	12" f/10
FOV(35mm film):	2.7° x 4.1°	1.50° x 2.26°	1.29° x 1.93°	1.0° x 1.50°	1.07° x 1.61°	0.68° x 1.02°	0.86° x 1.29°	0.54° x 0.81°	0.72° x 1.07°	0.45° x 0.68°

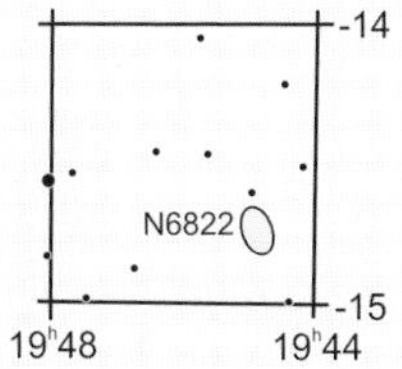

NGC 6822 (Barnard's Galaxy)

RA:	$19^h 44^m\ 59.5^s$	Con:	Sagittarius
Dec:	-14° 47' 45"	Type:	Peculiar Galaxy
Size:	10.0'	Mag:	9.0

NGC 6822 also known as "Barnard's Galaxy" is located in the northeastern region of the constellation.

Telescope Aperture:	4" f/5	4" f/9	6" f/7	6" f/9	8" f/6.3	8" f/10	10" f/6.3	10" f/10	12" f/6.3	12" f/10
FOV(35mm film):	2.7° x 4.1°	1.50° x 2.26°	1.29° x 1.93°	1.0° x 1.50°	1.07° x 1.61°	0.68° x 1.02°	0.86° x 1.29°	0.54° x 0.81°	0.72° x 1.07°	0.45° x 0.68°

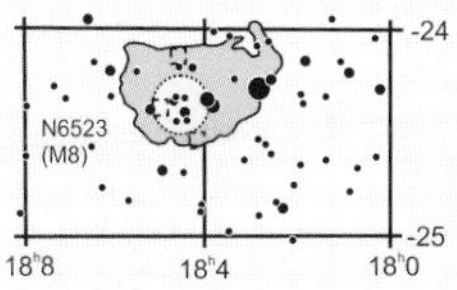

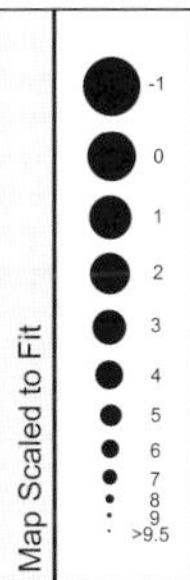

M8 (NGC 6523) "Lagoon Nebula"

RA:	$18^h 03^m\ 53.8^s$	Con:	Sagittarius
Dec:	-24° 22' 58"	Type:	Emission Nebula
Size:	50' x 40'	Mag:	5.8

M8 (NGC 6523) known as the Lagoon Nebula, is a large and impressive emission nebula.

Telescope Aperture:	4" f/5	4" f/9	6" f/7	6" f/9	8" f/6.3	8" f/10	10" f/6.3	10" f/10	12" f/6.3	12" f/10
FOV(35mm film):	2.7° x 4.1°	1.50° x 2.26°	1.29° x 1.93°	1.0° x 1.50°	1.07° x 1.61°	0.68° x 1.02°	0.86° x 1.29°	0.54° x 0.81°	0.72° x 1.07°	0.45° x 0.68°

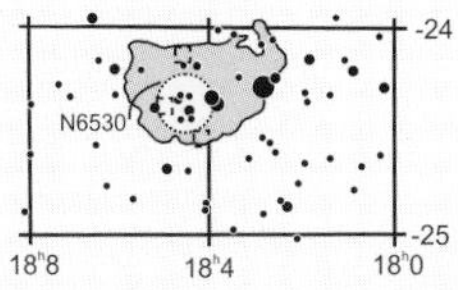

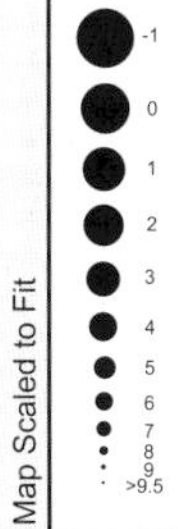

NGC 6530

RA:	$18^h\ 04^m\ 53.8^s$	Con:	Sagittarius
Dec:	-24° 19' 58"	Type:	Open Cluster
Size:	15.0'	Mag:	4.6

NGC 6530 is a rich open cluster inside the nebular complex M8.

Telescope Aperture:	4" f/5	4" f/9	6" f/7	6" f/9	8" f/6.3	8" f/10	10" f/6.3	10" f/10	12" f/6.3	12" f/10
FOV(35mm film):	2.7° x 4.1°	1.50° x 2.26°	1.29° x 1.93°	1.0° x 1.50°	1.07° x 1.61°	0.68° x 1.02°	0.86° x 1.29°	0.54° x 0.81°	0.72° x 1.07°	0.45° x 0.68°

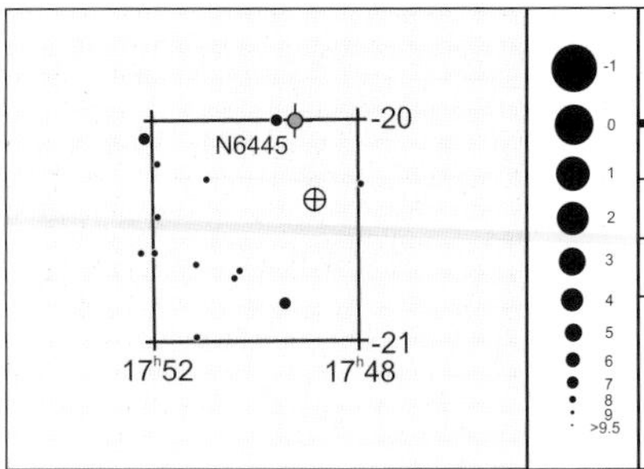

NGC 6445

RA:	$17^h\ 49^m\ 17.6^s$	Con:	Sagittarius
Dec:	-20° 01' 00"	Type:	Planetary Nebula
Size:	0.6'	Mag:	13.0

NGC 6445 is a small irregular planetary nebula.

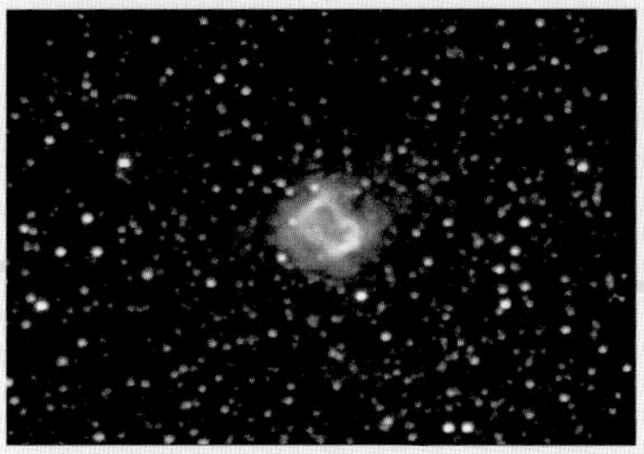

Telescope Aperture:	4" f/5	4" f/9	6" f/7	6" f/9	8" f/6.3	8" f/10	10" f/6.3	10" f/10	12" f/6.3	12" f/10
FOV(35mm film):	2.7° x 4.1°	1.50° x 2.26°	1.29° x 1.93°	1.0° x 1.50°	1.07° x 1.61°	0.68° x 1.02°	0.86° x 1.29°	0.54° x 0.81°	0.72° x 1.07°	0.45° x 0.68°

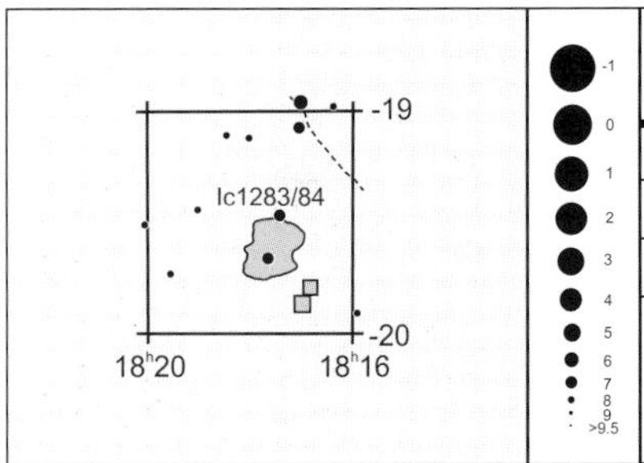

IC 1283/84

RA:	$18^h\ 17^m\ 23.6^s$	Con:	Sagittarius
Dec:	-19° 43' 56"	Type:	E & R Nebula
Size:	17' x 15'	Mag:	

IC 1283/84 is an emission and reflection nebula, irregularly shaped with many faint stars involved in the nebulosity.

Telescope Aperture:	4" f/5	4" f/9	6" f/7	6" f/9	8" f/6.3	8" f/10	10" f/6.3	10" f/10	12" f/6.3	12" f/10
FOV(35mm film):	2.7° x 4.1°	1.50° x 2.26°	1.29° x 1.93°	1.0° x 1.50°	1.07° x 1.61°	0.68° x 1.02°	0.86° x 1.29°	0.54° x 0.81°	0.72° x 1.07°	0.45° x 0.68°

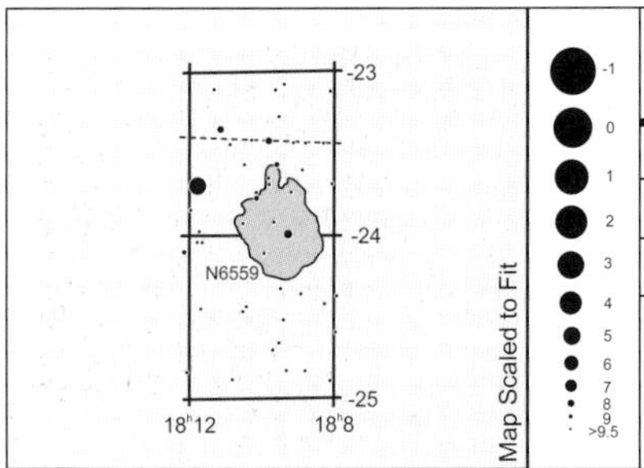

IC 4685

RA:	$18^h\ 09^m\ 23.8^s$	Con:	Sagittarius
Dec:	-23° 58' 58"	Type:	Reflection Nebula
Size:	10.0'	Mag:	

IC 4685 is a reflection nebula that is the largest member of a group of nebulae.

Telescope Aperture:	4" f/5	4" f/9	6" f/7	6" f/9	8" f/6.3	8" f/10	10" f/6.3	10" f/10	12" f/6.3	12" f/10
FOV(35mm film):	2.7° x 4.1°	1.50° x 2.26°	1.29° x 1.93°	1.0° x 1.50°	1.07° x 1.61°	0.68° x 1.02°	0.86° x 1.29°	0.54° x 0.81°	0.72° x 1.07°	0.45° x 0.68°

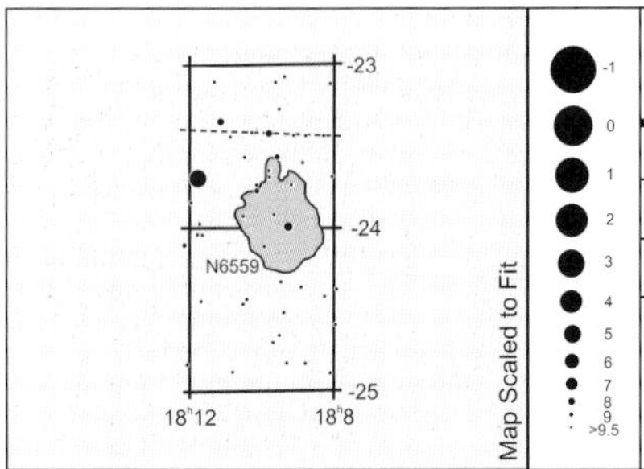

IC 1275

RA:	$18^h\ 10^m\ 5.8^s$	Con:	Sagittarius
Dec:	-23° 49' 58"	Type:	Emission Nebula
Size:	10.0'	Mag:	

IC 1275 is defined on its northern border by B-91. Object is a small round nebula in a grouping of gas objects.

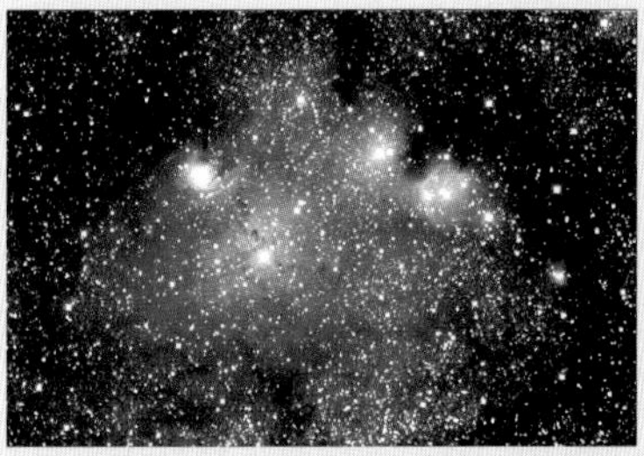

Telescope Aperture:	4" f/5	4" f/9	6" f/7	6" f/9	8" f/6.3	8" f/10	10" f/6.3	10" f/10	12" f/6.3	12" f/10
FOV(35mm film):	2.7° x 4.1°	1.50° x 2.26°	1.29° x 1.93°	1.0° x 1.50°	1.07° x 1.61°	0.68° x 1.02°	0.86° x 1.29°	0.54° x 0.81°	0.72° x 1.07°	0.45° x 0.68°

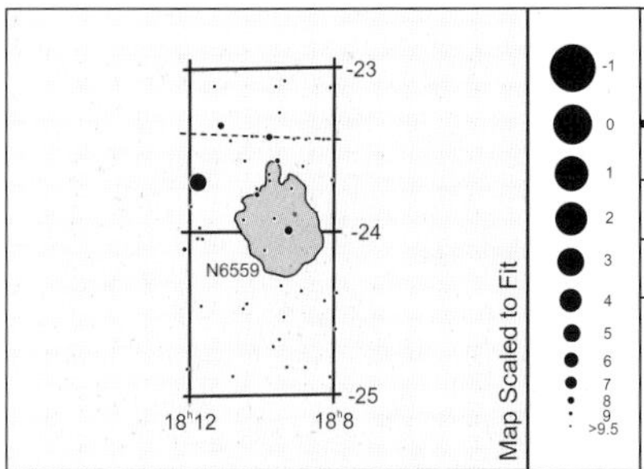

NGC 6559

RA:	$18^h\ 10^m\ 5.8^s$	Con:	Sagittarius
Dec:	-24° 05' 58"	Type:	Emission Nebula
Size:	8.0'	Mag:	

NGC 6559 is the largest nebular complex in a group of objects including IC 1275, IC 1284 and IC 1283.

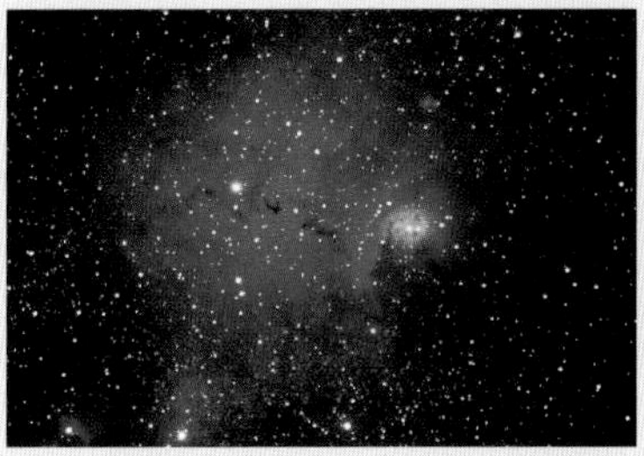

Telescope Aperture:	4" f/5	4" f/9	6" f/7	6" f/9	8" f/6.3	8" f/10	10" f/6.3	10" f/10	12" f/6.3	12" f/10
FOV(35mm film):	2.7° x 4.1°	1.50° x 2.26°	1.29° x 1.93°	1.0° x 1.50°	1.07° x 1.61°	0.68° x 1.02°	0.86° x 1.29°	0.54° x 0.81°	0.72° x 1.07°	0.45° x 0.68°

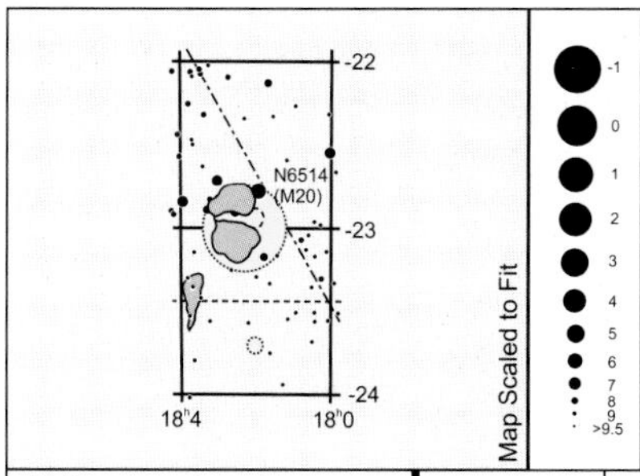

M20 (NGC 6514) "Trifid Nebula"

RA:	$18^h\ 02^m\ 23.7^s$	Con:	Sagittarius
Dec:	-23° 01' 58"	Type:	E & R Nebula
Size:	29.0'	Mag:	6.3

M20 (NGC 6514) known as the Trifid Nebula is located 1° northwest of M8.

Telescope Aperture:	4" f/5	4" f/9	6" f/7	6" f/9	8" f/6.3	8" f/10	10" f/6.3	10" f/10	12" f/6.3	12" f/10
FOV(35mm film):	2.7° x 4.1°	1.50° x 2.26°	1.29° x 1.93°	1.0° x 1.50°	1.07° x 1.61°	0.68° x 1.02°	0.86° x 1.29°	0.54° x 0.81°	0.72° x 1.07°	0.45° x 0.68°

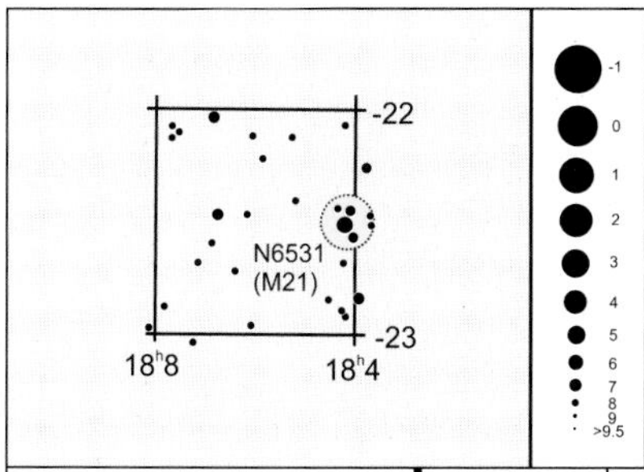

M21 (NGC 6531)

RA:	$18^h\ 04^m\ 41.7^s$	Con:	Sagittarius
Dec:	-22° 29' 58"	Type:	Open Cluster
Size:	13.0'	Mag:	5.9

M21 (NGC 6531) is located less than 1° northeast of the Trifid Nebula. Object is a rich open cluster comprised of 50 stars.

Telescope Aperture:	4" f/5	4" f/9	6" f/7	6" f/9	8" f/6.3	8" f/10	10" f/6.3	10" f/10	12" f/6.3	12" f/10
FOV(35mm film):	2.7° x 4.1°	1.50° x 2.26°	1.29° x 1.93°	1.0° x 1.50°	1.07° x 1.61°	0.68° x 1.02°	0.86° x 1.29°	0.54° x 0.81°	0.72° x 1.07°	0.45° x 0.68°

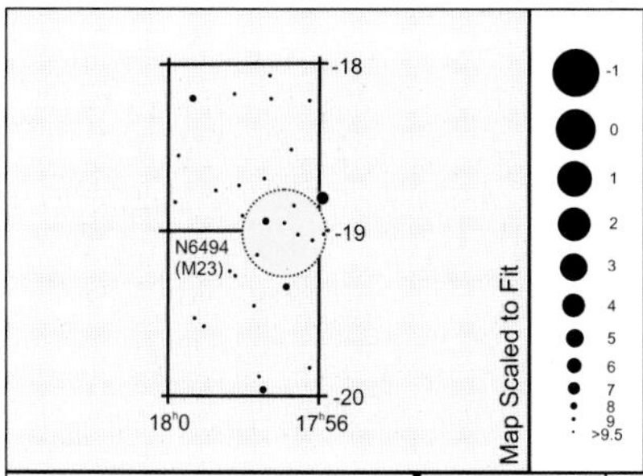

M23 (NGC 6494)

RA:	$17^h\ 56^m\ 53.6^s$	Con:	Sagittarius
Dec:	-19° 00' 59"	Type:	Open Cluster
Size:	27.0'	Mag:	5.5

M23 (NGC 6494) is a dense open cluster containing 120 stars.

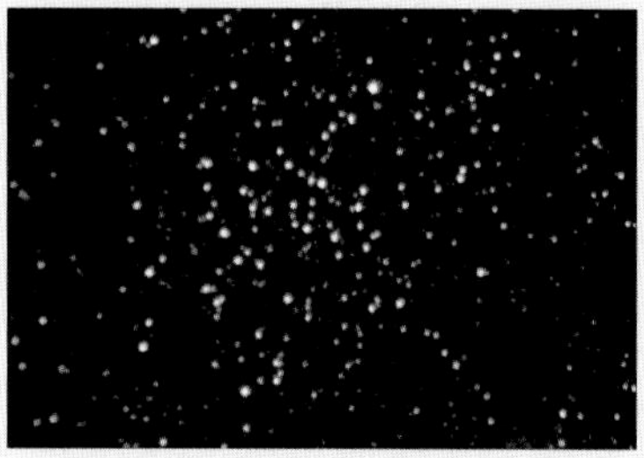

Telescope Aperture:	4" f/5	4" f/9	6" f/7	6" f/9	8" f/6.3	8" f/10	10" f/6.3	10" f/10	12" f/6.3	12" f/10
FOV(35mm film):	2.7° x 4.1°	1.50° x 2.26°	1.29° x 1.93°	1.0° x 1.50°	1.07° x 1.61°	0.68° x 1.02°	0.86° x 1.29°	0.54° x 0.81°	0.72° x 1.07°	0.45° x 0.68°

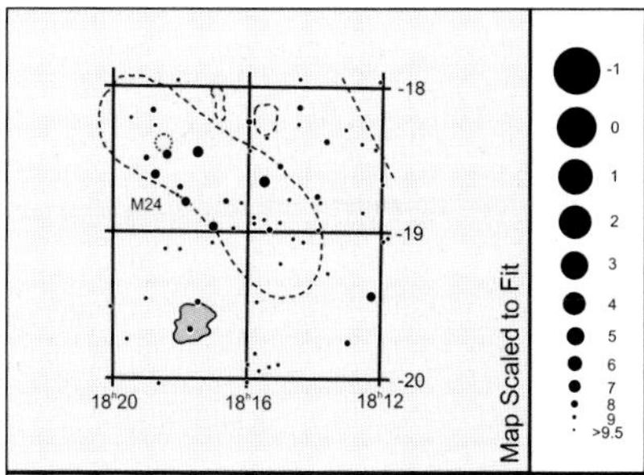

M24 (NGC 6603) "Sag. Star Cloud"

RA:	$18^h\ 18^m\ 29.5^s$	Con:	Sagittarius
Dec:	-18° 24' 56"	Type:	Open Cluster
Size:	5.0'	Mag:	11.0

M24 (NGC 6603), known as the Sagittarius Star Cloud, is a dense open cluster found east of M23.

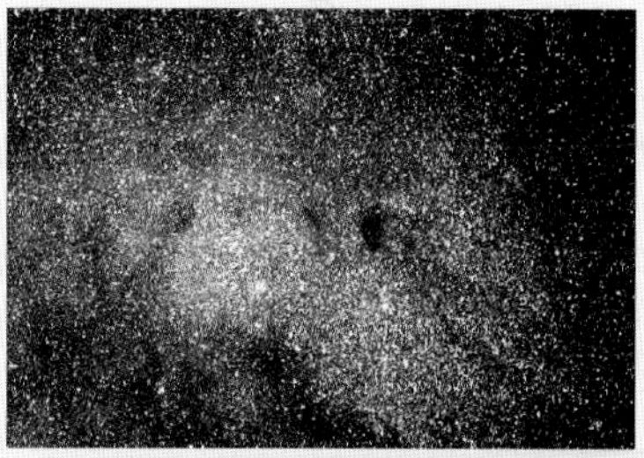

Telescope Aperture:	4" f/5	4" f/9	6" f/7	6" f/9	8" f/6.3	8" f/10	10" f/6.3	10" f/10	12" f/6.3	12" f/10
FOV(35mm film):	2.7° x 4.1°	1.50° x 2.26°	1.29° x 1.93°	1.0° x 1.50°	1.07° x 1.61°	0.68° x 1.02°	0.86° x 1.29°	0.54° x 0.81°	0.72° x 1.07°	0.45° x 0.68°

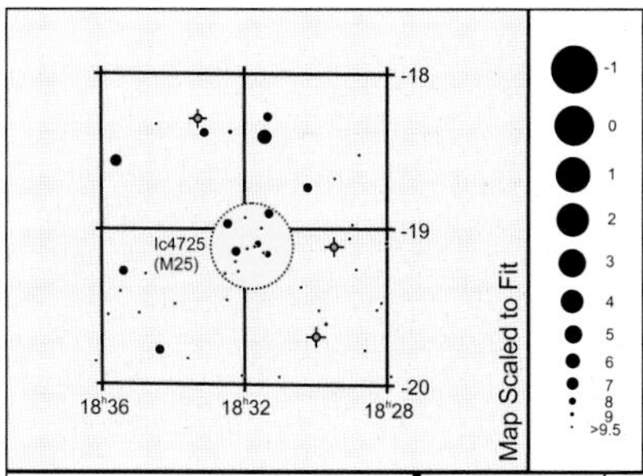

M25 (IC 4725)

RA:	$18^h\ 31^m\ 41.6^s$	Con:	Sagittarius
Dec:	-19° 14' 54"	Type:	Open Cluster
Size:	32.0'	Mag:	4.0

M25 (IC 4725) is a large, bright open cluster. Object is comprised of 80 stars.

Telescope Aperture:	4" f/5	4" f/9	6" f/7	6" f/9	8" f/6.3	8" f/10	10" f/6.3	10" f/10	12" f/6.3	12" f/10
FOV(35mm film):	2.7° x 4.1°	1.50° x 2.26°	1.29° x 1.93°	1.0° x 1.50°	1.07° x 1.61°	0.68° x 1.02°	0.86° x 1.29°	0.54° x 0.81°	0.72° x 1.07°	0.45° x 0.68°

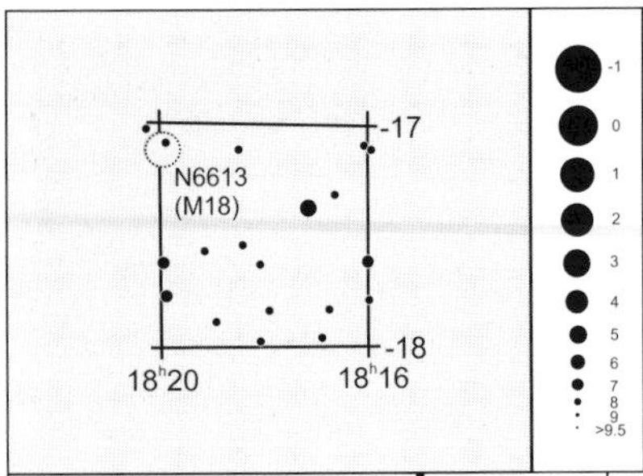

M18 (NGC 6613)

RA:	$18^h\ 19^m\ 59.5^s$	Con:	Sagittarius
Dec:	-17° 07' 55"	Type:	Open Cluster
Size:	9.0'	Mag:	6.9

M18 (NGC 6613) is a bright, scattered open cluster. Object is located north of M24, containing 18 stars.

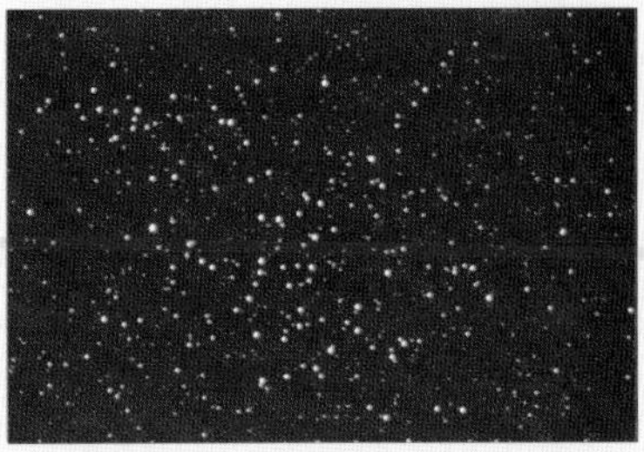

Telescope Aperture:	4" f/5	4" f/9	6" f/7	6" f/9	8" f/6.3	8" f/10	10" f/6.3	10" f/10	12" f/6.3	12" f/10
FOV(35mm film):	2.7° x 4.1°	1.50° x 2.26°	1.29° x 1.93°	1.0° x 1.50°	1.07° x 1.61°	0.68° x 1.02°	0.86° x 1.29°	0.54° x 0.81°	0.72° x 1.07°	0.45° x 0.68°

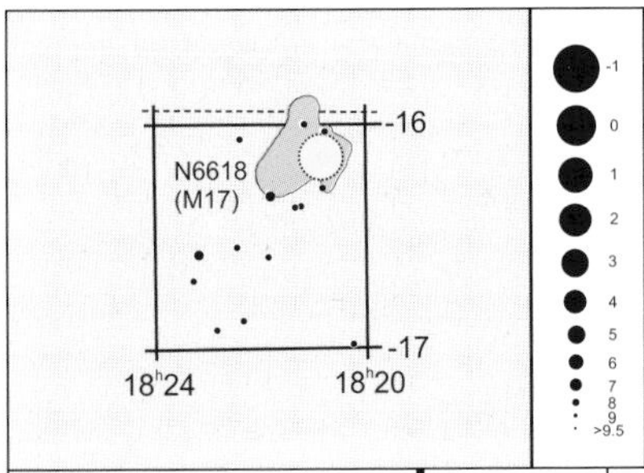

M17 (NGC 6618) "Omega Nebula"

RA:	$18^h\ 20^m\ 53.5^s$	Con:	Sagittarius
Dec:	-16° 10' 55"	Type:	Nebula & Cluster
Size:	46.0'	Mag:	6.0

M17 (NGC 6618) known as the Omega Nebula, is a bright emission nebula and its associated open cluster are a grouping of loosely arranged stars.

Telescope Aperture:	4" f/5	4" f/9	6" f/7	6" f/9	8" f/6.3	8" f/10	10" f/6.3	10" f/10	12" f/6.3	12" f/10
FOV(35mm film):	2.7° x 4.1°	1.50° x 2.26°	1.29° x 1.93°	1.0° x 1.50°	1.07° x 1.61°	0.68° x 1.02°	0.86° x 1.29°	0.54° x 0.81°	0.72° x 1.07°	0.45° x 0.68°

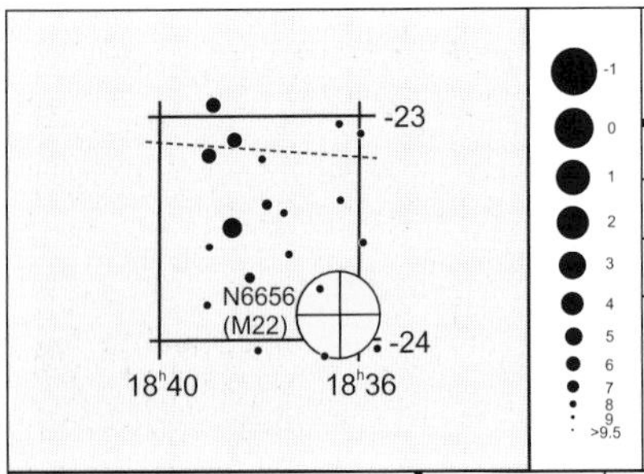

M22 (NGC 6656)

RA:	$18^h\ 36^m\ 29.8^s$	Con:	Sagittarius
Dec:	-23° 53' 54"	Type:	Globular Cluster
Size:	24.0'	Mag:	5.1

M22 (NGC 6656) is a highly resolved globular cluster. Object contains approximately 500,000 stars.

Telescope Aperture:	4" f/5	4" f/9	6" f/7	6" f/9	8" f/6.3	8" f/10	10" f/6.3	10" f/10	12" f/6.3	12" f/10
FOV(35mm film):	2.7° x 4.1°	1.50° x 2.26°	1.29° x 1.93°	1.0° x 1.50°	1.07° x 1.61°	0.68° x 1.02°	0.86° x 1.29°	0.54° x 0.81°	0.72° x 1.07°	0.45° x 0.68°

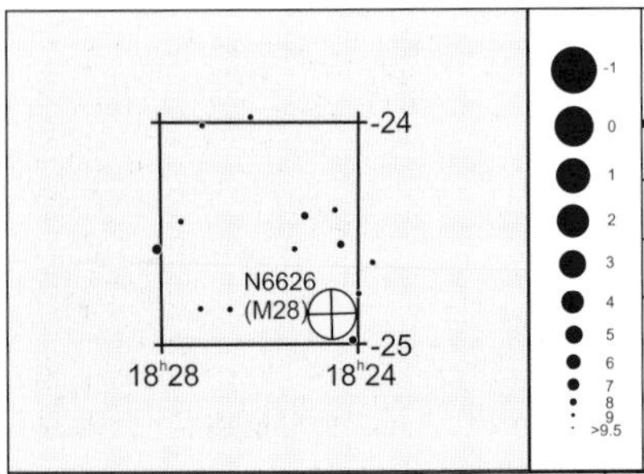

M28 (NGC 6626)

RA:	$18^h\ 24^m\ 35.9^s$	Con:	Sagittarius
Dec:	-24° 51' 56"	Type:	Globular Cluster
Size:	11.2'	Mag:	6.9

M28 (NGC 6626) is a highly resolved globular cluster. Object is smaller than M22, but impressive none-the-less.

Telescope Aperture:	4" f/5	4" f/9	6" f/7	6" f/9	8" f/6.3	8" f/10	10" f/6.3	10" f/10	12" f/6.3	12" f/10
FOV(35mm film):	2.7° x 4.1°	1.50° x 2.26°	1.29° x 1.93°	1.0° x 1.50°	1.07° x 1.61°	0.68° x 1.02°	0.86° x 1.29°	0.54° x 0.81°	0.72° x 1.07°	0.45° x 0.68°

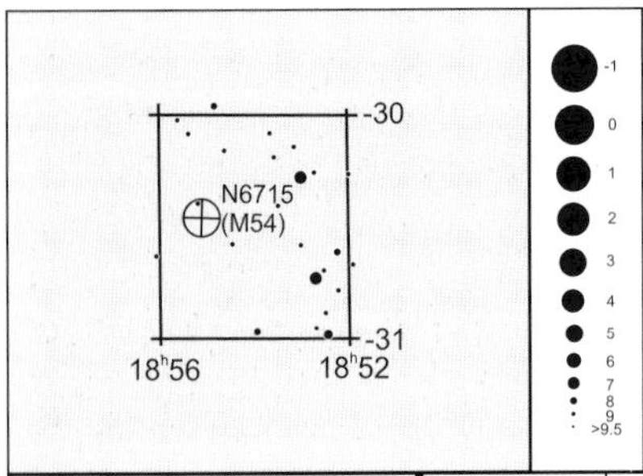

M54 (NGC 6715)

RA:	$18^h\ 55^m\ 12.1^s$	Con:	Sagittarius
Dec:	-30° 28' 52"	Type:	Globular Cluster
Size:	9.1'	Mag:	7.7

M54 (NGC 6715) is a small, mottled globular cluster, found in the Teapot asterism of stars that defines the constellation's outline.

Telescope Aperture:	4" f/5	4" f/9	6" f/7	6" f/9	8" f/6.3	8" f/10	10" f/6.3	10" f/10	12" f/6.3	12" f/10
FOV(35mm film):	2.7° x 4.1°	1.50° x 2.26°	1.29° x 1.93°	1.0° x 1.50°	1.07° x 1.61°	0.68° x 1.02°	0.86° x 1.29°	0.54° x 0.81°	0.72° x 1.07°	0.45° x 0.68°

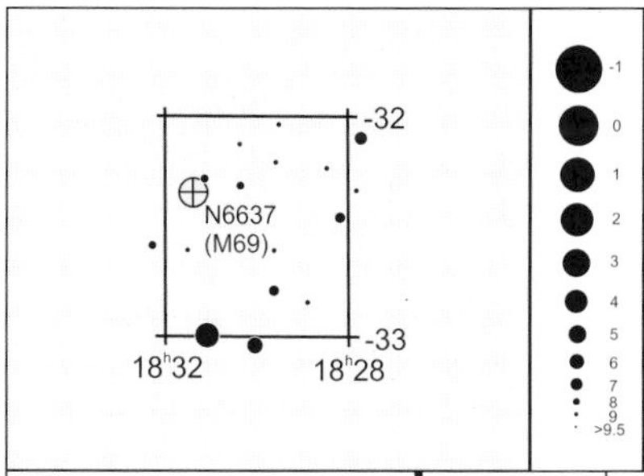

M69 (NGC 6637)

RA:	18h 31m 30.2s	Con:	Sagittarius
Dec:	-32° 20' 55"	Type:	Globular Cluster
Size:	7.1'	Mag:	7.7

M69 (NGC 6637) like M54 is located in the constellation's Teapot. Object is a globular cluster.

Telescope Aperture:	4" f/5	4" f/9	6" f/7	6" f/9	8" f/6.3	8" f/10	10" f/6.3	10" f/10	12" f/6.3	12" f/10
FOV(35mm film):	2.7° x 4.1°	1.50° x 2.26°	1.29° x 1.93°	1.0° x 1.50°	1.07° x 1.61°	0.68° x 1.02°	0.86° x 1.29°	0.54° x 0.81°	0.72° x 1.07°	0.45° x 0.68°

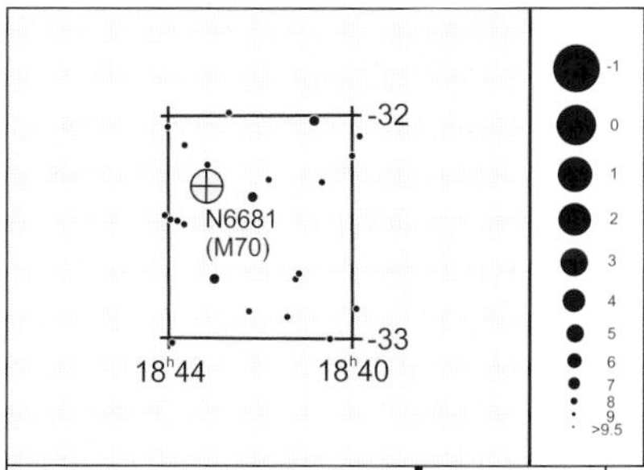

M70 (NGC 6681)

RA:	18h 43m 18.2s	Con:	Sagittarius
Dec:	-32° 17' 53"	Type:	Globular Cluster
Size:	7.8'	Mag:	8.1

M70 (NGC 6681) is another globular cluster located in the constellation's Teapot.

Telescope Aperture:	4" f/5	4" f/9	6" f/7	6" f/9	8" f/6.3	8" f/10	10" f/6.3	10" f/10	12" f/6.3	12" f/10
FOV(35mm film):	2.7° x 4.1°	1.50° x 2.26°	1.29° x 1.93°	1.0° x 1.50°	1.07° x 1.61°	0.68° x 1.02°	0.86° x 1.29°	0.54° x 0.81°	0.72° x 1.07°	0.45° x 0.68°

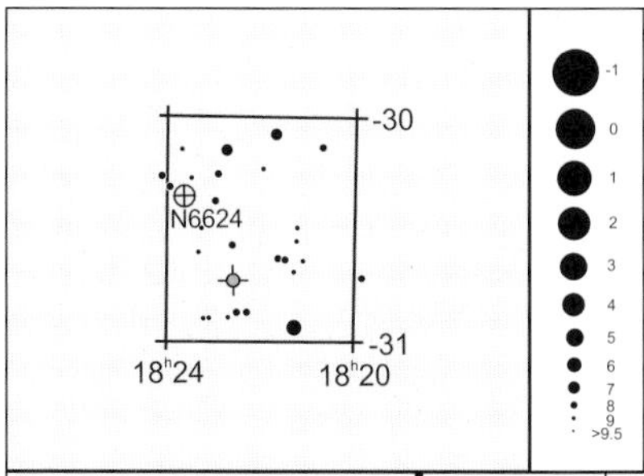

NGC 6624

RA:	18h 23m 48.1s	Con:	Sagittarius
Dec:	-30° 21' 56"	Type:	Globular Cluster
Size:	5.9'	Mag:	8.3

NGC 6624 is the fourth globular cluster located in the Teapot.

Telescope Aperture:	4" f/5	4" f/9	6" f/7	6" f/9	8" f/6.3	8" f/10	10" f/6.3	10" f/10	12" f/6.3	12" f/10
FOV(35mm film):	2.7° x 4.1°	1.50° x 2.26°	1.29° x 1.93°	1.0° x 1.50°	1.07° x 1.61°	0.68° x 1.02°	0.86° x 1.29°	0.54° x 0.81°	0.72° x 1.07°	0.45° x 0.68°

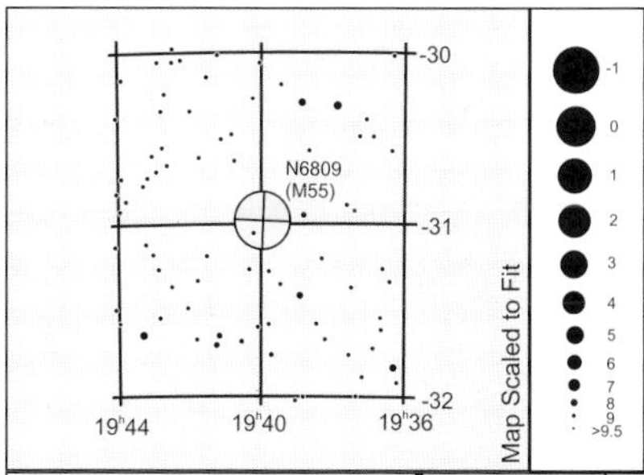

M55 (NGC 6809)

RA:	19h 40m 6.1s	Con:	Sagittarius
Dec:	-30° 57' 45"	Type:	Globular Cluster
Size:	19.0'	Mag:	7.0

M55 (NGC 6809) is a highly resolved globular cluster, located east of the Teapot asterism.

Telescope Aperture:	4" f/5	4" f/9	6" f/7	6" f/9	8" f/6.3	8" f/10	10" f/6.3	10" f/10	12" f/6.3	12" f/10
FOV(35mm film):	2.7° x 4.1°	1.50° x 2.26°	1.29° x 1.93°	1.0° x 1.50°	1.07° x 1.61°	0.68° x 1.02°	0.86° x 1.29°	0.54° x 0.81°	0.72° x 1.07°	0.45° x 0.68°

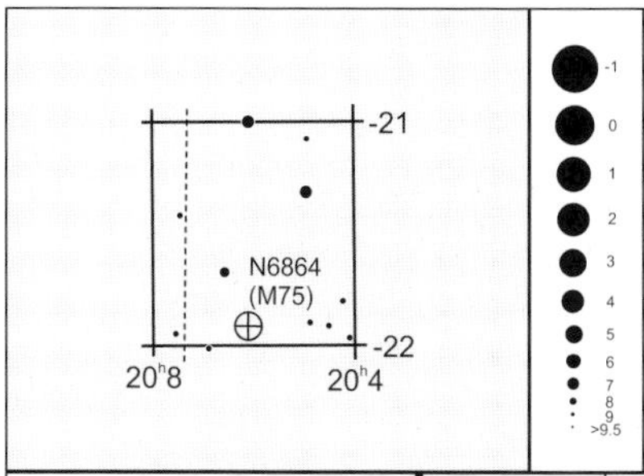

M75 (NGC 6864)

RA:	20h 06m 11.7s	Con:	Sagittarius
Dec:	-21° 54' 42"	Type:	Globular Cluster
Size:	6.0'	Mag:	8.6

M75 (NGC 6864) is a bright globular cluster that is difficult to resolve into individual stars.

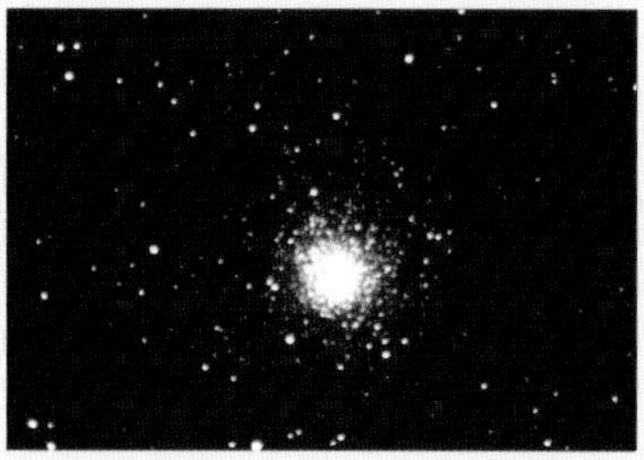

Telescope Aperture:	4" f/5	4" f/9	6" f/7	6" f/9	8" f/6.3	8" f/10	10" f/6.3	10" f/10	12" f/6.3	12" f/10
FOV(35mm film):	2.7° x 4.1°	1.50° x 2.26°	1.29° x 1.93°	1.0° x 1.50°	1.07° x 1.61°	0.68° x 1.02°	0.86° x 1.29°	0.54° x 0.81°	0.72° x 1.07°	0.45° x 0.68°

Sco

Crosses Prime Meridian:
June thru July

SPRING

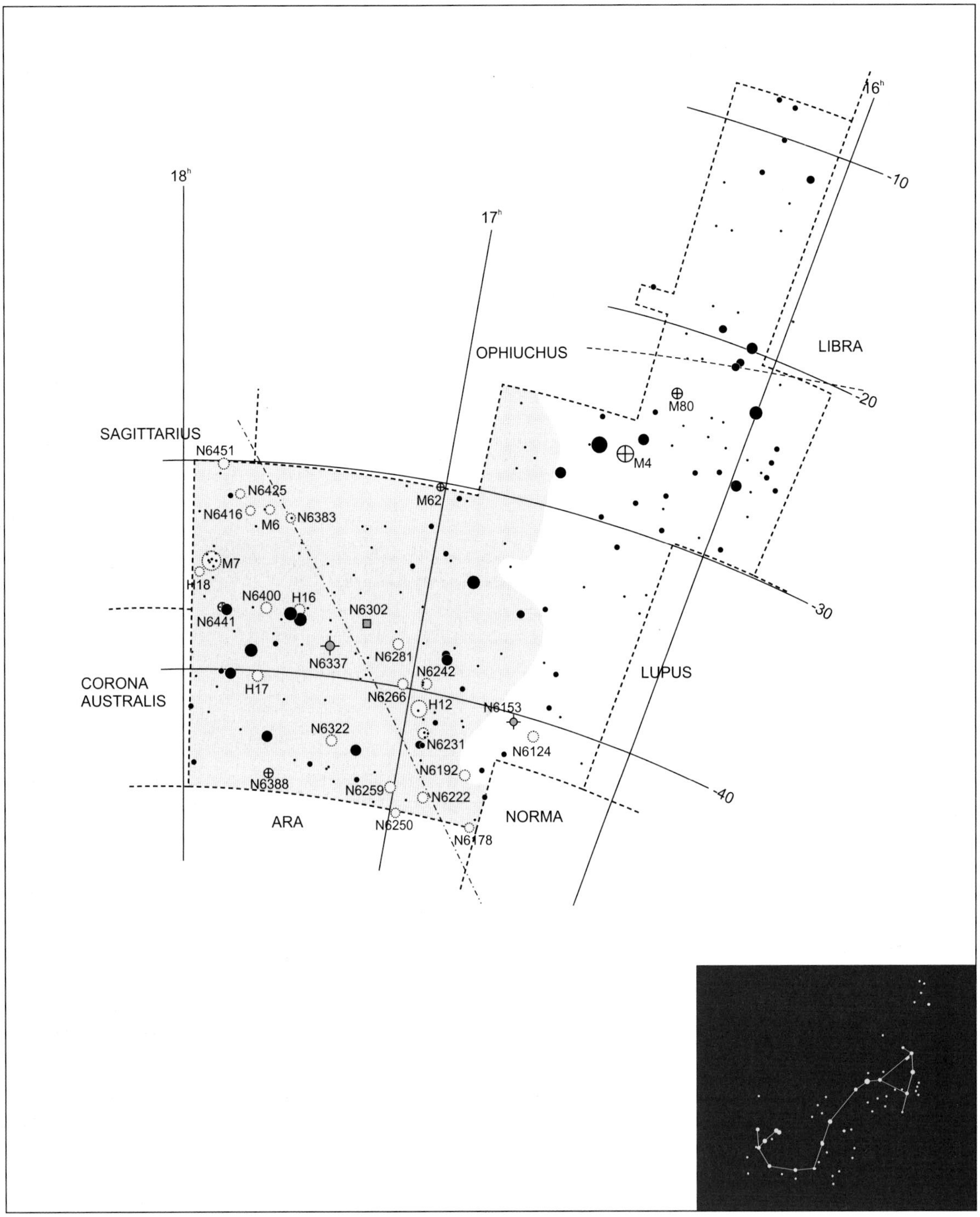

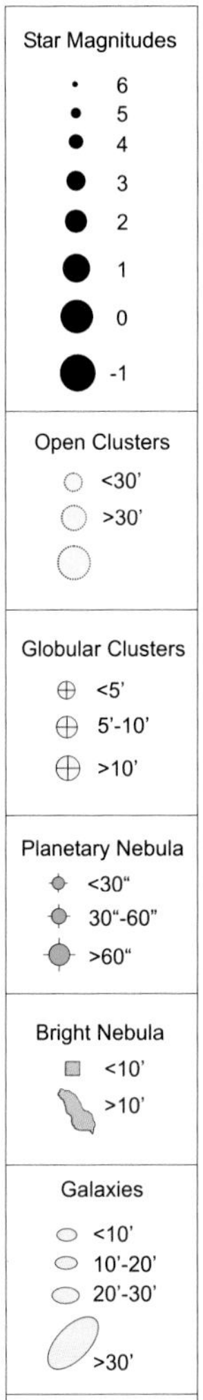

SCORPIUS

Constellation Facts:

Scorpius; (SKOR-pee-us)

Scorpius, the Scorpion.
This constellation contains several of the most southerly of the bright stars seen from the northern hemisphere. Because of the location of the stars, Scorpius is seen near the horizon as it moves from the southeast to the southwest.
The constellation covers 497 square degrees.

Constellation is visible from 44° N to 900° S. Partially visible from 44° N to 70° N.

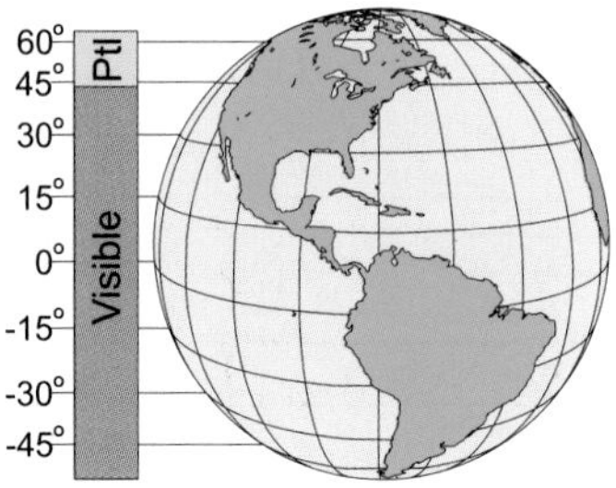

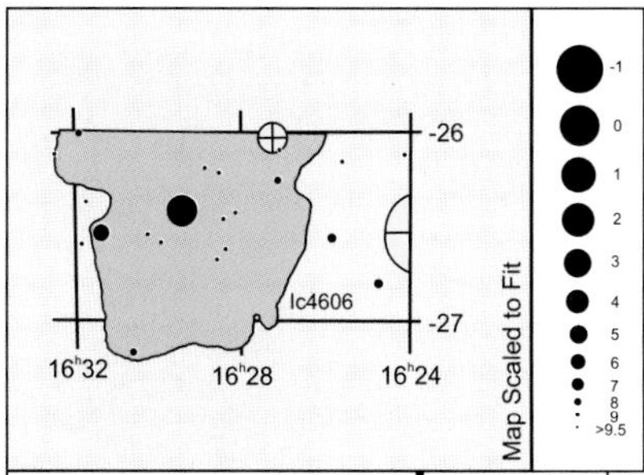

IC 4606

RA:	$16^h\ 31^m\ 40.2^s$	Con:	Scorpius
Dec:	-26° 03' 09"	Type:	Reflection Nebula
Size:	60' x 40'	Mag:	

IC 4606 is a large reflection nebula that surrounds Antares.

Telescope Aperture:	4" f/5	4" f/9	6" f/7	6" f/9	8" f/6.3	8" f/10	10" f/6.3	10" f/10	12" f/6.3	12" f/10
FOV(35mm film):	2.7° x 4.1°	1.50° x 2.26°	1.29° x 1.93°	1.0° x 1.50°	1.07° x 1.61°	0.68° x 1.02°	0.86° x 1.29°	0.54° x 0.81°	0.72° x 1.07°	0.45° x 0.68°

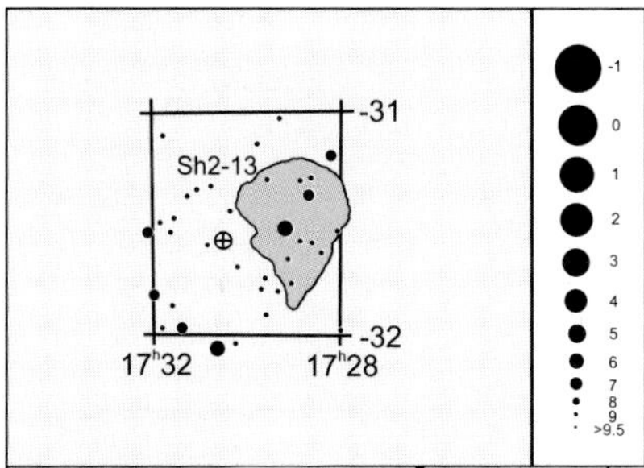

Sh2-13

RA:	$17^h\ 29^m\ 6.0^s$	Con:	Scorpius
Dec:	-31° 33' 0.0"	Type:	Emission Nebula
Size:	40' x 35'	Mag:	

Sh2-13 is a large emission nebula that is associated with many other nebulosities in the region. Object is part of a dense star field.

Telescope Aperture:	4" f/5	4" f/9	6" f/7	6" f/9	8" f/6.3	8" f/10	10" f/6.3	10" f/10	12" f/6.3	12" f/10
FOV(35mm film):	2.7° x 4.1°	1.50° x 2.26°	1.29° x 1.93°	1.0° x 1.50°	1.07° x 1.61°	0.68° x 1.02°	0.86° x 1.29°	0.54° x 0.81°	0.72° x 1.07°	0.45° x 0.68°

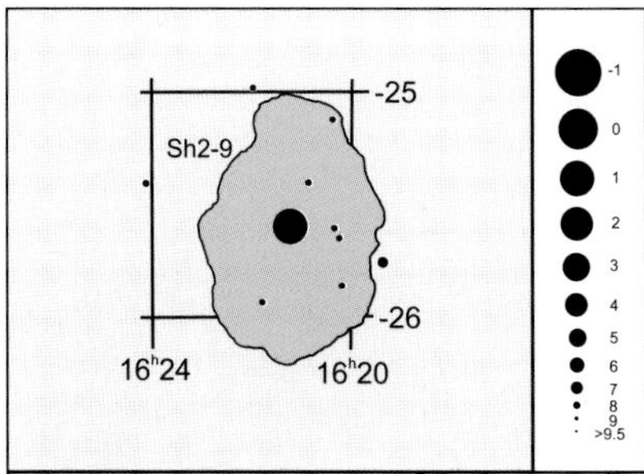

Sh2-9

RA:	$16^h\ 31^m\ 6.0^s$	Con:	Scorpius
Dec:	-25° 35' 0.0"	Type:	E & R Nebula
Size:	60.0' x 15.0'	Mag:	

Sh2-9 is a diffuse emission and reflection nebula.

Telescope Aperture:	4" f/5	4" f/9	6" f/7	6" f/9	8" f/6.3	8" f/10	10" f/6.3	10" f/10	12" f/6.3	12" f/10
FOV(35mm film):	2.7° x 4.1°	1.50° x 2.26°	1.29° x 1.93°	1.0° x 1.50°	1.07° x 1.61°	0.68° x 1.02°	0.86° x 1.29°	0.54° x 0.81°	0.72° x 1.07°	0.45° x 0.68°

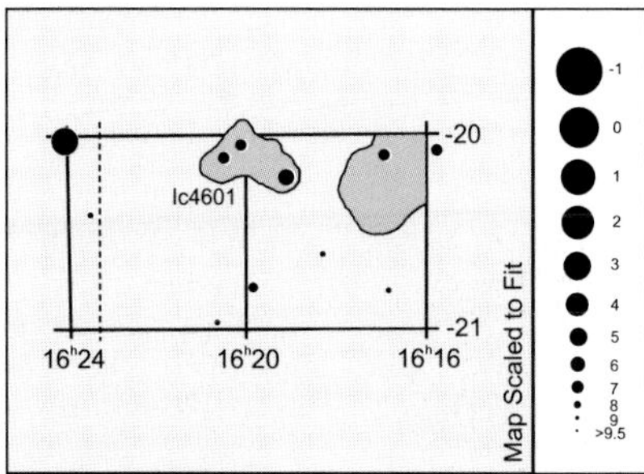

IC 4601

RA:	$16^h\ 20^m\ 3.9^s$	Con:	Scorpius
Dec:	-20° 02' 10"	Type:	Reflection Nebula
Size:	20.0' x 10.0'	Mag:	

IC 4601 is a reflection nebula, where the nebula is most prominent around the 7th magnitude star.

Telescope Aperture:	4" f/5	4" f/9	6" f/7	6" f/9	8" f/6.3	8" f/10	10" f/6.3	10" f/10	12" f/6.3	12" f/10
FOV(35mm film):	2.7° x 4.1°	1.50° x 2.26°	1.29° x 1.93°	1.0° x 1.50°	1.07° x 1.61°	0.68° x 1.02°	0.86° x 1.29°	0.54° x 0.81°	0.72° x 1.07°	0.45° x 0.68°

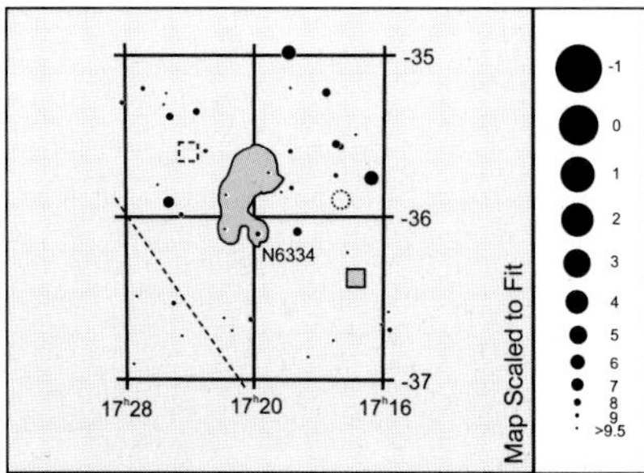

NGC 6334 "Cat Paw Nebula"

RA:	$17^h\ 20^m\ 34.6^s$	Con:	Scorpius
Dec:	-35° 43' 05"	Type:	Emission Nebula
Size:	35.0' x 20.0'	Mag:	

NGC 6334 known as the Cat Paw Nebula is a large emission nebula that is bisected several times by dark lanes divides object into 4 lobes.

Telescope Aperture:	4" f/5	4" f/9	6" f/7	6" f/9	8" f/6.3	8" f/10	10" f/6.3	10" f/10	12" f/6.3	12" f/10
FOV(35mm film):	2.7° x 4.1°	1.50° x 2.26°	1.29° x 1.93°	1.0° x 1.50°	1.07° x 1.61°	0.68° x 1.02°	0.86° x 1.29°	0.54° x 0.81°	0.72° x 1.07°	0.45° x 0.68°

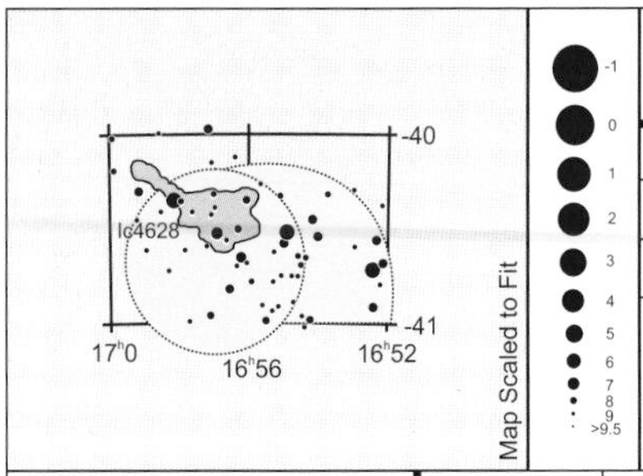

IC4628

RA:	16h 57m 4.7s	Con:	Scorpius
Dec:	-40° 20' 07"	Type:	Emission Nebula
Size:	90'x60'	Mag:	

Ic4628 appears elongated, and is irregular in shape.

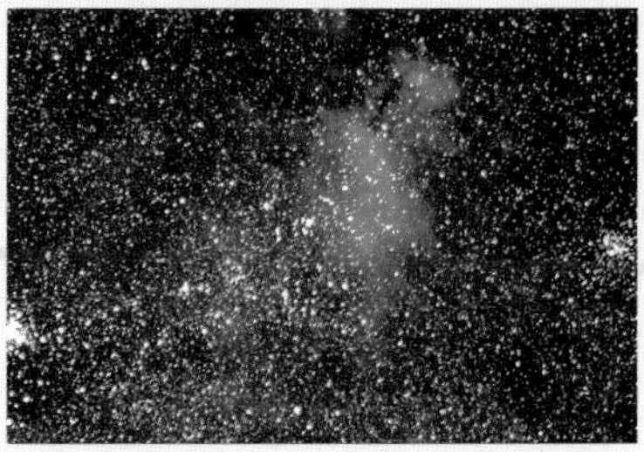

Telescope Aperture:	4" f/5	4" f/9	6" f/7	6" f/9	8" f/6.3	8" f/10	10" f/6.3	10" f/10	12" f/6.3	12" f/10
FOV(35mm film):	2.7°x 4.1°	1.50°x 2.26°	1.29° x 1.93°	1.0° x 1.50°	1.07° x 1.61°	0.68°x 1.02°	0.86° x 1.29°	0.54° x 0.81°	0.72° x 1.07°	0.45°x 0.68°

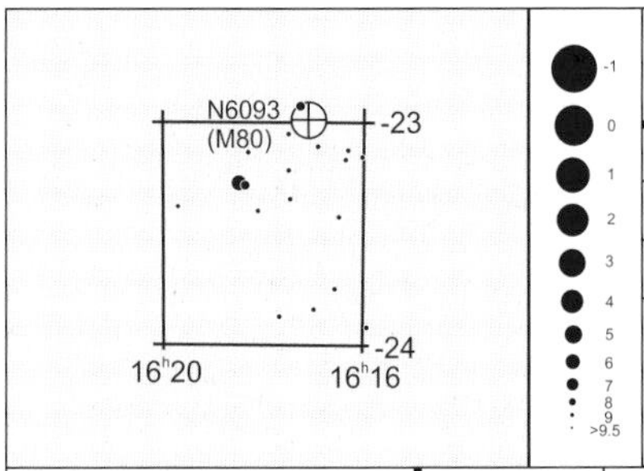

M80 (NGC6093)

RA:	16h 17m 4.1s	Con:	Scorpius
Dec:	-22° 59' 10"	Type:	Globular Cluster
Size:	8.9'	Mag:	7.2

M80 (NGC-6093) is a mottled globular cluster.

Telescope Aperture:	4" f/5	4" f/9	6" f/7	6" f/9	8" f/6.3	8" f/10	10" f/6.3	10" f/10	12" f/6.3	12" f/10
FOV(35mm film):	2.7°x 4.1°	1.50°x 2.26°	1.29° x 1.93°	1.0° x 1.50°	1.07° x 1.61°	0.68°x 1.02°	0.86° x 1.29°	0.54° x 0.81°	0.72° x 1.07°	0.45°x 0.68°

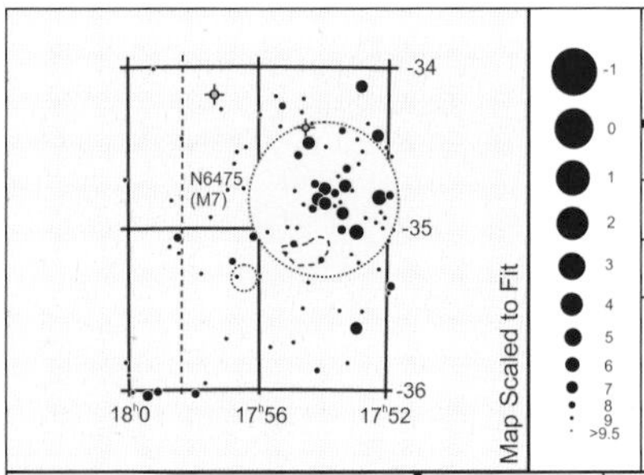

M7 (NGC 6475)

RA:	17h 53m 58.6s	Con:	Scorpius
Dec:	-34° 49' 02"	Type:	Open Cluster
Size:	80.0'	Mag:	3.3

M7 (NGC 6475) is a bright, scattered open cluster.
Object contains about 130 stars.

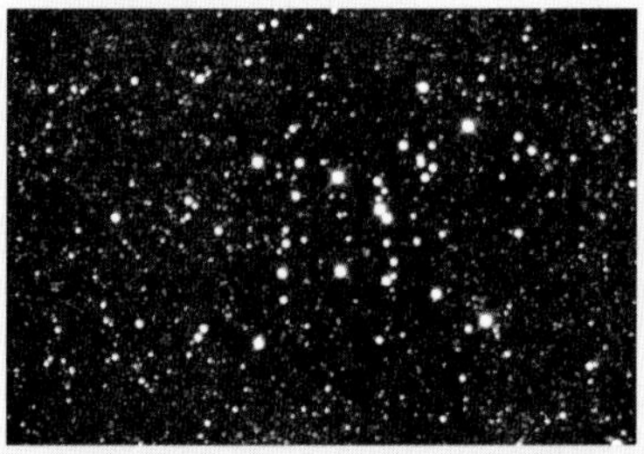

Telescope Aperture:	4" f/5	4" f/9	6" f/7	6" f/9	8" f/6.3	8" f/10	10" f/6.3	10" f/10	12" f/6.3	12" f/10
FOV(35mm film):	2.7°x 4.1°	1.50°x 2.26°	1.29° x 1.93°	1.0° x 1.50°	1.07° x 1.61°	0.68°x 1.02°	0.86° x 1.29°	0.54° x 0.81°	0.72° x 1.07°	0.45°x 0.68°

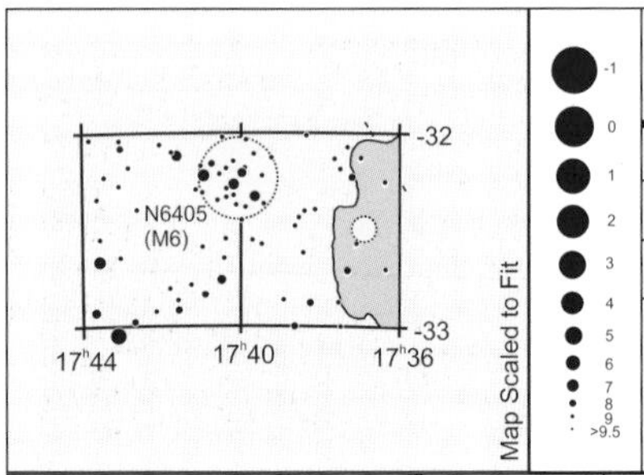

M6 (NGC 6405) "Butterfly Cluster"

RA:	17h 40m 10.4s	Con:	Scorpius
Dec:	-32° 13' 03"	Type:	Open Cluster
Size:	15.0'	Mag:	4.2'

M6 (NGC 6405) known as the Butterfly Cluster, is a rich open cluster containing 132 stars and is found 3.5° northwest of M7.

Telescope Aperture:	4" f/5	4" f/9	6" f/7	6" f/9	8" f/6.3	8" f/10	10" f/6.3	10" f/10	12" f/6.3	12" f/10
FOV(35mm film):	2.7°x 4.1°	1.50°x 2.26°	1.29° x 1.93°	1.0° x 1.50°	1.07° x 1.61°	0.68°x 1.02°	0.86° x 1.29°	0.54° x 0.81°	0.72° x 1.07°	0.45°x 0.68°

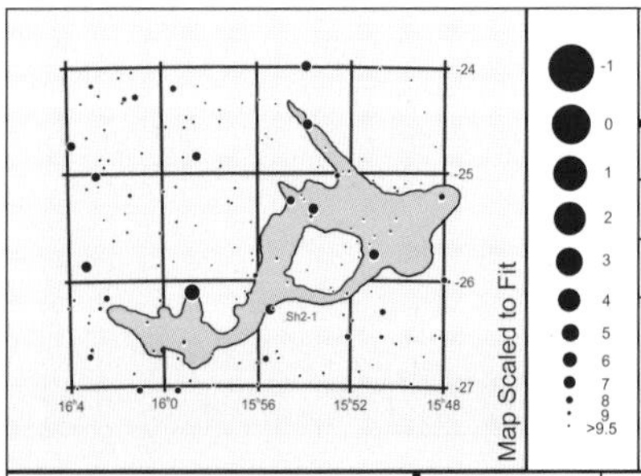

Sh2-1

RA:	15h 58m 50.0s	Con:	Scorpius
Dec:	-26° 09' 0.0"	Type:	Reflection Nebula
Size:	90' x 10'	Mag:	

Telescope Aperture:	4" f/5	4" f/9	6" f/7	6" f/9	8" f/6.3	8" f/10	10" f/6.3	10" f/10	12" f/6.3	12" f/10
FOV(35mm film):	2.7°x 4.1°	1.50°x 2.26°	1.29° x 1.93°	1.0° x 1.50°	1.07° x 1.61°	0.68°x 1.02°	0.86° x 1.29°	0.54° x 0.81°	0.72° x 1.07°	0.45°x 0.68°

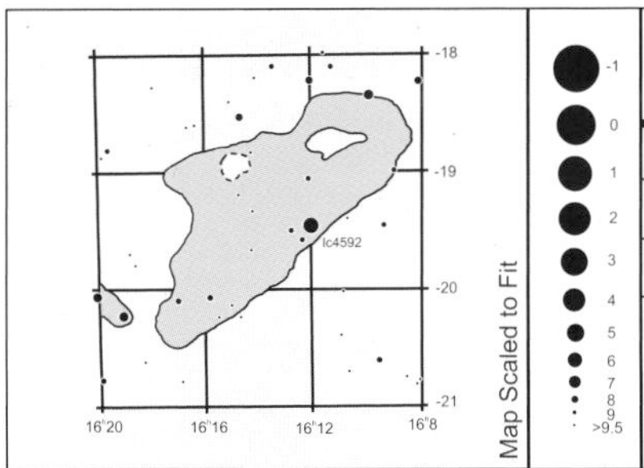

IC 4592

RA:	16h 12m 4.0s	Con:	Scorpius
Dec:	-19° 28' 10"	Type:	Reflection Nebula
Size:	150' x 60'	Mag:	

IC 4592 is a large reflection nebula, brightest in the region surrounding 14 Scorpii.

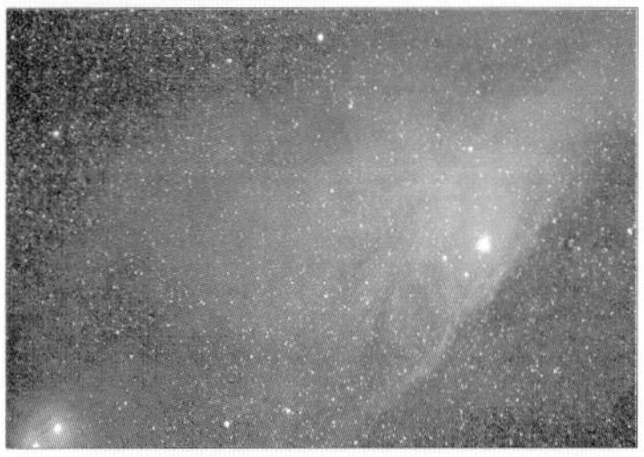

Telescope Aperture:	4" f/5	4" f/9	6" f/7	6" f/9	8" f/6.3	8" f/10	10" f/6.3	10" f/10	12" f/6.3	12" f/10
FOV(35mm film):	2.7° x 4.1°	1.50° x 2.26°	1.29° x 1.93°	1.0° x 1.50°	1.07° x 1.61°	0.68° x 1.02°	0.86° x 1.29°	0.54° x 0.81°	0.72° x 1.07°	0.45° x 0.68°

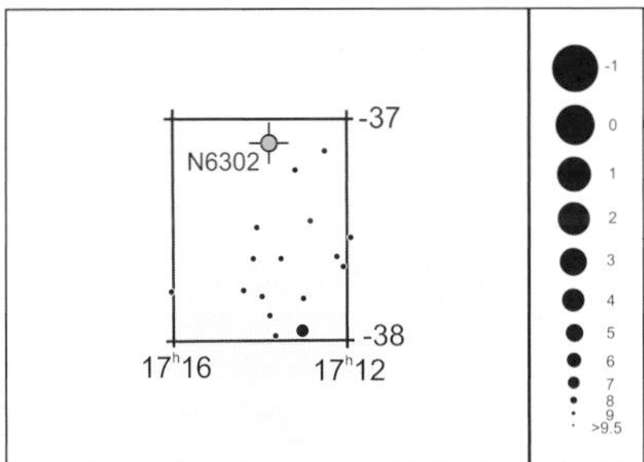

NGC 6302 "Bug Nebula"

RA:	17h 13m 46.5s	Con:	Scorpius
Dec:	-37° 06' 05"	Type:	Planetary Nebula
Size:	0.8'	Mag:	13.0

NGC 6302 known as the Bug Nebula, is an irregular planetary nebula displaying a bipolar lobe structure.

Telescope Aperture:	4" f/5	4" f/9	6" f/7	6" f/9	8" f/6.3	8" f/10	10" f/6.3	10" f/10	12" f/6.3	12" f/10
FOV(35mm film):	2.7° x 4.1°	1.50° x 2.26°	1.29° x 1.93°	1.0° x 1.50°	1.07° x 1.61°	0.68° x 1.02°	0.86° x 1.29°	0.54° x 0.81°	0.72° x 1.07°	0.45° x 0.68°

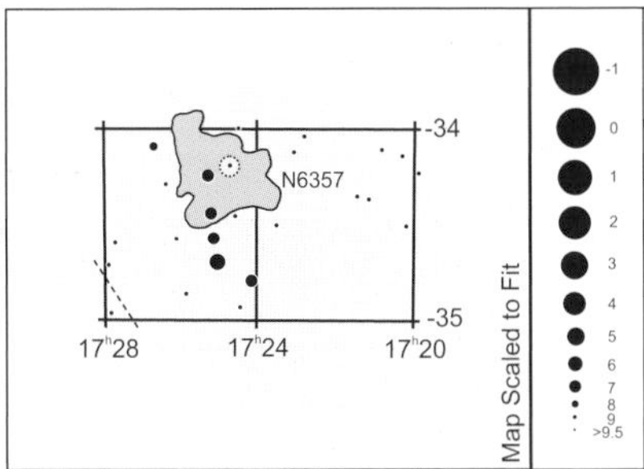

NGC 6357

RA:	17h 24m 40.5s	Con:	Scorpius
Dec:	-34° 10' 03"	Type:	Emission Nebula
Size:	50.0'	Mag:	

NGC 6357 is very irregular in shape and quite mottled.

Telescope Aperture:	4" f/5	4" f/9	6" f/7	6" f/9	8" f/6.3	8" f/10	10" f/6.3	10" f/10	12" f/6.3	12" f/10
FOV(35mm film):	2.7° x 4.1°	1.50° x 2.26°	1.29° x 1.93°	1.0° x 1.50°	1.07° x 1.61°	0.68° x 1.02°	0.86° x 1.29°	0.54° x 0.81°	0.72° x 1.07°	0.45° x 0.68°

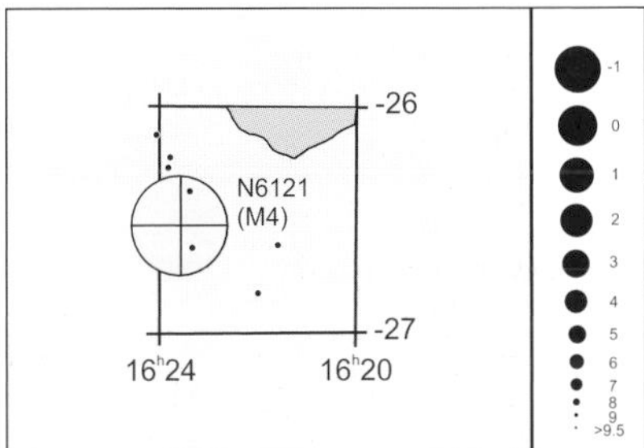

M4 (NGC 6121)

RA:	16h 23m 40.2s	Con:	Scorpius
Dec:	-26° 32' 09"	Type:	Globular Cluster
Size:	26.3'	Mag:	5.9

M4 (NGC 6121) is a highly resolved globular cluster found near Antares. Object has a loose structure.

Telescope Aperture:	4" f/5	4" f/9	6" f/7	6" f/9	8" f/6.3	8" f/10	10" f/6.3	10" f/10	12" f/6.3	12" f/10
FOV(35mm film):	2.7° x 4.1°	1.50° x 2.26°	1.29° x 1.93°	1.0° x 1.50°	1.07° x 1.61°	0.68° x 1.02°	0.86° x 1.29°	0.54° x 0.81°	0.72° x 1.07°	0.45° x 0.68°

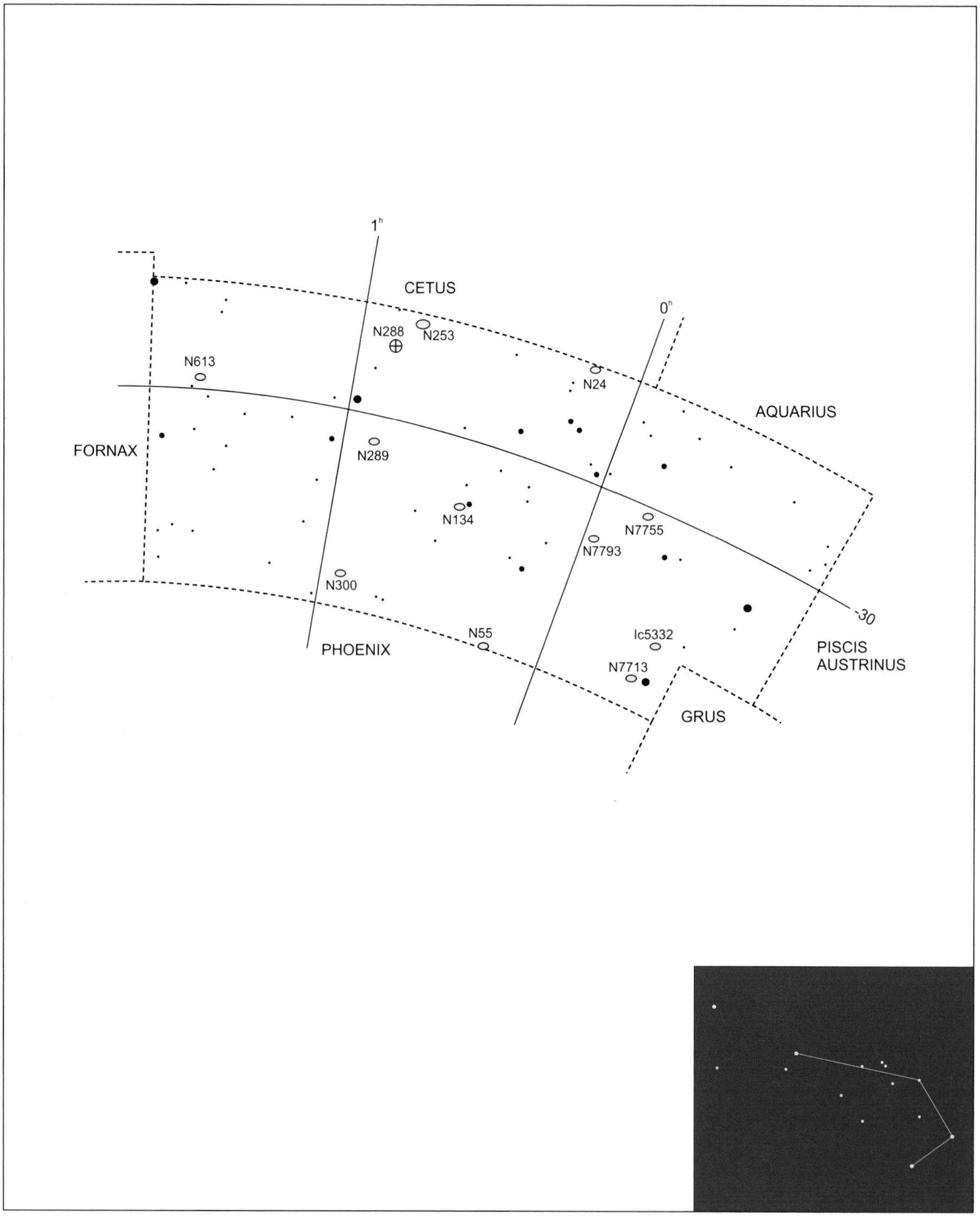

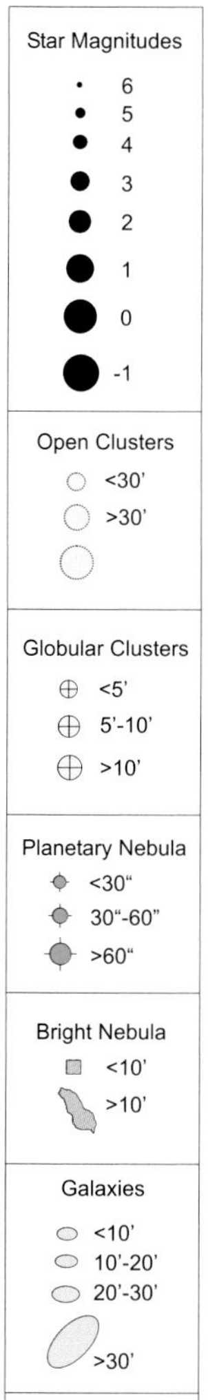

SCULPTOR

Constellation Facts:

Sculptor; (SKULPT-tor)

Constellation remains very low on the horizon, rising in the southeast and setting in the southwest. The constellation covers 475 square degrees.

Constellation is visible from 50° N to 90° S. Partially visible from 50° N to 60° N.

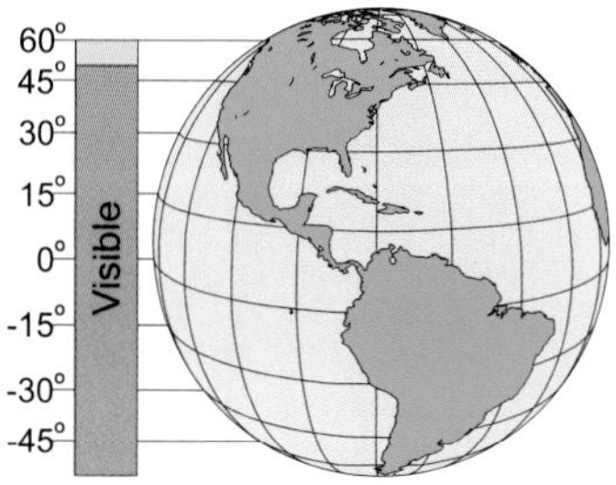

NGC 253

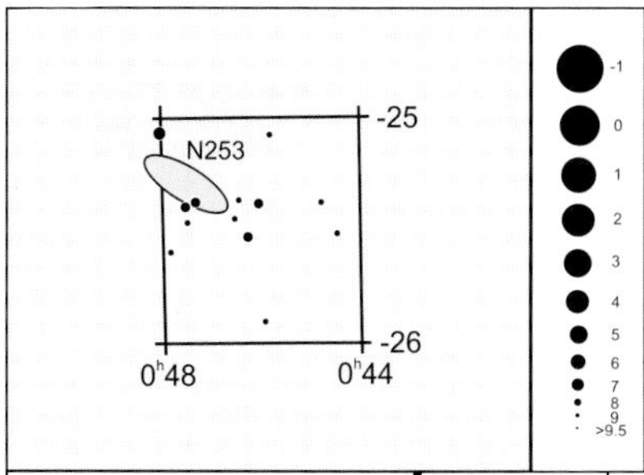

RA:	$00^h\ 47^m\ 39.8^s$	Con:	Sculptor
Dec:	-25° 16' 25"	Type:	Spiral Galaxy
Size:	25.0' x 5.0'	Mag:	7.1

NGC 253 is the brightest member of its local group called the Sculptor group. Object is a large spiral galaxy the appears elongated with visible dust lanes and bright knots.

Telescope Aperture:	4" f/5	4" f/9	6" f/7	6" f/9	8" f/6.3	8" f/10	10" f/6.3	10" f/10	12" f/6.3	12" f/10
FOV(35mm film):	2.7° x 4.1°	1.50° x 2.26°	1.29° x 1.93°	1.0° x 1.50°	1.07° x 1.61°	0.68° x 1.02°	0.86° x 1.29°	0.54° x 0.81°	0.72° x 1.07°	0.45° x 0.68°

NGC 288

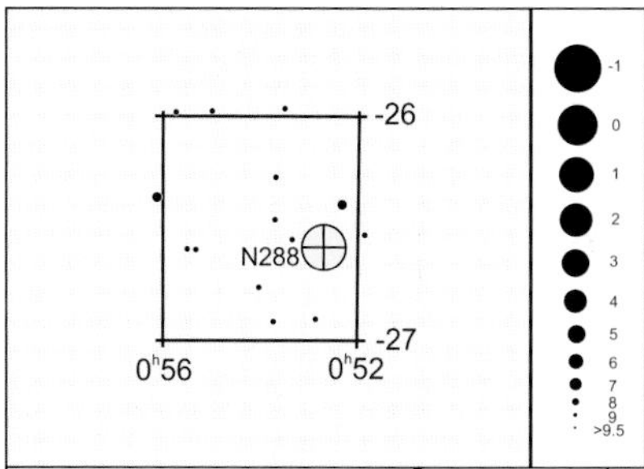

RA:	$00^h\ 52^m\ 51.7^s$	Con:	Sculptor
Dec:	-26° 34' 25"	Type:	Globular Cluster
Size:	13.8'	Mag:	8.1

NGC 288 is a highly resolved globular cluster. It's located 2° southeast of NGC 253. Object is a loosely arranged cluster.

Telescope Aperture:	4" f/5	4" f/9	6" f/7	6" f/9	8" f/6.3	8" f/10	10" f/6.3	10" f/10	12" f/6.3	12" f/10
FOV(35mm film):	2.7° x 4.1°	1.50° x 2.26°	1.29° x 1.93°	1.0° x 1.50°	1.07° x 1.61°	0.68° x 1.02°	0.86° x 1.29°	0.54° x 0.81°	0.72° x 1.07°	0.45° x 0.68°

NGC 613

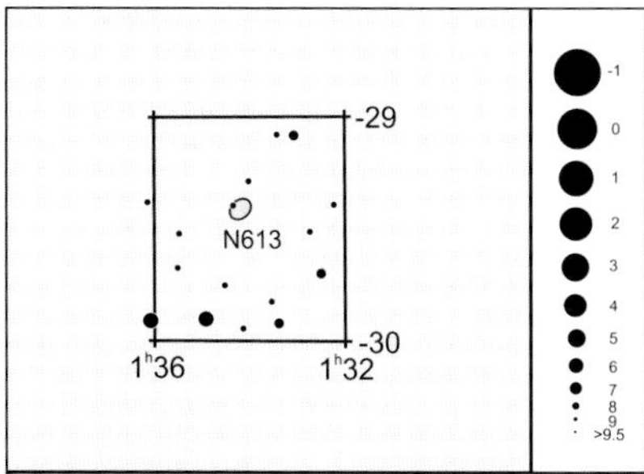

RA:	$01^h\ 34^m\ 21.3^s$	Con:	Sculptor
Dec:	-29° 24' 26"	Type:	Spiral Galaxy
Size:	5.0'	Mag:	10.0

NGC 613 is a bright barred spiral galaxy located in the northeastern region of the constellation.

Telescope Aperture:	4" f/5	4" f/9	6" f/7	6" f/9	8" f/6.3	8" f/10	10" f/6.3	10" f/10	12" f/6.3	12" f/10
FOV(35mm film):	2.7° x 4.1°	1.50° x 2.26°	1.29° x 1.93°	1.0° x 1.50°	1.07° x 1.61°	0.68° x 1.02°	0.86° x 1.29°	0.54° x 0.81°	0.72° x 1.07°	0.45° x 0.68°

NGC 300

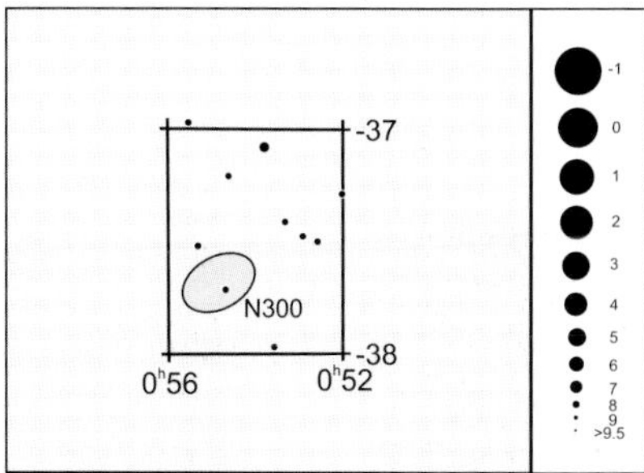

RA:	$00^h\ 54^m\ 57.6^s$	Con:	Sculptor
Dec:	-37° 40' 22"	Type:	Spiral Galaxy
Size:	20.0' x 14.8'	Mag:	9.0

NGC 300 is a large spiral galaxy found near the border with Phoenix.

Telescope Aperture:	4" f/5	4" f/9	6" f/7	6" f/9	8" f/6.3	8" f/10	10" f/6.3	10" f/10	12" f/6.3	12" f/10
FOV(35mm film):	2.7° x 4.1°	1.50° x 2.26°	1.29° x 1.93°	1.0° x 1.50°	1.07° x 1.61°	0.68° x 1.02°	0.86° x 1.29°	0.54° x 0.81°	0.72° x 1.07°	0.45° x 0.68°

NGC 55

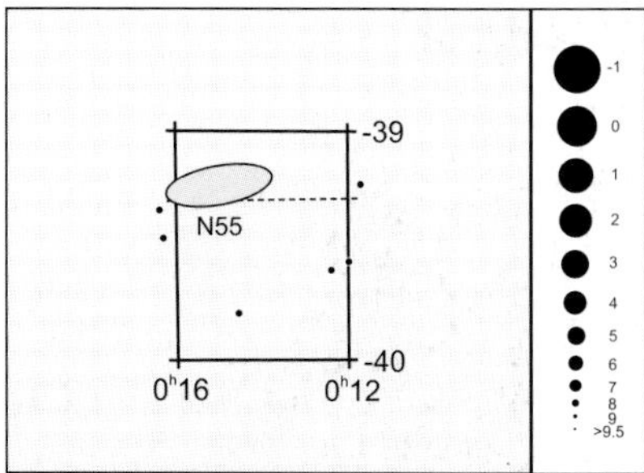

RA:	$00^h\ 14^m\ 58.1^s$	Con:	Sculptor
Dec:	-39° 10' 21"	Type:	Spiral Galaxy
Size:	32.4' x 6.5'	Mag:	8.0

NGC 55 is a large barred spiral galaxy, oriented edge-on to our line of sight. Object is the largest galaxy in the constellation.

Telescope Aperture:	4" f/5	4" f/9	6" f/7	6" f/9	8" f/6.3	8" f/10	10" f/6.3	10" f/10	12" f/6.3	12" f/10
FOV(35mm film):	2.7° x 4.1°	1.50° x 2.26°	1.29° x 1.93°	1.0° x 1.50°	1.07° x 1.61°	0.68° x 1.02°	0.86° x 1.29°	0.54° x 0.81°	0.72° x 1.07°	0.45° x 0.68°

Sct

Crosses Prime Meridian: July thru August

SUMMER

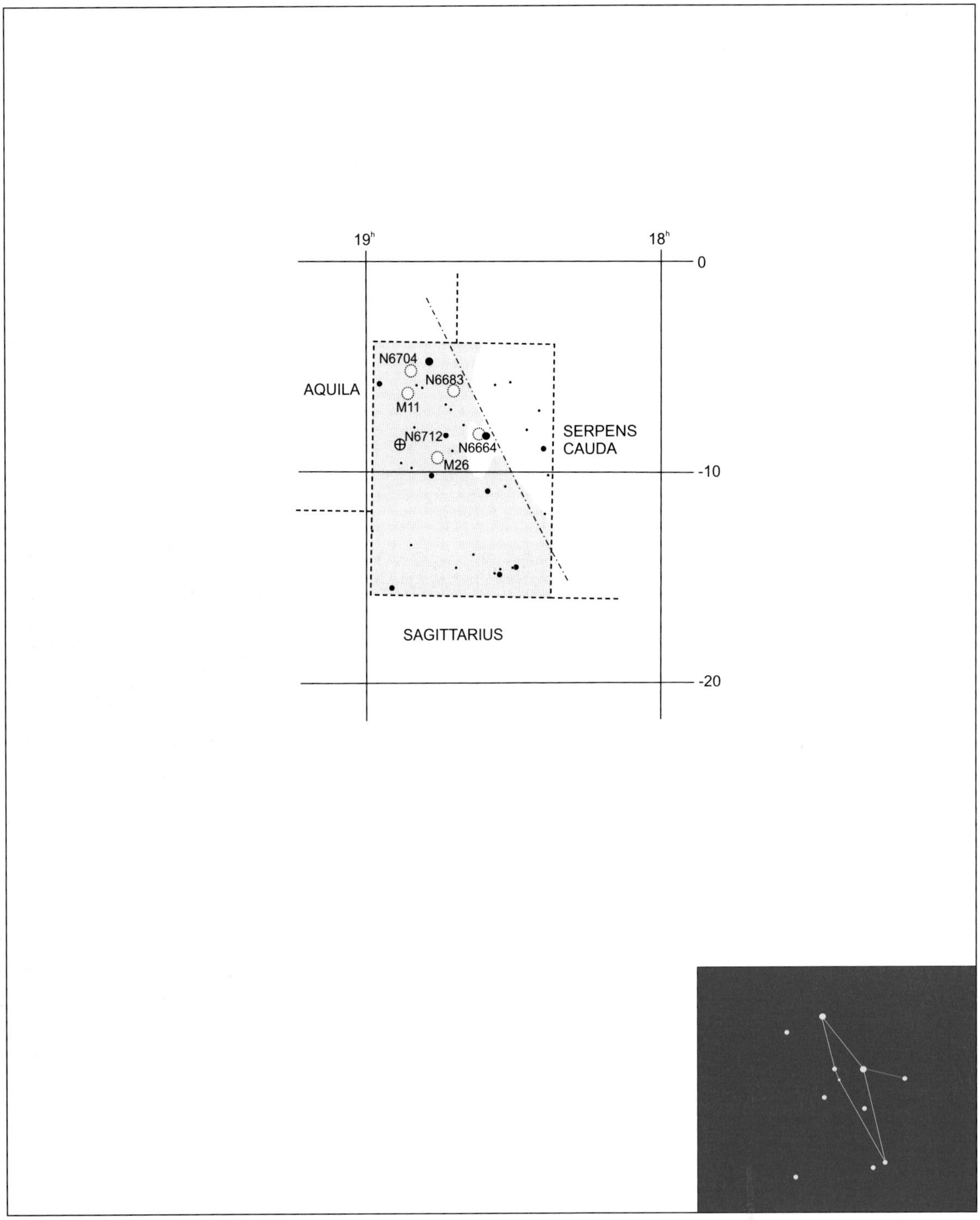

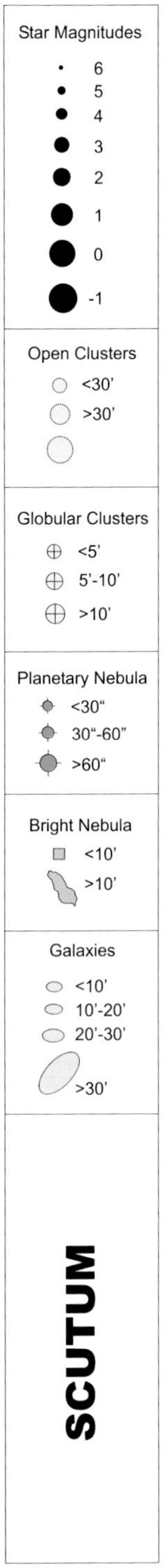

Constellation Facts:

Scutum; (SKU-tum)

Scutum, the Shield.
Scutum is a small constellation that is located just south of the celestial equator in a rich area of the Milky Way. The stars rise close to the eastern point on the horizon, pass the meridian about halfway between the horizon and the zenith, and set towards the west.
The constellation covers 109 square degrees.

Constellation is visible from 74° N to 90° S. Partially visible from 74° N to 80° N.

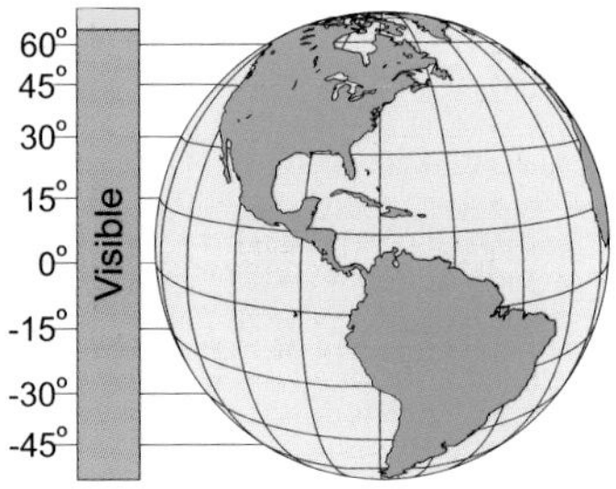

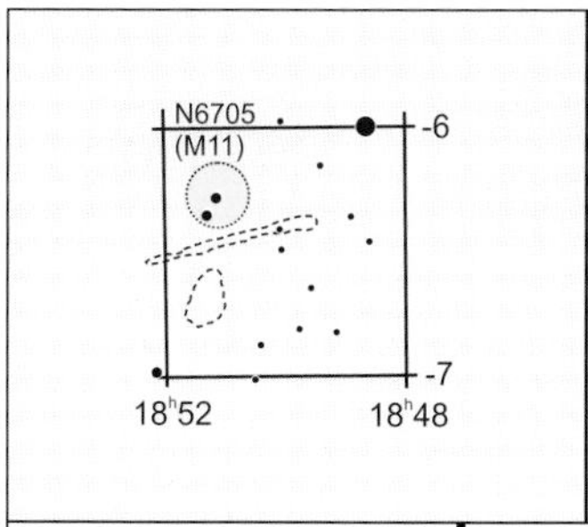

M11 (NGC 6705) "Wild Duck Cluster"

RA:	18h 51m 11.1s	Con:	Scutum
Dec:	-06° 15' 51"	Type:	Open Cluster
Size:	14.0'	Mag:	5.8

M11 (NGC 6705) is a dense, rich open cluster known as the Wild Duck Cluster. Object is found on the northern edge of the Scutum Star Cloud.

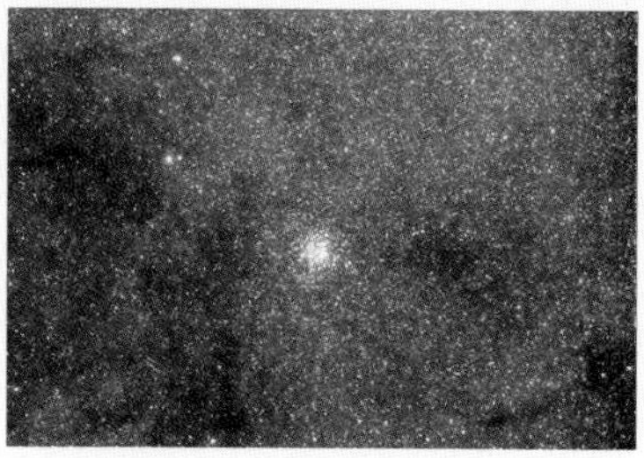

Telescope Aperture:	4" f/5	4" f/9	6" f/7	6" f/9	8" f/6.3	8" f/10	10" f/6.3	10" f/10	12" f/6.3	12" f/10
FOV(35mm film):	2.7° x 4.1°	1.50° x 2.26°	1.29° x 1.93°	1.0° x 1.50°	1.07° x 1.61°	0.68° x 1.02°	0.86° x 1.29°	0.54° x 0.81°	0.72° x 1.07°	0.45° x 0.68°

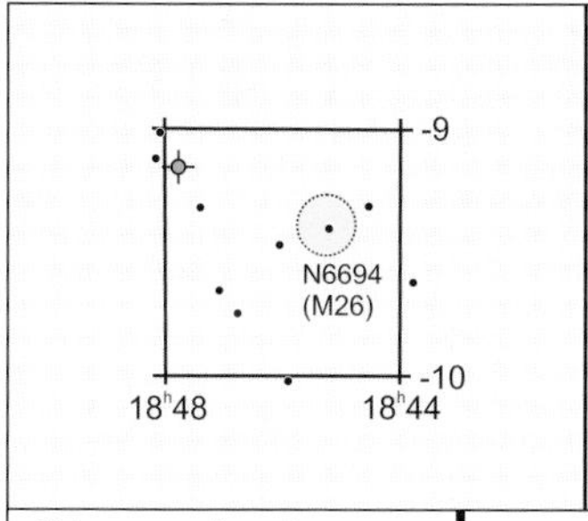

M26 (NGC 6694)

RA:	18h 45m 17.3s	Con:	Scutum
Dec:	-09° 23' 51"	Type:	Open Cluster
Size:	15.0'	Mag:	8.0

M26 (NGC 6694) is a rich open cluster containing 30 stars.

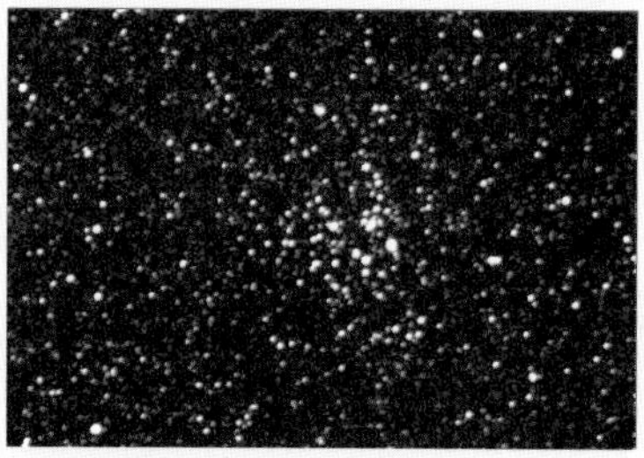

Telescope Aperture:	4" f/5	4" f/9	6" f/7	6" f/9	8" f/6.3	8" f/10	10" f/6.3	10" f/10	12" f/6.3	12" f/10
FOV(35mm film):	2.7° x 4.1°	1.50° x 2.26°	1.29° x 1.93°	1.0° x 1.50°	1.07° x 1.61°	0.68° x 1.02°	0.86° x 1.29°	0.54° x 0.81°	0.72° x 1.07°	0.45° x 0.68°

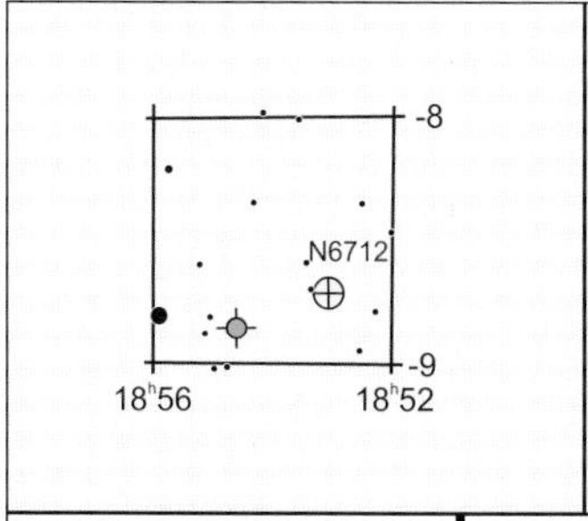

NGC 6712

RA:	18h 53m 11.3s	Con:	Scutum
Dec:	-08° 41' 50"	Type:	Globular Cluster
Size:	7.2'	Mag:	8.2

NGC 6712 is a highly resolved globular cluster. Object is found 2° east and north of M26.

Telescope Aperture:	4" f/5	4" f/9	6" f/7	6" f/9	8" f/6.3	8" f/10	10" f/6.3	10" f/10	12" f/6.3	12" f/10
FOV(35mm film):	2.7° x 4.1°	1.50° x 2.26°	1.29° x 1.93°	1.0° x 1.50°	1.07° x 1.61°	0.68° x 1.02°	0.86° x 1.29°	0.54° x 0.81°	0.72° x 1.07°	0.45° x 0.68°

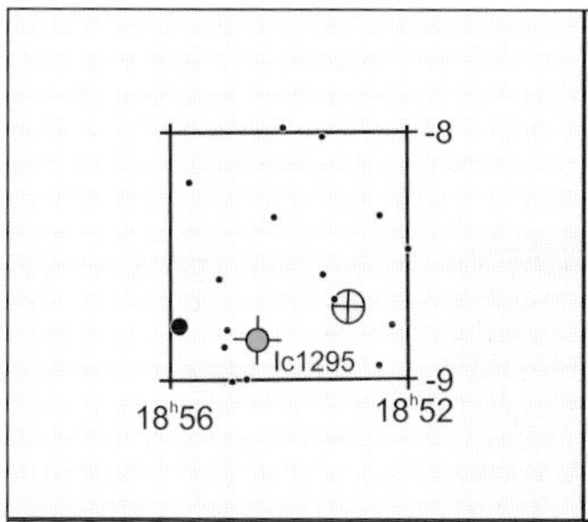

IC 1295

RA:	18h 54m 41.2s	Con:	Scutum
Dec:	-08° 49' 50"	Type:	Planetary Nebula
Size:	1.4'	Mag:	15.0

IC 1295 is a small, faint planetary nebula.

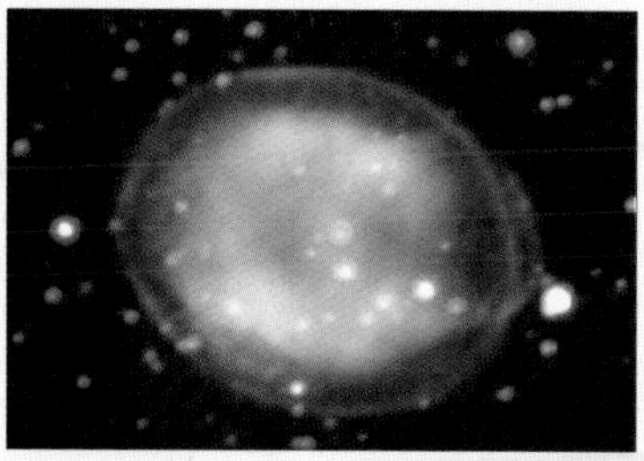

Telescope Aperture:	4" f/5	4" f/9	6" f/7	6" f/9	8" f/6.3	8" f/10	10" f/6.3	10" f/10	12" f/6.3	12" f/10
FOV(35mm film):	2.7° x 4.1°	1.50° x 2.26°	1.29° x 1.93°	1.0° x 1.50°	1.07° x 1.61°	0.68° x 1.02°	0.86° x 1.29°	0.54° x 0.81°	0.72° x 1.07°	0.45° x 0.68°

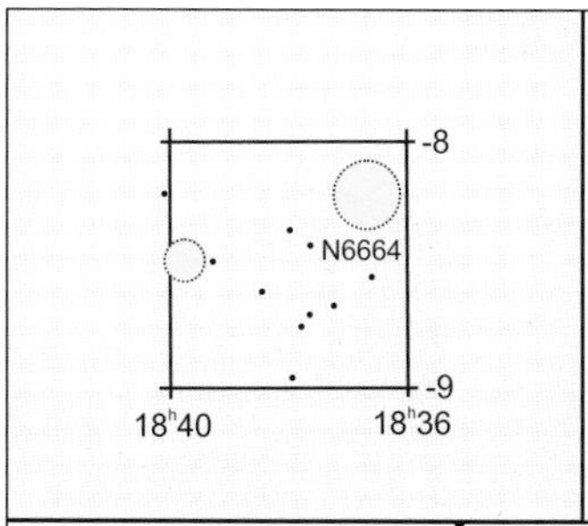

NGC 6664

RA:	18h 36m 47.3s	Con:	Scutum
Dec:	-08° 12' 52"	Type:	Open Cluster
Size:	16.0'	Mag:	7.8

NGC 6664 is a bright, scattered open cluster. Object contains 50 stars.

Telescope Aperture:	4" f/5	4" f/9	6" f/7	6" f/9	8" f/6.3	8" f/10	10" f/6.3	10" f/10	12" f/6.3	12" f/10
FOV(35mm film):	2.7° x 4.1°	1.50° x 2.26°	1.29° x 1.93°	1.0° x 1.50°	1.07° x 1.61°	0.68° x 1.02°	0.86° x 1.29°	0.54° x 0.81°	0.72° x 1.07°	0.45° x 0.68°

Ser **Crosses Prime Meridian:** June thru August **SPRING**

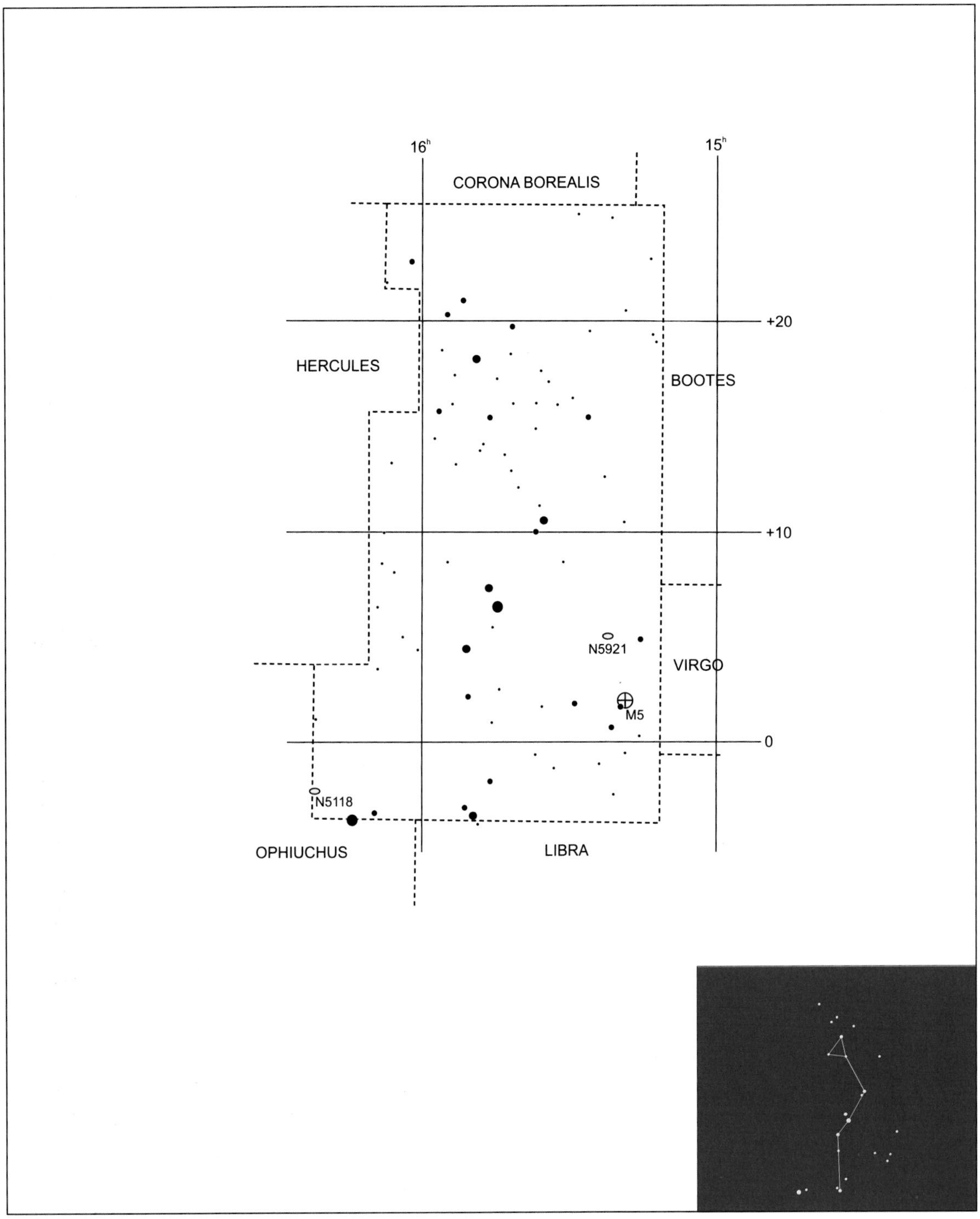

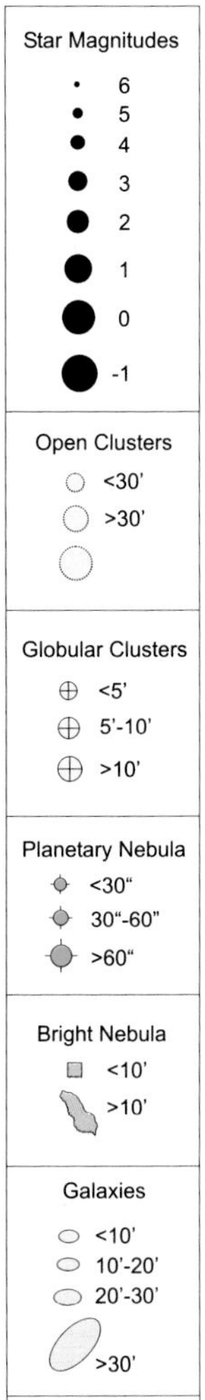

SERPENS CAPUT

Constellation Facts:

Serpens; (SIR-pens)

Serpens, the Serpent.
The constellation rises around the eastern point of the horizon, passes the meridian halfway between the horizon and the zenith, and sets towards the west.
The constellation covers 637 square degrees.

Constellation is visible from 74° N to 64° S. Partially visible from 74° N to 90° N.

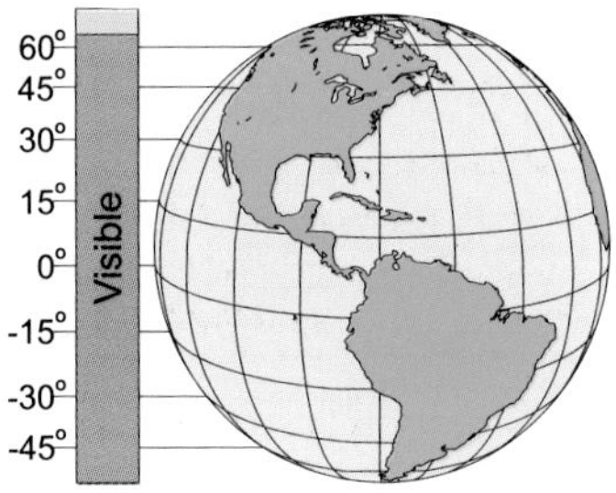

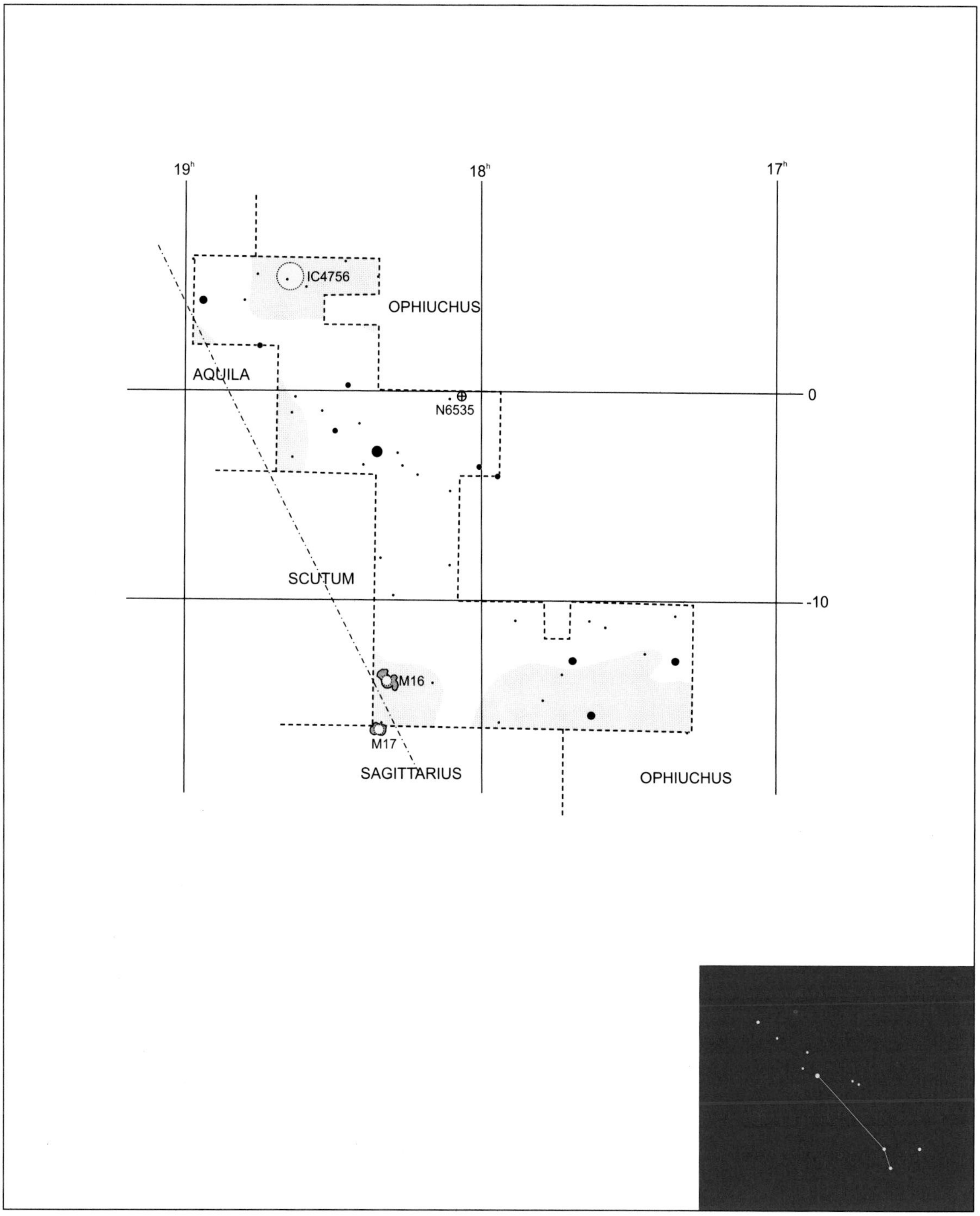

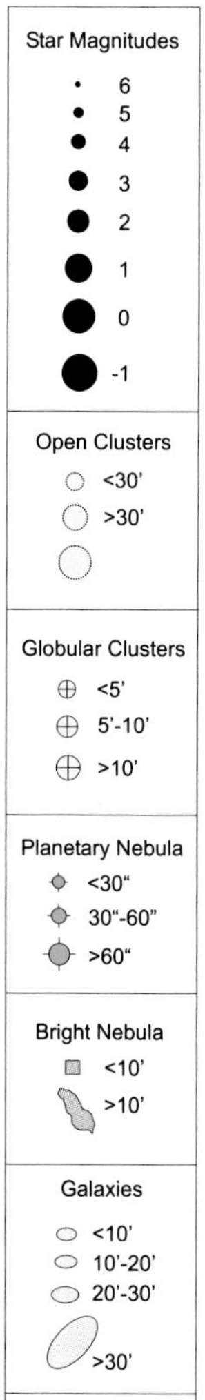

SERPENS CAUDA

Constellation Facts:

Serpens; (SIR-pens)

Serpens, the Serpent.
The constellation rises around the eastern point of the horizon, passes the meridian halfway between the horizon and the zenith, and sets towards the west.
The constellation covers 637 square degrees.

Constellation is visible from 74° N to 64° S. Partially visible from 74° N to 90° N.

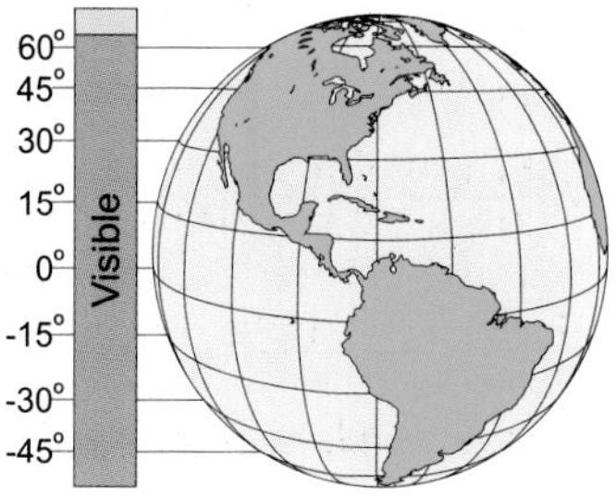

NGC 6118

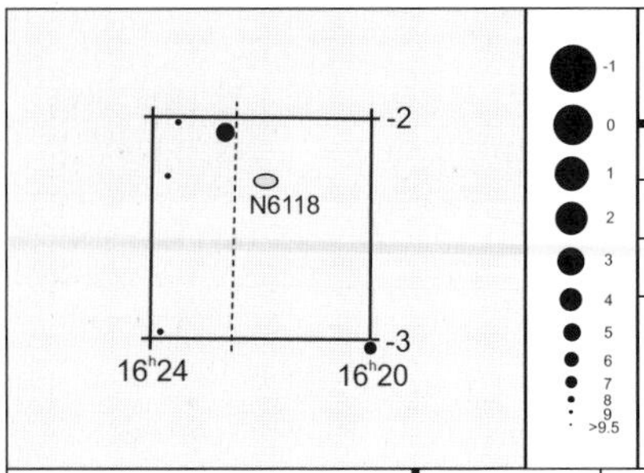

RA:	$16^h\ 21^m\ 52.7^s$	Con:	Serpens Caput
Dec:	-02° 17' 06"	Type:	Spiral Galaxy
Size:	4.9' x 2.3'	Mag:	12.0

NGC 6118 is an Sb-type spiral galaxy that, in the eyepiece appears very large and has low surface brightness.

Telescope Aperture:	4" f/5	4" f/9	6" f/7	6" f/9	8" f/6.3	8" f/10	10" f/6.3	10" f/10	12" f/6.3	12" f/10
FOV(35mm film):	2.7° x 4.1°	1.50° x 2.26°	1.29° x 1.93°	1.0° x 1.50°	1.07° x 1.61°	0.68° x 1.02°	0.86° x 1.29°	0.54° x 0.81°	0.72° x 1.07°	0.45° x 0.68°

M5 (NGC 5904)

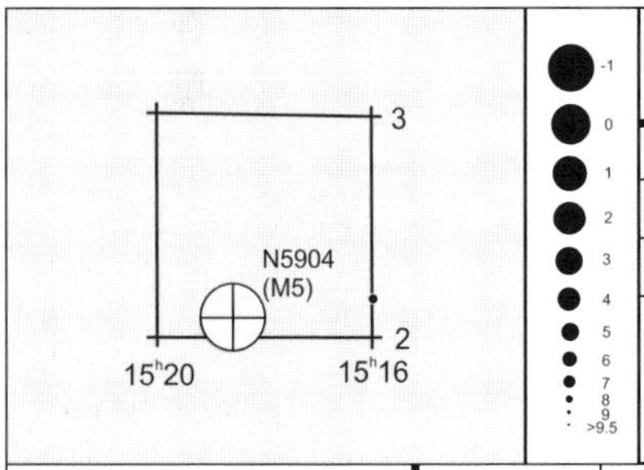

RA:	$15^h\ 18^m\ 40.3^s$	Con:	Serpens Caput
Dec:	02° 04' 49"	Type:	Globular Cluster
Size:	17.4'	Mag:	5.8

M5 (NGC 5904) is a highly resolved globular cluster.

Telescope Aperture:	4" f/5	4" f/9	6" f/7	6" f/9	8" f/6.3	8" f/10	10" f/6.3	10" f/10	12" f/6.3	12" f/10
FOV(35mm film):	2.7° x 4.1°	1.50° x 2.26°	1.29° x 1.93°	1.0° x 1.50°	1.07° x 1.61°	0.68° x 1.02°	0.86° x 1.29°	0.54° x 0.81°	0.72° x 1.07°	0.45° x 0.68°

M16 (NGC 6611) "Eagle Nebula"

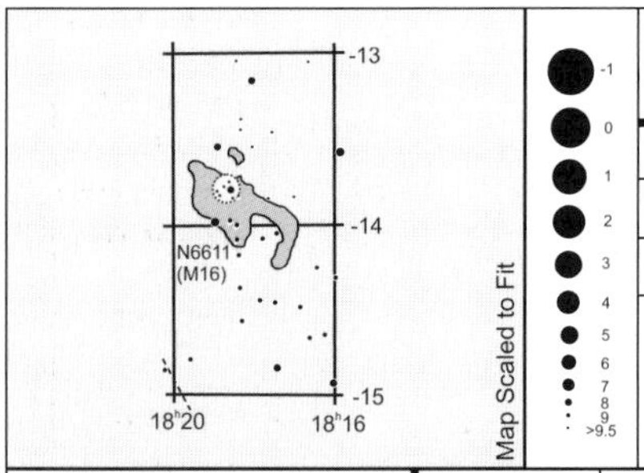

RA:	$18^h\ 18^m\ 53.4^s$	Con:	Serpens Cauda
Dec:	-13° 46' 55"	Type:	Nebula & Cluster
Size:	35.0'	Mag:	6.0

M16 (NGC 6611) is known as the Eagle Nebula.
Large emission nebula with associated open cluster.

Telescope Aperture:	4" f/5	4" f/9	6" f/7	6" f/9	8" f/6.3	8" f/10	10" f/6.3	10" f/10	12" f/6.3	12" f/10
FOV(35mm film):	2.7° x 4.1°	1.50° x 2.26°	1.29° x 1.93°	1.0° x 1.50°	1.07° x 1.61°	0.68° x 1.02°	0.86° x 1.29°	0.54° x 0.81°	0.72° x 1.07°	0.45° x 0.68°

NGC 6539

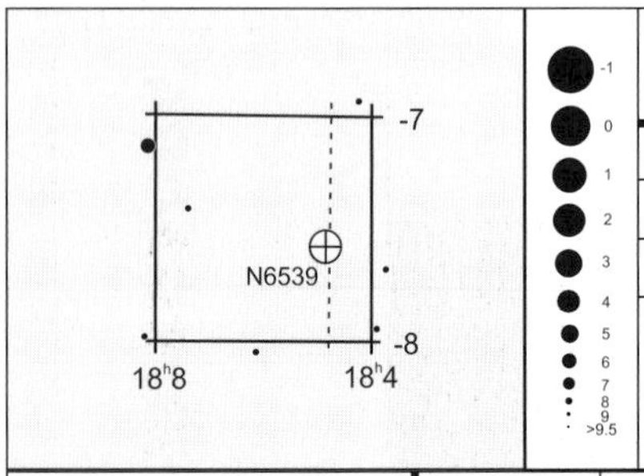

RA:	$18^h\ 04^m\ 53.2^s$	Con:	Serpens Cauda
Dec:	-07° 34' 56"	Type:	Globular Cluster
Size:	6.9'	Mag:	9.6

NGC 6539 is an unresolved globular cluster.

Telescope Aperture:	4" f/5	4" f/9	6" f/7	6" f/9	8" f/6.3	8" f/10	10" f/6.3	10" f/10	12" f/6.3	12" f/10
FOV(35mm film):	2.7° x 4.1°	1.50° x 2.26°	1.29° x 1.93°	1.0° x 1.50°	1.07° x 1.61°	0.68° x 1.02°	0.86° x 1.29°	0.54° x 0.81°	0.72° x 1.07°	0.45° x 0.68°

NGC 6604

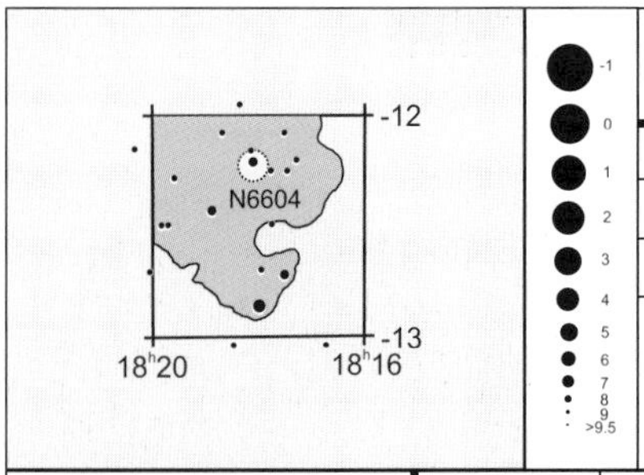

RA:	$18^h\ 18^m\ 11.3^s$	Con:	Serpens Cauda
Dec:	-12° 13' 55"	Type:	Cluster & Nebula
Size:	60.0'	Mag:	6.5

NGC 6604 is a large open cluster with associated nebulosity.

Telescope Aperture:	4" f/5	4" f/9	6" f/7	6" f/9	8" f/6.3	8" f/10	10" f/6.3	10" f/10	12" f/6.3	12" f/10
FOV(35mm film):	2.7° x 4.1°	1.50° x 2.26°	1.29° x 1.93°	1.0° x 1.50°	1.07° x 1.61°	0.68° x 1.02°	0.86° x 1.29°	0.54° x 0.81°	0.72° x 1.07°	0.45° x 0.68°

Tau

Crosses Prime Meridian:
December thru January

AUTUMN

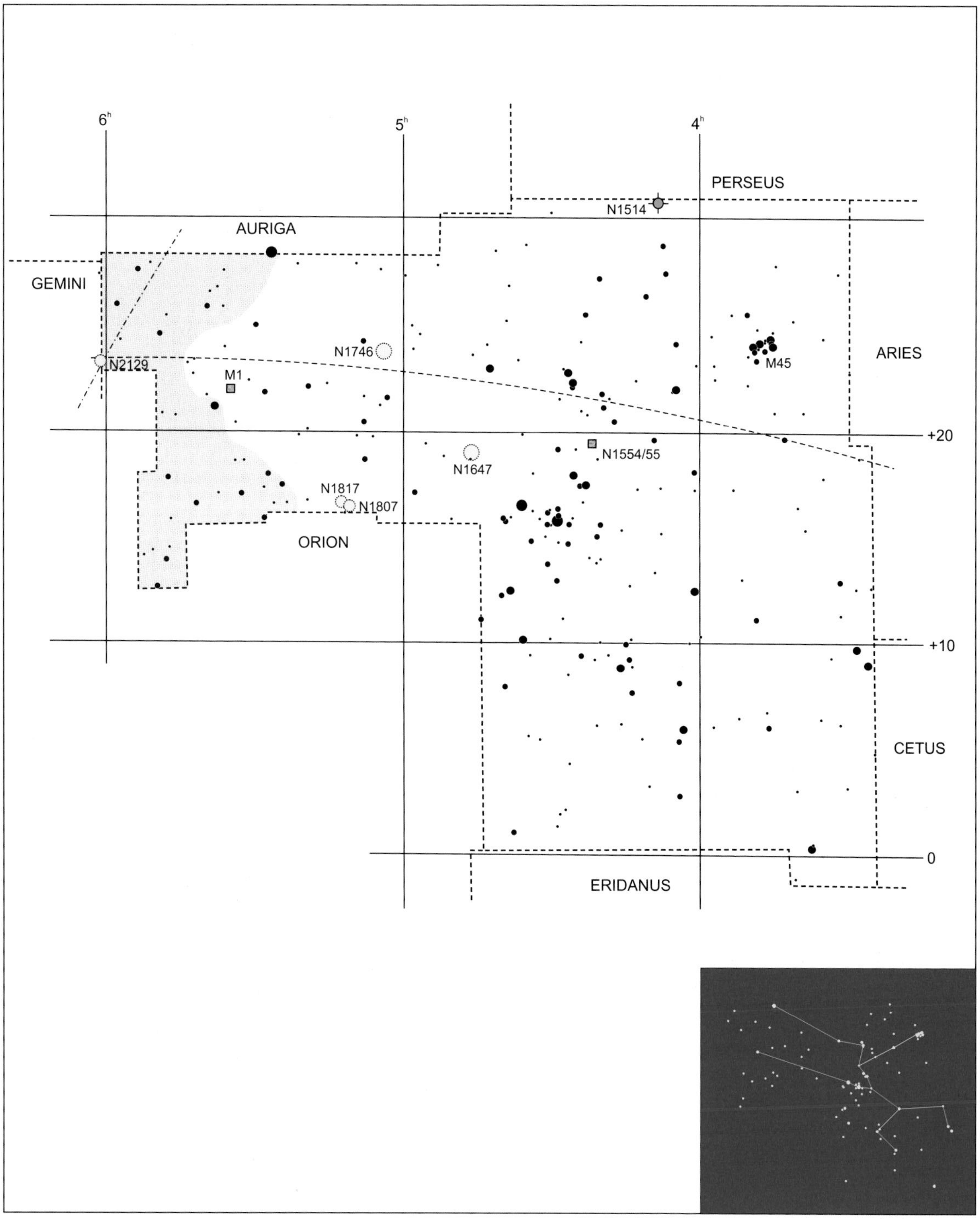

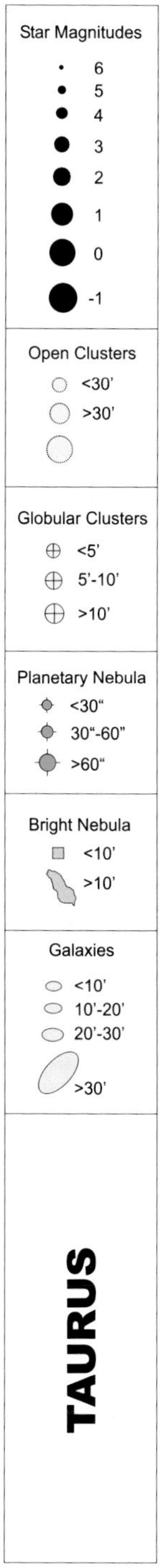

Constellation Facts:

Taurus; (TAW-rus)

Taurus, the Bull.
This constellation rises in the northeast, crosses the meridian high in the southern sky, and sets in the northwest.
The constellation covers 797 square degrees.

Constellation is visible from 88° N to 58° S. Partially visible from 58° S to 90° S.

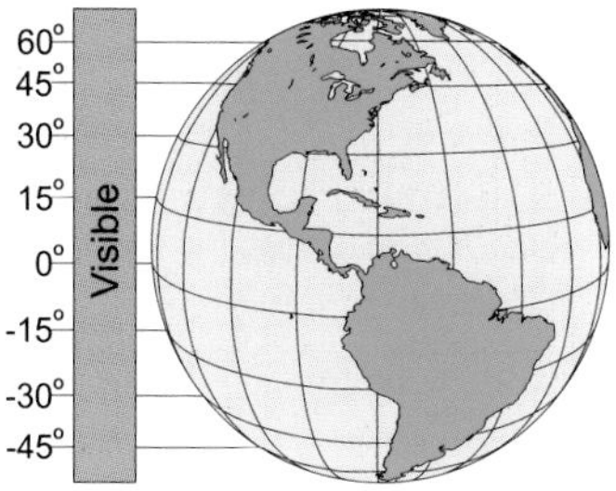

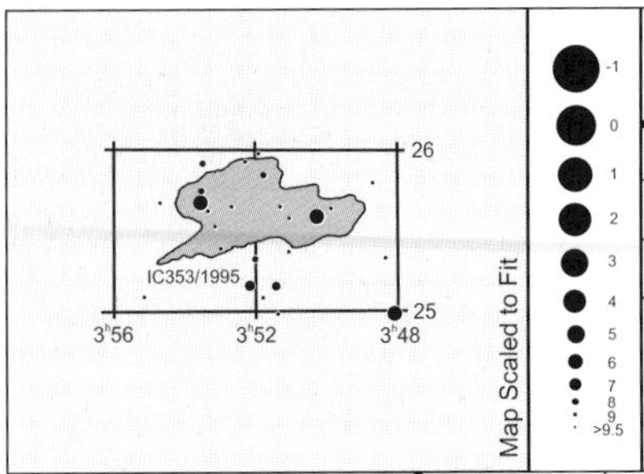

IC 353

RA:	03h 55m 3.5s	Con:	Taurus
Dec:	25° 29' 08"	Type:	Reflection Nebula
Size:	180.0'	Mag:	

IC 353 is a large reflection nebula that is part of a greater nebula complex.

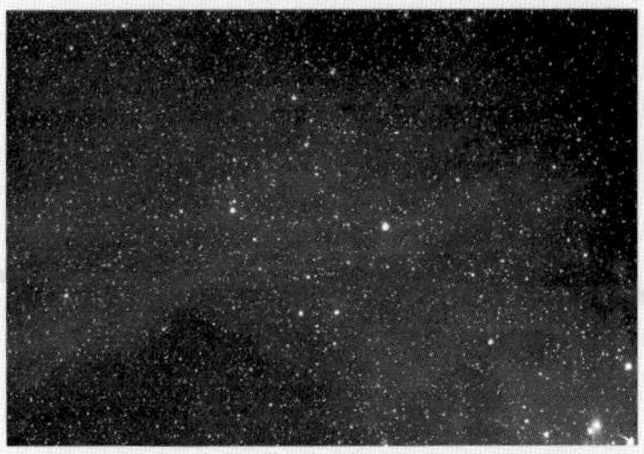

Telescope Aperture:	4" f/5	4" f/9	6" f/7	6" f/9	8" f/6.3	8" f/10	10" f/6.3	10" f/10	12" f/6.3	12" f/10
FOV(35mm film):	2.7° x 4.1°	1.50° x 2.26°	1.29° x 1.93°	1.0° x 1.50°	1.07° x 1.61°	0.68° x 1.02°	0.86° x 1.29°	0.54° x 0.81°	0.72° x 1.07°	0.45° x 0.68°

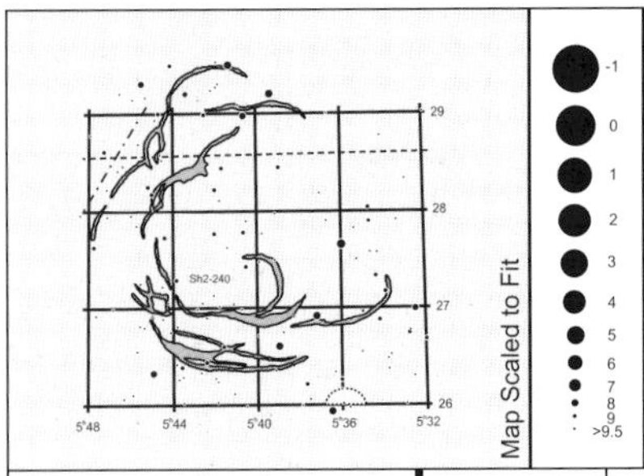

Sh2-240

RA:	05h 39m 6.0s	Con:	Taurus
Dec:	28° 00' 00"	Type:	Emission Nebula
Size:	200' x 180'	Mag:	

Sh2-240 is a super nova remnant that displays filamentary structure.

Telescope Aperture:	4" f/5	4" f/9	6" f/7	6" f/9	8" f/6.3	8" f/10	10" f/6.3	10" f/10	12" f/6.3	12" f/10
FOV(35mm film):	2.7° x 4.1°	1.50° x 2.26°	1.29° x 1.93°	1.0° x 1.50°	1.07° x 1.61°	0.68° x 1.02°	0.86° x 1.29°	0.54° x 0.81°	0.72° x 1.07°	0.45° x 0.68°

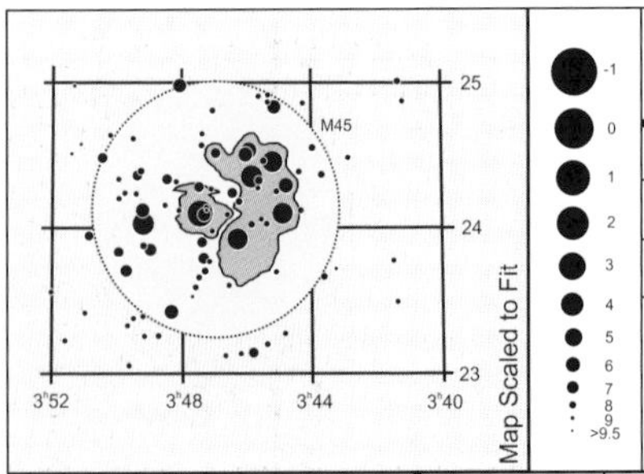

M45 "The Pleiades"

RA:	03h 47m 3.5s	Con:	Taurus
Dec:	24° 07' 09"	Type:	Nebula & Cluster
Size:	110.0'	Mag:	1.2

M45 known as the Pleiades. The brightest stars are called the "Seven Sisters". Some of the stars are surrounded by a blue reflection nebula.

Telescope Aperture:	4" f/5	4" f/9	6" f/7	6" f/9	8" f/6.3	8" f/10	10" f/6.3	10" f/10	12" f/6.3	12" f/10
FOV(35mm film):	2.7° x 4.1°	1.50° x 2.26°	1.29° x 1.93°	1.0° x 1.50°	1.07° x 1.61°	0.68° x 1.02°	0.86° x 1.29°	0.54° x 0.81°	0.72° x 1.07°	0.45° x 0.68°

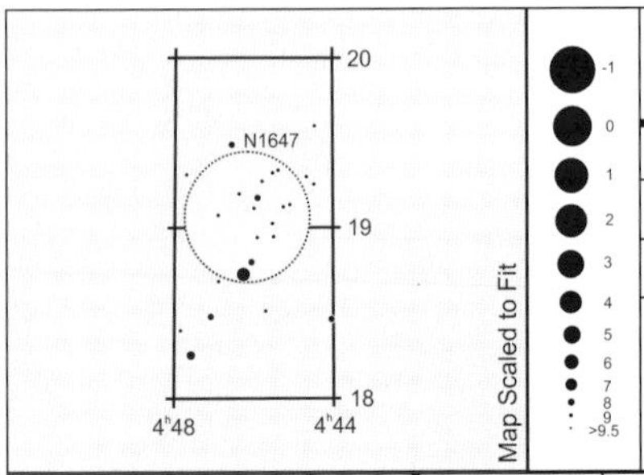

NGC 1647

RA:	04h 46m 3.2s	Con:	Taurus
Dec:	19° 04' 05"	Type:	Open Cluster
Size:	45.0'	Mag:	6.4

NGC 1647 is a bright, scattered open cluster. Object contains about 200 stars.

Telescope Aperture:	4" f/5	4" f/9	6" f/7	6" f/9	8" f/6.3	8" f/10	10" f/6.3	10" f/10	12" f/6.3	12" f/10
FOV(35mm film):	2.7° x 4.1°	1.50° x 2.26°	1.29° x 1.93°	1.0° x 1.50°	1.07° x 1.61°	0.68° x 1.02°	0.86° x 1.29°	0.54° x 0.81°	0.72° x 1.07°	0.45° x 0.68°

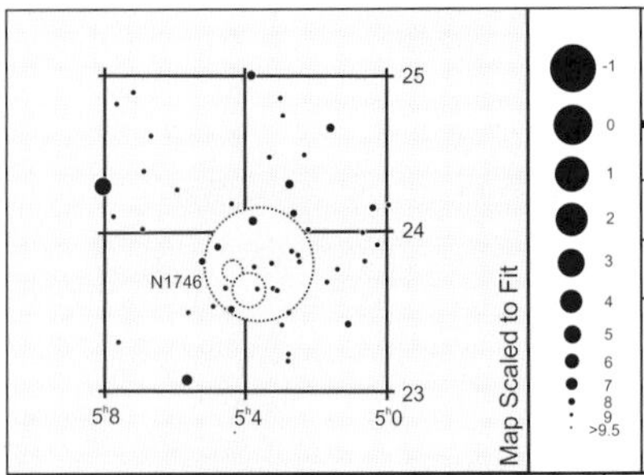

NGC 1746

RA:	05h 03m 39.2s	Con:	Taurus
Dec:	23° 49' 03"	Type:	Open Cluster
Size:	42.0'	Mag:	6.0

NGC 1746 is a bright, scattered open cluster. Object contains about 20 stars.

Telescope Aperture:	4" f/5	4" f/9	6" f/7	6" f/9	8" f/6.3	8" f/10	10" f/6.3	10" f/10	12" f/6.3	12" f/10
FOV(35mm film):	2.7° x 4.1°	1.50° x 2.26°	1.29° x 1.93°	1.0° x 1.50°	1.07° x 1.61°	0.68° x 1.02°	0.86° x 1.29°	0.54° x 0.81°	0.72° x 1.07°	0.45° x 0.68°

NGC 1807/17

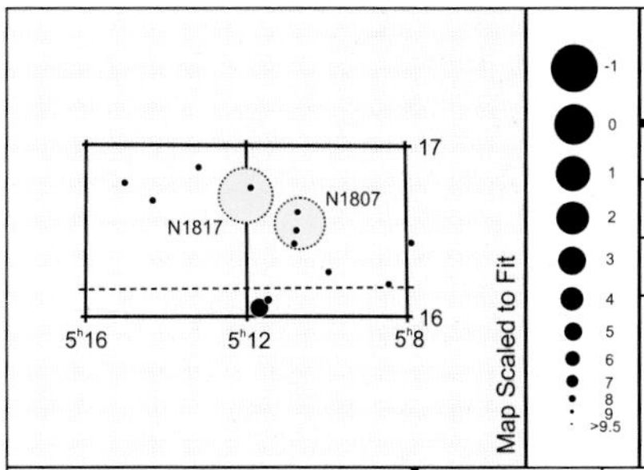

RA:	$05^h\ 10^m\ 45.1^s$	Con:	Taurus
Dec:	16° 32' 04"	Type:	Open Cluster
Size:	17.0'	Mag:	7.0'

NGC 1807/17 are two bright, scattered open clusters, separated by only a few degrees.

Telescope Aperture:	4" f/5	4" f/9	6" f/7	6" f/9	8" f/6.3	8" f/10	10" f/6.3	10" f/10	12" f/6.3	12" f/10
FOV(35mm film):	2.7° x 4.1°	1.50° x 2.26°	1.29° x 1.93°	1.0° x 1.50°	1.07° x 1.61°	0.68° x 1.02°	0.86° x 1.29°	0.54° x 0.81°	0.72° x 1.07°	0.45° x 0.68°

NGC 1514

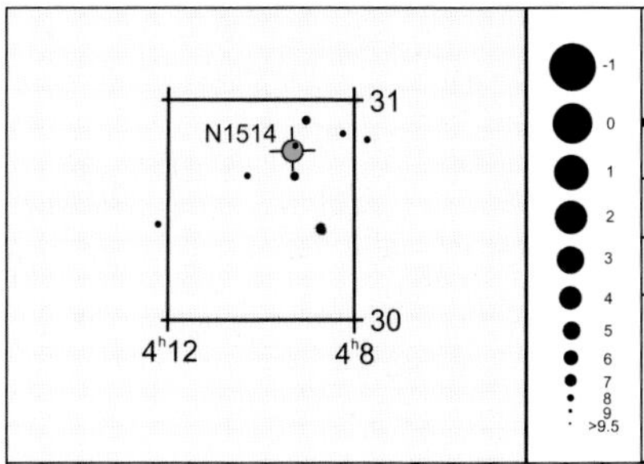

RA:	$04^h\ 09^m\ 15.6^s$	Con:	Taurus
Dec:	30° 47' 05"	Type:	Planetary Nebula
Size:	1.9'	Mag:	10.0'

NGC 1514 is a bright planetary nebula, with a central star.

Telescope Aperture:	4" f/5	4" f/9	6" f/7	6" f/9	8" f/6.3	8" f/10	10" f/6.3	10" f/10	12" f/6.3	12" f/10
FOV(35mm film):	2.7° x 4.1°	1.50° x 2.26°	1.29° x 1.93°	1.0° x 1.50°	1.07° x 1.61°	0.68° x 1.02°	0.86° x 1.29°	0.54° x 0.81°	0.72° x 1.07°	0.45° x 0.68°

M1 (NGC 1952) "Crab Nebula"

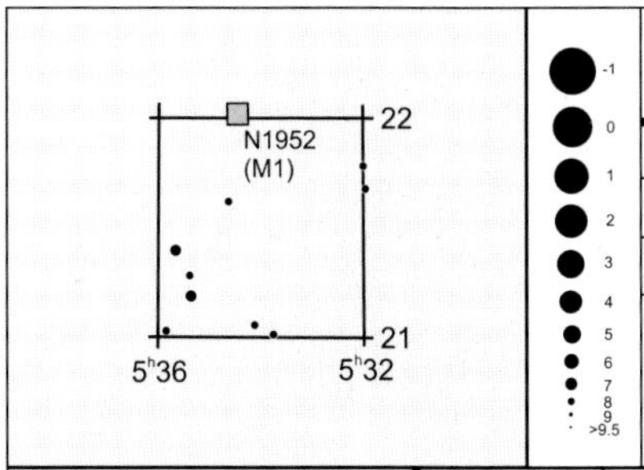

RA:	$05^h\ 34^m\ 33.1^s$	Con:	Taurus
Dec:	22° 01' 01"	Type:	SNR Nebula
Size:	6.0'	Mag:	8.4

M1 (NGC 1952) known as the Crab Nebula which is a super nova remnant.

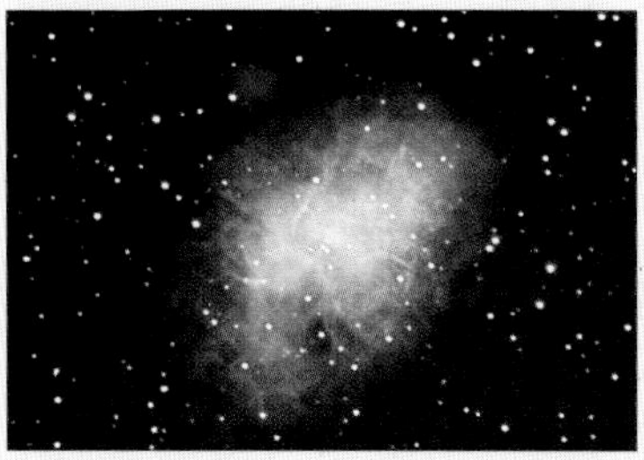

Telescope Aperture:	4" f/5	4" f/9	6" f/7	6" f/9	8" f/6.3	8" f/10	10" f/6.3	10" f/10	12" f/6.3	12" f/10
FOV(35mm film):	2.7° x 4.1°	1.50° x 2.26°	1.29° x 1.93°	1.0° x 1.50°	1.07° x 1.61°	0.68° x 1.02°	0.86° x 1.29°	0.54° x 0.81°	0.72° x 1.07°	0.45° x 0.68°

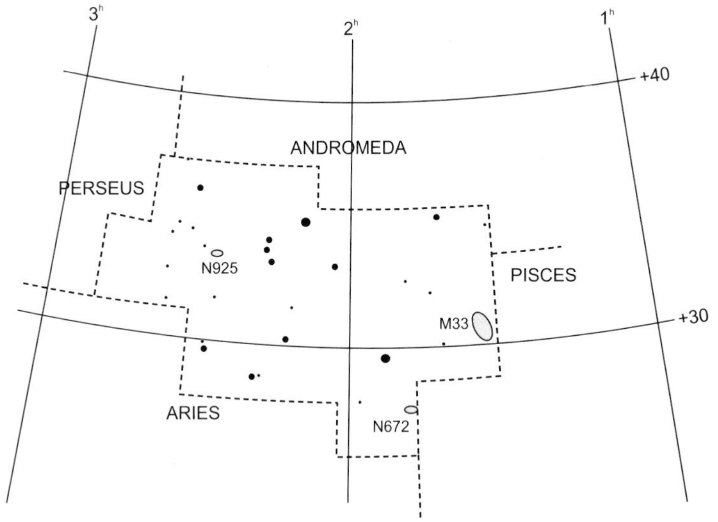

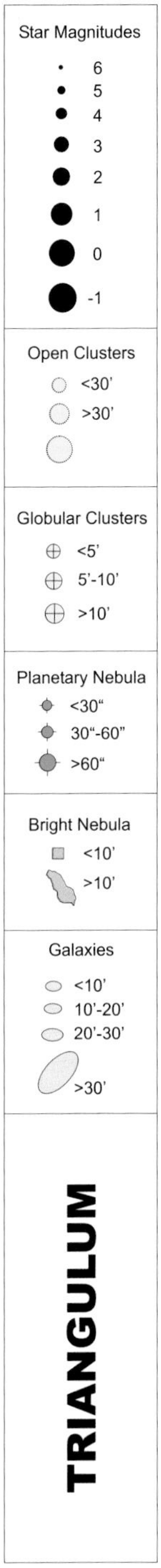

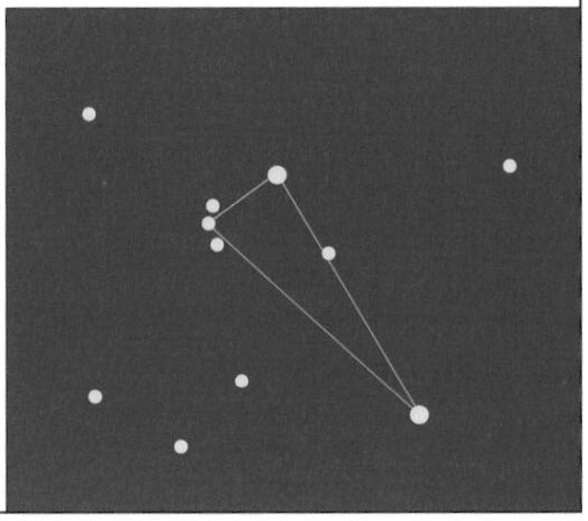

Constellation Facts:

Triangulum; (tri-AN-gue-lum)

Triangulum, the Triangle.
This small constellation is located between Andromeda and Aries.
The constellation covers 132 square degrees.

Constellation is visible from 90° N to 52° S. Partially visible from 52° S to 90° S.

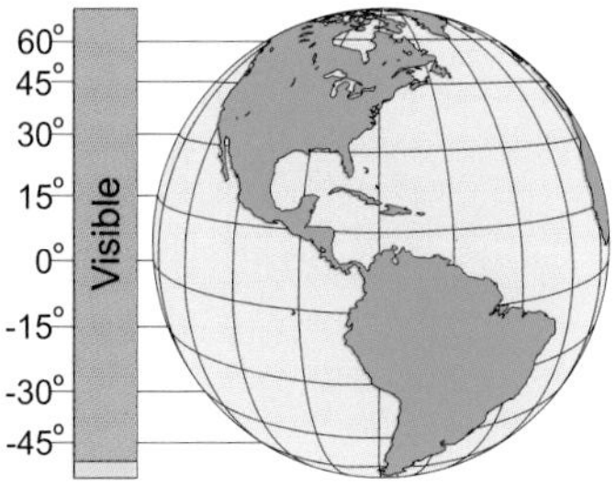

NGC 672

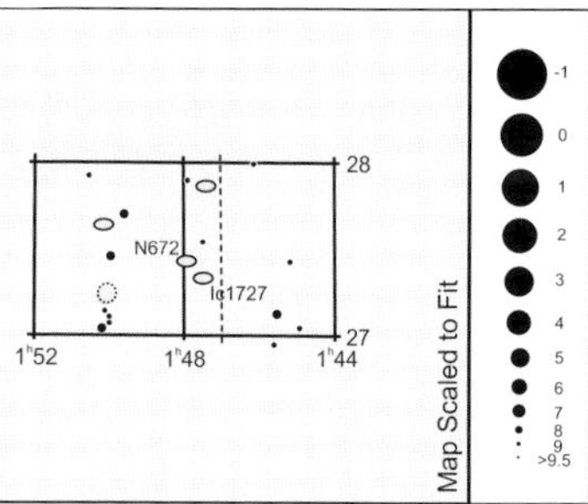

RA:	01^h 47^m 58.0^s	Con:	Triangulum
Dec:	27° 26' 15"	Type:	Barred Spiral
Size:	7.0' x 2.2'	Mag:	10.8

NGC 672 is a bright barred spiral galaxy, which is the brightest galaxy in its local group.

Telescope Aperture:	4" f/5	4" f/9	6" f/7	6" f/9	8" f/6.3	8" f/10	10" f/6.3	10" f/10	12" f/6.3	12" f/10
FOV(35mm film):	2.7° x 4.1°	1.50° x 2.26°	1.29° x 1.93°	1.0° x 1.50°	1.07° x 1.61°	0.68° x 1.02°	0.86° x 1.29°	0.54° x 0.81°	0.72° x 1.07°	0.45° x 0.68°

IC 1727

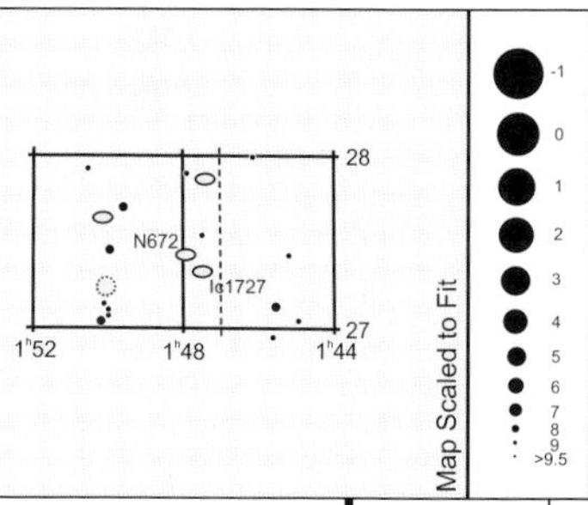

RA:	01^h 47^m 34.0^s	Con:	Triangulum
Dec:	27° 20' 15"	Type:	Galaxy
Size:	6.0' x 3.0'	Mag:	11.0

IC 1727 is a close companion galaxy of NGC 672.

Telescope Aperture:	4" f/5	4" f/9	6" f/7	6" f/9	8" f/6.3	8" f/10	10" f/6.3	10" f/10	12" f/6.3	12" f/10
FOV(35mm film):	2.7° x 4.1°	1.50° x 2.26°	1.29° x 1.93°	1.0° x 1.50°	1.07° x 1.61°	0.68° x 1.02°	0.86° x 1.29°	0.54° x 0.81°	0.72° x 1.07°	0.45° x 0.68°

M33 (NGC 598) "Triangulum Galaxy"

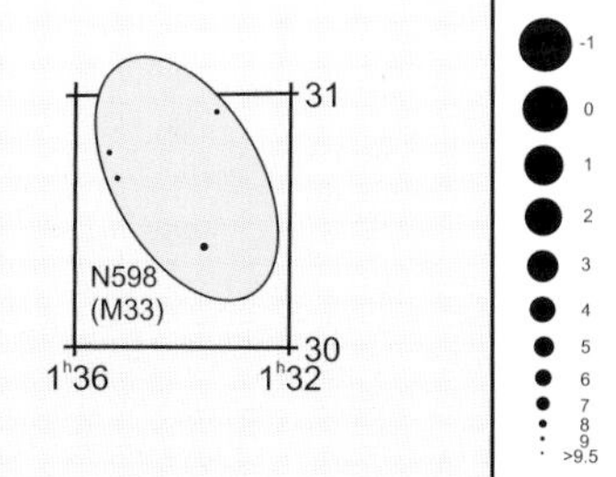

RA:	01^h 33^m 58.1^s	Con:	Triangulum
Dec:	30° 39' 14"	Type:	Spiral Galaxy
Size:	64.0' x 35.0'	Mag:	5.7

M33 (NGC 598) known as the Triangulum Galaxy is a large face-on spiral galaxy.

Telescope Aperture:	4" f/5	4" f/9	6" f/7	6" f/9	8" f/6.3	8" f/10	10" f/6.3	10" f/10	12" f/6.3	12" f/10
FOV(35mm film):	2.7° x 4.1°	1.50° x 2.26°	1.29° x 1.93°	1.0° x 1.50°	1.07° x 1.61°	0.68° x 1.02°	0.86° x 1.29°	0.54° x 0.81°	0.72° x 1.07°	0.45° x 0.68°

NGC 925

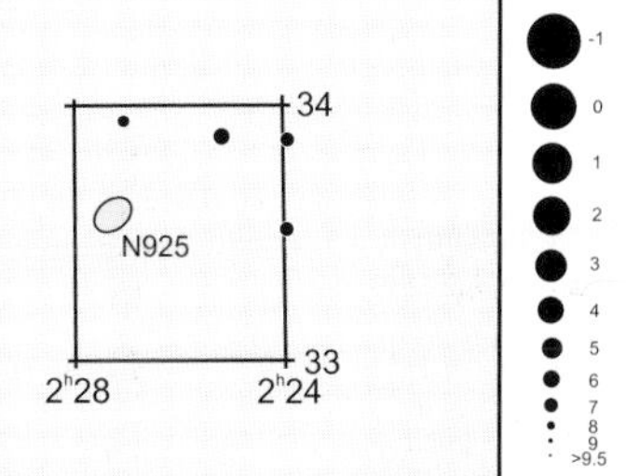

RA:	02^h 27^m 22.0^s	Con:	Triangulum
Dec:	33° 35' 11"	Type:	Spiral Galaxy
Size:	10.0' x 5.0'	Mag:	10.0

NGC 925 is a large Sb-type spiral galaxy. Object is faint and is difficult to observe.

Telescope Aperture:	4" f/5	4" f/9	6" f/7	6" f/9	8" f/6.3	8" f/10	10" f/6.3	10" f/10	12" f/6.3	12" f/10
FOV(35mm film):	2.7° x 4.1°	1.50° x 2.26°	1.29° x 1.93°	1.0° x 1.50°	1.07° x 1.61°	0.68° x 1.02°	0.86° x 1.29°	0.54° x 0.81°	0.72° x 1.07°	0.45° x 0.68°

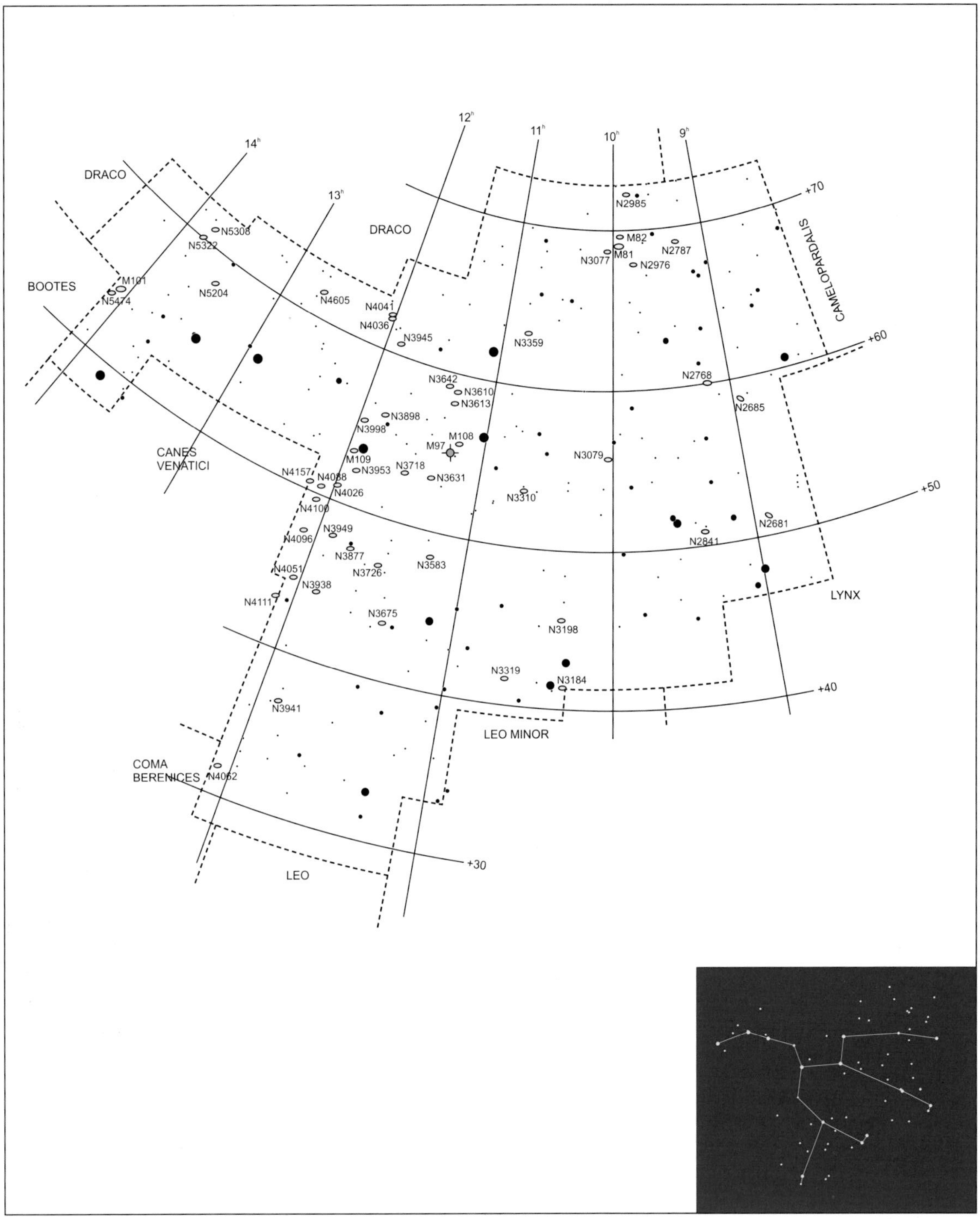

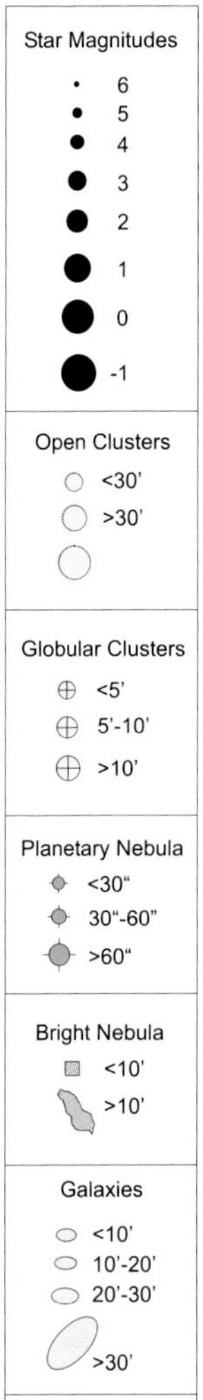

URSA MAJOR

Constellation Facts:

Ursa Major; (URR-sah MAY-jer)

Ursa Major, the Great Bear.
This constellation is a circumpolar constellation. It follows a counterclockwise path around the north celestial pole and does not set beneath the horizon.
The constellation covers 1280 square degrees.

Constellation is visible from 90° N to 16° S. Partially visible from 16° S to 90° S.

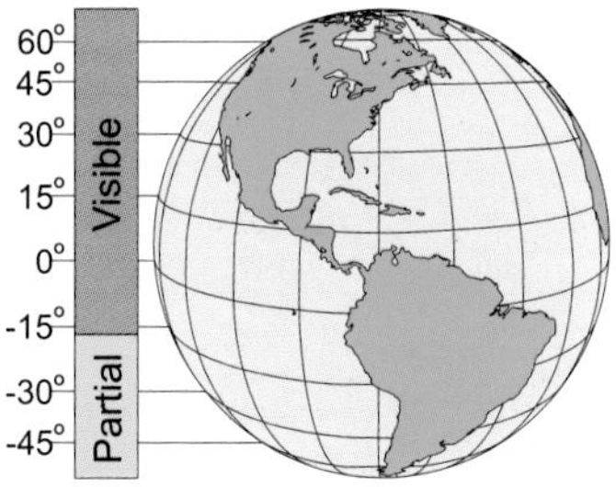

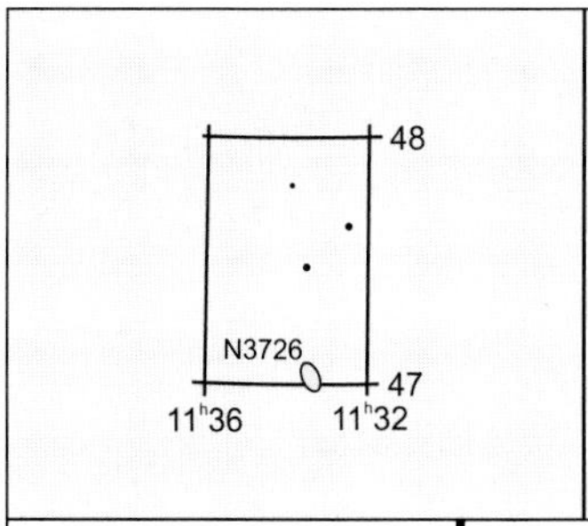

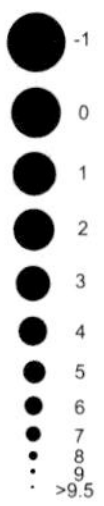

NGC 3726

RA:	$11^h\ 33^m\ 20.9^s$	Con:	Ursa Major
Dec:	47° 01’ 51”	Type:	Spiral Galaxy
Size:	6.0’ x 4.0’	Mag:	10.4

NGC 3726 is a spiral galaxy that when viewed appears an elongated band of light, with a bright core.

Telescope Aperture:	4” f/5	4” f/9	6” f/7	6” f/9	8” f/6.3	8” f/10	10” f/6.3	10” f/10	12” f/6.3	12” f/10
FOV(35mm film):	2.7° x 4.1°	1.50° x 2.26°	1.29° x 1.93°	1.0° x 1.50°	1.07° x 1.61°	0.68° x 1.02°	0.86° x 1.29°	0.54° x 0.81°	0.72° x 1.07°	0.45° x 0.68°

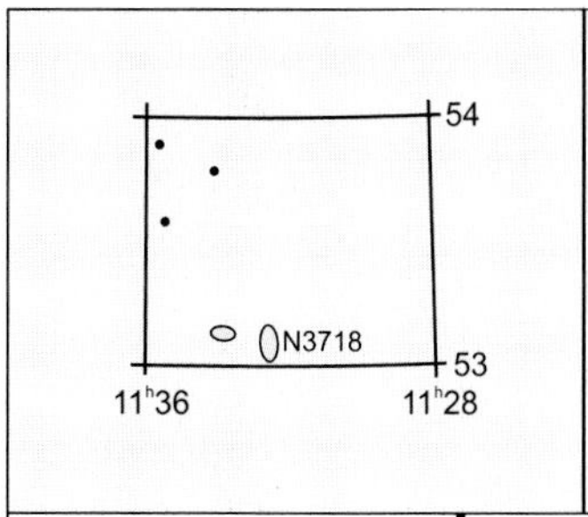

NGC 3718

RA:	$11^h\ 32^m\ 38.8^s$	Con:	Ursa Major
Dec:	53° 03’ 52”	Type:	Spiral Galaxy
Size:	8.0’ x 2.5’	Mag:	10.5

NGC 3718 is a small Sb spiral galaxy that displays a bright nucleus that is partially hidden by a prominent dark lane running laterally.

Telescope Aperture:	4” f/5	4” f/9	6” f/7	6” f/9	8” f/6.3	8” f/10	10” f/6.3	10” f/10	12” f/6.3	12” f/10
FOV(35mm film):	2.7° x 4.1°	1.50° x 2.26°	1.29° x 1.93°	1.0° x 1.50°	1.07° x 1.61°	0.68° x 1.02°	0.86° x 1.29°	0.54° x 0.81°	0.72° x 1.07°	0.45° x 0.68°

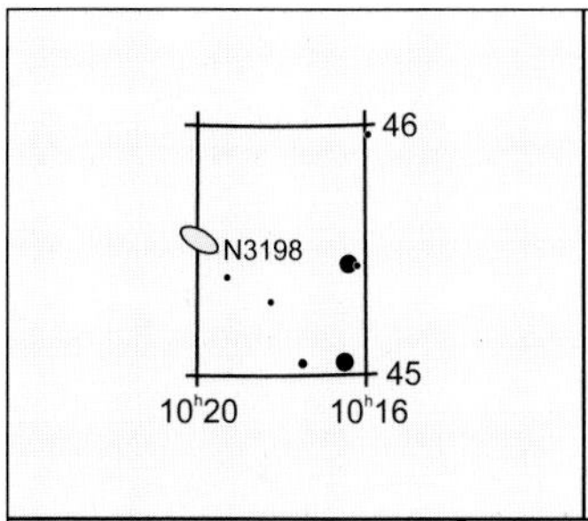

NGC 3198

RA:	$10^h\ 19^m\ 57.0^s$	Con:	Ursa Major
Dec:	45° 32’ 50”	Type:	Galaxy
Size:	6.0’ x 2.0’	Mag:	10.4

NGC 3198 appears as a very elongated galaxy thru the eyepiece.

Telescope Aperture:	4” f/5	4” f/9	6” f/7	6” f/9	8” f/6.3	8” f/10	10” f/6.3	10” f/10	12” f/6.3	12” f/10
FOV(35mm film):	2.7° x 4.1°	1.50° x 2.26°	1.29° x 1.93°	1.0° x 1.50°	1.07° x 1.61°	0.68° x 1.02°	0.86° x 1.29°	0.54° x 0.81°	0.72° x 1.07°	0.45° x 0.68°

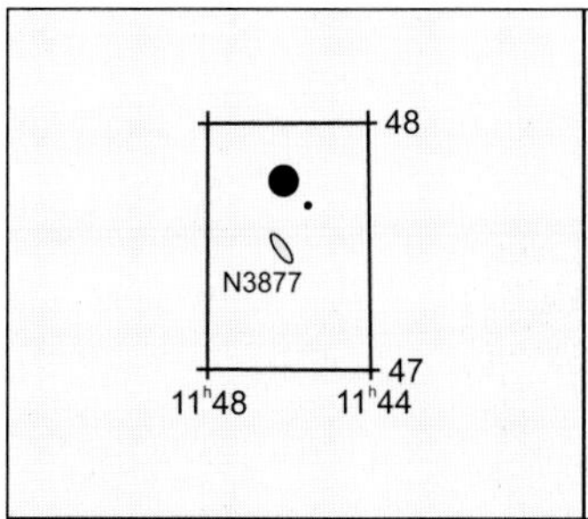

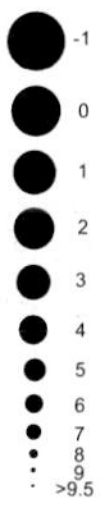

NGC 3877

RA:	$11^h\ 46^m\ 8.9^s$	Con:	Ursa Major
Dec:	47° 29’ 51”	Type:	Spiral Galaxy
Size:	5.0’ x 1.2’	Mag:	12.0

NGC 3877 appears very elongated with a bright core.

Telescope Aperture:	4” f/5	4” f/9	6” f/7	6” f/9	8” f/6.3	8” f/10	10” f/6.3	10” f/10	12” f/6.3	12” f/10
FOV(35mm film):	2.7° x 4.1°	1.50° x 2.26°	1.29° x 1.93°	1.0° x 1.50°	1.07° x 1.61°	0.68° x 1.02°	0.86° x 1.29°	0.54° x 0.81°	0.72° x 1.07°	0.45° x 0.68°

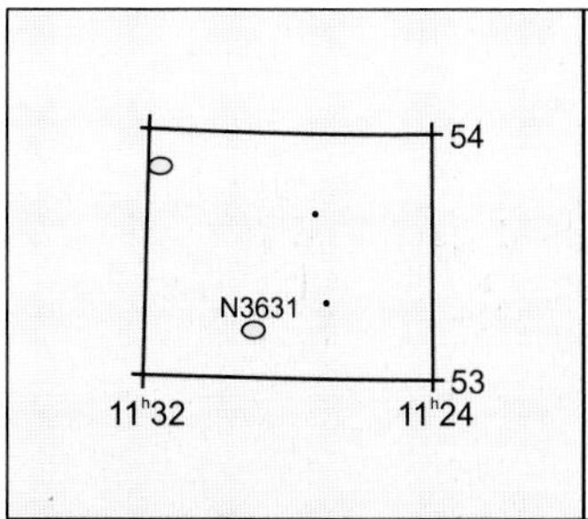

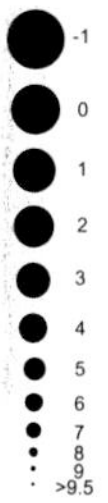

NGC 3631

RA:	$11^h\ 21^m\ 2.9^s$	Con:	Ursa Major
Dec:	53° 59’ 52”	Type:	Galaxy
Size:	4.5’ x 0.3’	Mag:	10.4

NGC 3631 appears as a round galaxy with a bright core.

Telescope Aperture:	4” f/5	4” f/9	6” f/7	6” f/9	8” f/6.3	8” f/10	10” f/6.3	10” f/10	12” f/6.3	12” f/10
FOV(35mm film):	2.7° x 4.1°	1.50° x 2.26°	1.29° x 1.93°	1.0° x 1.50°	1.07° x 1.61°	0.68° x 1.02°	0.86° x 1.29°	0.54° x 0.81°	0.72° x 1.07°	0.45° x 0.68°

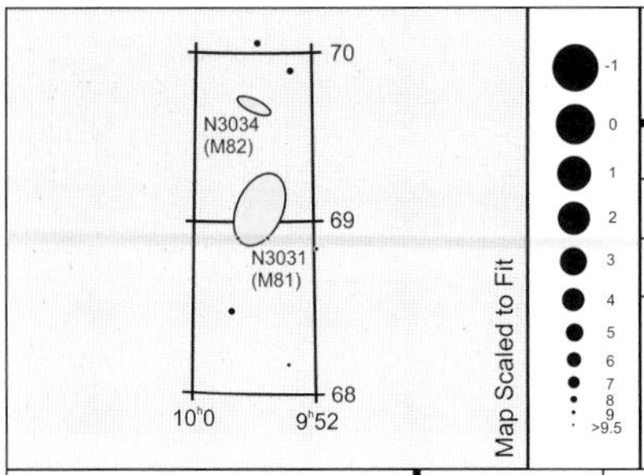

M81 (NGC 3031)

RA:	$09^h\ 20^m\ 21.0^s$	Con:	Ursa Major
Dec:	69° 03' 51"	Type:	Spiral Galaxy
Size:	20.0' x 9.0'	Mag:	6.9

M81 (NGC 3031) is a large Sb-type spiral galaxy. M81 is the largest galaxy in a group of galaxies.

Telescope Aperture:	4" f/5	4" f/9	6" f/7	6" f/9	8" f/6.3	8" f/10	10" f/6.3	10" f/10	12" f/6.3	12" f/10
FOV(35mm film):	2.7° x 4.1°	1.50° x 2.26°	1.29° x 1.93°	1.0° x 1.50°	1.07° x 1.61°	0.68° x 1.02°	0.86° x 1.29°	0.54° x 0.81°	0.72° x 1.07°	0.45° x 0.68°

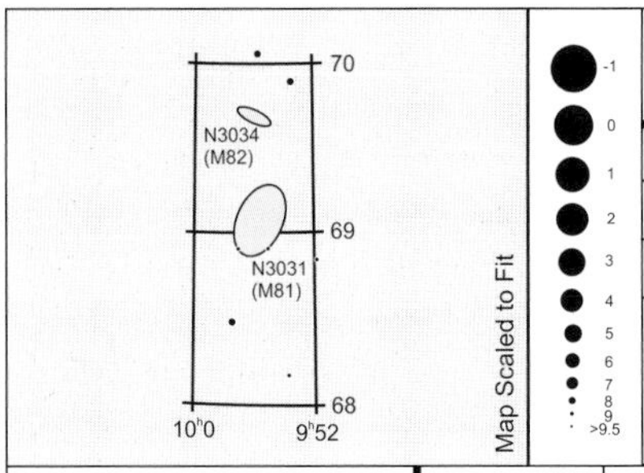

M82 (NGC 3034)

RA:	$09^h\ 55^m\ 50.8^s$	Con:	Ursa Major
Dec:	69° 40' 51"	Type:	Irregular Galaxy
Size:	8.4' x 3.4'	Mag:	8.4

M82 (NGC 3034) is a large edge-on irregular galaxy. Object displays visible dust lanes and bright knots.

Telescope Aperture:	4" f/5	4" f/9	6" f/7	6" f/9	8" f/6.3	8" f/10	10" f/6.3	10" f/10	12" f/6.3	12" f/10
FOV(35mm film):	2.7° x 4.1°	1.50° x 2.26°	1.29° x 1.93°	1.0° x 1.50°	1.07° x 1.61°	0.68° x 1.02°	0.86° x 1.29°	0.54° x 0.81°	0.72° x 1.07°	0.45° x 0.68°

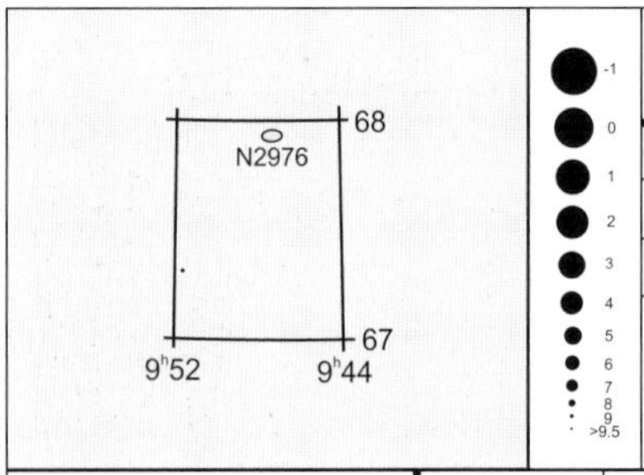

NGC 2976

RA:	$09^h\ 47^m\ 20.9^s$	Con:	Ursa Major
Dec:	67° 54' 51"	Type:	Spiral Galaxy
Size:	5.1' x 2.2'	Mag:	10.2

NGC 2976 is a small, bright Sc-type spiral galaxy, with an irregular spiral arm pattern. Object is part of the M81 galaxy group.

Telescope Aperture:	4" f/5	4" f/9	6" f/7	6" f/9	8" f/6.3	8" f/10	10" f/6.3	10" f/10	12" f/6.3	12" f/10
FOV(35mm film):	2.7° x 4.1°	1.50° x 2.26°	1.29° x 1.93°	1.0° x 1.50°	1.07° x 1.61°	0.68° x 1.02°	0.86° x 1.29°	0.54° x 0.81°	0.72° x 1.07°	0.45° x 0.68°

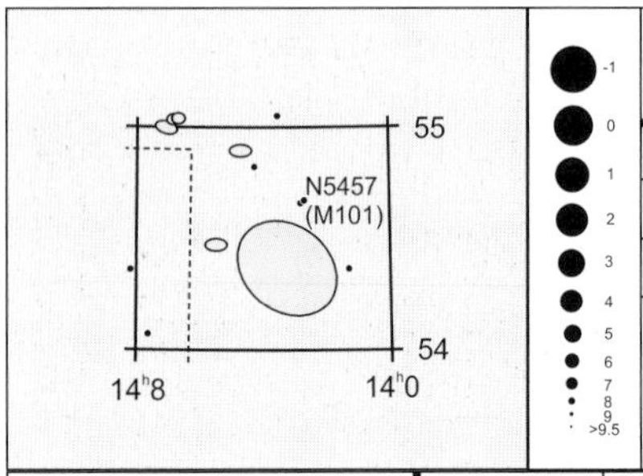

M101 (NGC 5457)

RA:	$14^h\ 03^m\ 14.8^s$	Con:	Ursa Major
Dec:	54° 20' 58"	Type:	Spiral Galaxy
Size:	22.0' x 20.0'	Mag:	7.7

M101 (NGC 5457) is a large Sc-type spiral galaxy that is oriented face-on to our line of sight.

Telescope Aperture:	4" f/5	4" f/9	6" f/7	6" f/9	8" f/6.3	8" f/10	10" f/6.3	10" f/10	12" f/6.3	12" f/10
FOV(35mm film):	2.7° x 4.1°	1.50° x 2.26°	1.29° x 1.93°	1.0° x 1.50°	1.07° x 1.61°	0.68° x 1.02°	0.86° x 1.29°	0.54° x 0.81°	0.72° x 1.07°	0.45° x 0.68°

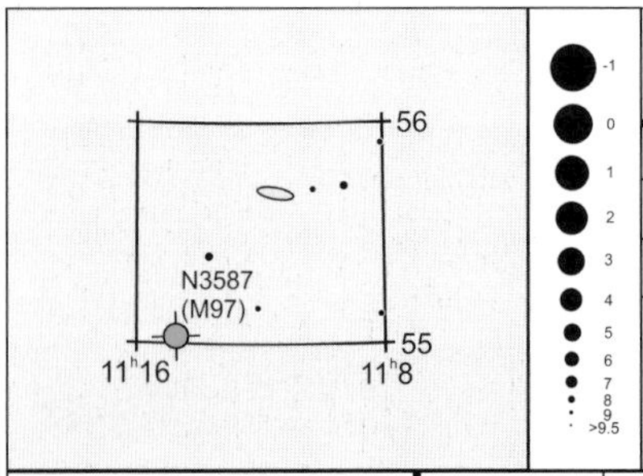

M97 (NGC 3587) "Owl Nebula"

RA:	$11^h\ 14^m\ 50.8^s$	Con:	Ursa Major
Dec:	55° 00' 52"	Type:	Planetary Nebula
Size:	3.2'	Mag:	11.2

M97 (NGC 3587) known as the Owl Nebula is an irregular planetary nebula, easily observed thru medium to large telescopes.

Telescope Aperture:	4" f/5	4" f/9	6" f/7	6" f/9	8" f/6.3	8" f/10	10" f/6.3	10" f/10	12" f/6.3	12" f/10
FOV(35mm film):	2.7° x 4.1°	1.50° x 2.26°	1.29° x 1.93°	1.0° x 1.50°	1.07° x 1.61°	0.68° x 1.02°	0.86° x 1.29°	0.54° x 0.81°	0.72° x 1.07°	0.45° x 0.68°

M108 (NGC 3556)

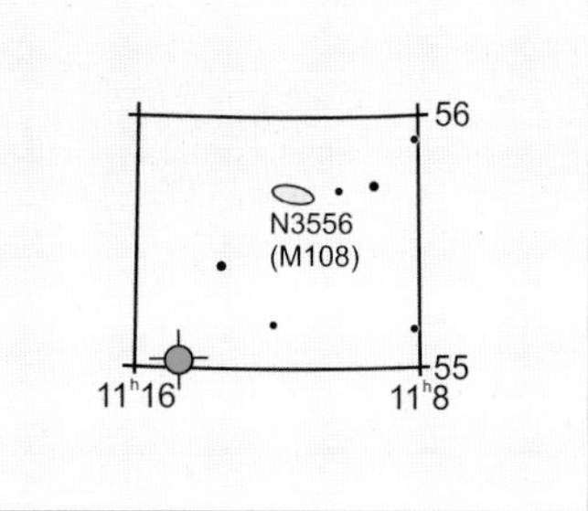

RA:	$11^h\ 11^m\ 32.8^s$	Con:	Ursa Major
Dec:	55° 39′ 52″	Type:	Spiral Galaxy
Size:	8.0′ x 2.0′	Mag:	10.1

M108 (NGC 3556) is found 48′ northwest of the Owl Nebula. Object is an edge-on spiral galaxy with visible dust lanes and bright knots.

Telescope Aperture:	4″ f/5	4″ f/9	6″ f/7	6″ f/9	8″ f/6.3	8″ f/10	10″ f/6.3	10″ f/10	12″ f/6.3	12″ f/10
FOV(35mm film):	2.7° x 4.1°	1.50° x 2.26°	1.29° x 1.93°	1.0° x 1.50°	1.07° x 1.61°	0.68° x 1.02°	0.86° x 1.29°	0.54° x 0.81°	0.72° x 1.07°	0.45° x 0.68°

M109 (NGC 3992)

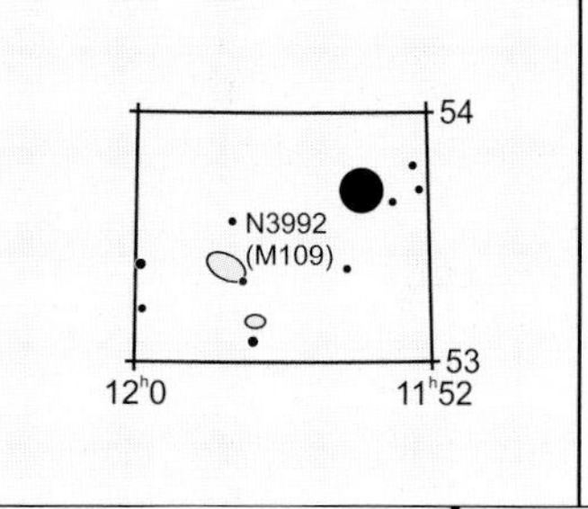

RA:	$11^h\ 57^m\ 38.8^s$	Con:	Ursa Major
Dec:	53° 22′ 53″	Type:	Spiral Galaxy
Size:	7.0′ x 4.0′	Mag:	9.8

M109 (NGC 3992) appears as an elongated galaxy with a bright core.

Telescope Aperture:	4″ f/5	4″ f/9	6″ f/7	6″ f/9	8″ f/6.3	8″ f/10	10″ f/6.3	10″ f/10	12″ f/6.3	12″ f/10
FOV(35mm film):	2.7° x 4.1°	1.50° x 2.26°	1.29° x 1.93°	1.0° x 1.50°	1.07° x 1.61°	0.68° x 1.02°	0.86° x 1.29°	0.54° x 0.81°	0.72° x 1.07°	0.45° x 0.68°

NGC 3953

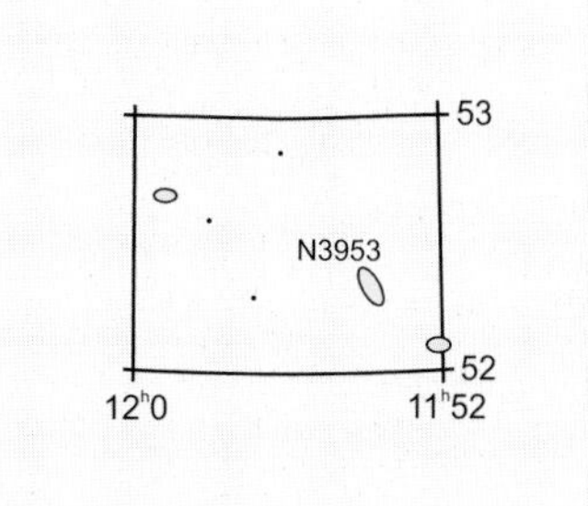

RA:	$11^h\ 53^m\ 50.8^s$	Con:	Ursa Major
Dec:	52° 19′ 52″	Type:	Spiral Galaxy
Size:	7.0′ x 2.8′	Mag:	10.1

NGC 3953 is a spiral galaxy located southeast of M109.

Telescope Aperture:	4″ f/5	4″ f/9	6″ f/7	6″ f/9	8″ f/6.3	8″ f/10	10″ f/6.3	10″ f/10	12″ f/6.3	12″ f/10
FOV(35mm film):	2.7° x 4.1°	1.50° x 2.26°	1.29° x 1.93°	1.0° x 1.50°	1.07° x 1.61°	0.68° x 1.02°	0.86° x 1.29°	0.54° x 0.81°	0.72° x 1.07°	0.45° x 0.68°

NGC 4051

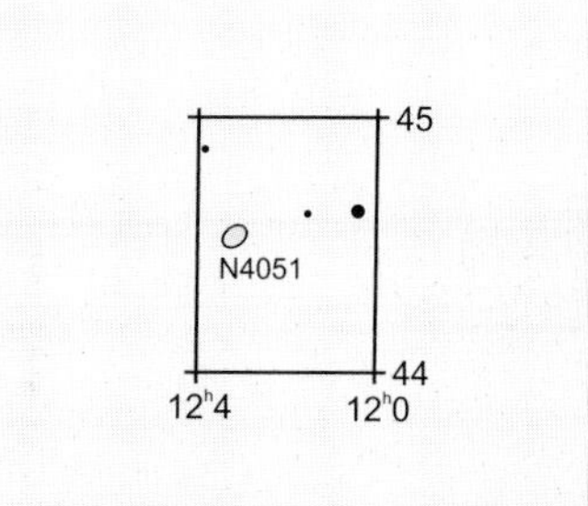

RA:	$12^h\ 03^m\ 14.9^s$	Con:	Ursa Major
Dec:	44° 31′ 51″	Type:	Spiral Galaxy
Size:	4.2′ x 4.0′	Mag:	10.3

NGC 4051 is an Sb-type spiral galaxy with large spiral arms.

Telescope Aperture:	4″ f/5	4″ f/9	6″ f/7	6″ f/9	8″ f/6.3	8″ f/10	10″ f/6.3	10″ f/10	12″ f/6.3	12″ f/10
FOV(35mm film):	2.7° x 4.1°	1.50° x 2.26°	1.29° x 1.93°	1.0° x 1.50°	1.07° x 1.61°	0.68° x 1.02°	0.86° x 1.29°	0.54° x 0.81°	0.72° x 1.07°	0.45° x 0.68°

NGC 2841

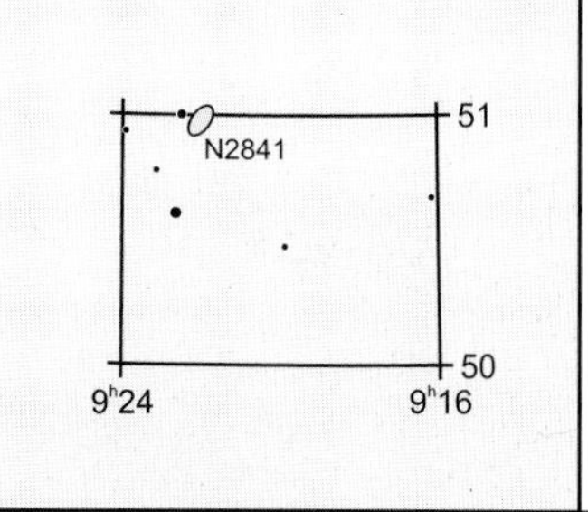

RA:	$09^h\ 22^m\ 3.1^s$	Con:	Ursa Major
Dec:	50° 57′ 50″	Type:	Spiral Galaxy
Size:	7.0′ x 3.0′	Mag:	9.3

NGC 2841 is an Sb-type spiral galaxy which is large, bright and appears elongated with visible dust lanes in the eyepiece.

Telescope Aperture:	4″ f/5	4″ f/9	6″ f/7	6″ f/9	8″ f/6.3	8″ f/10	10″ f/6.3	10″ f/10	12″ f/6.3	12″ f/10
FOV(35mm film):	2.7° x 4.1°	1.50° x 2.26°	1.29° x 1.93°	1.0° x 1.50°	1.07° x 1.61°	0.68° x 1.02°	0.86° x 1.29°	0.54° x 0.81°	0.72° x 1.07°	0.45° x 0.68°

Vir | **Crosses Prime Meridian:** April thru June | **SPRING**

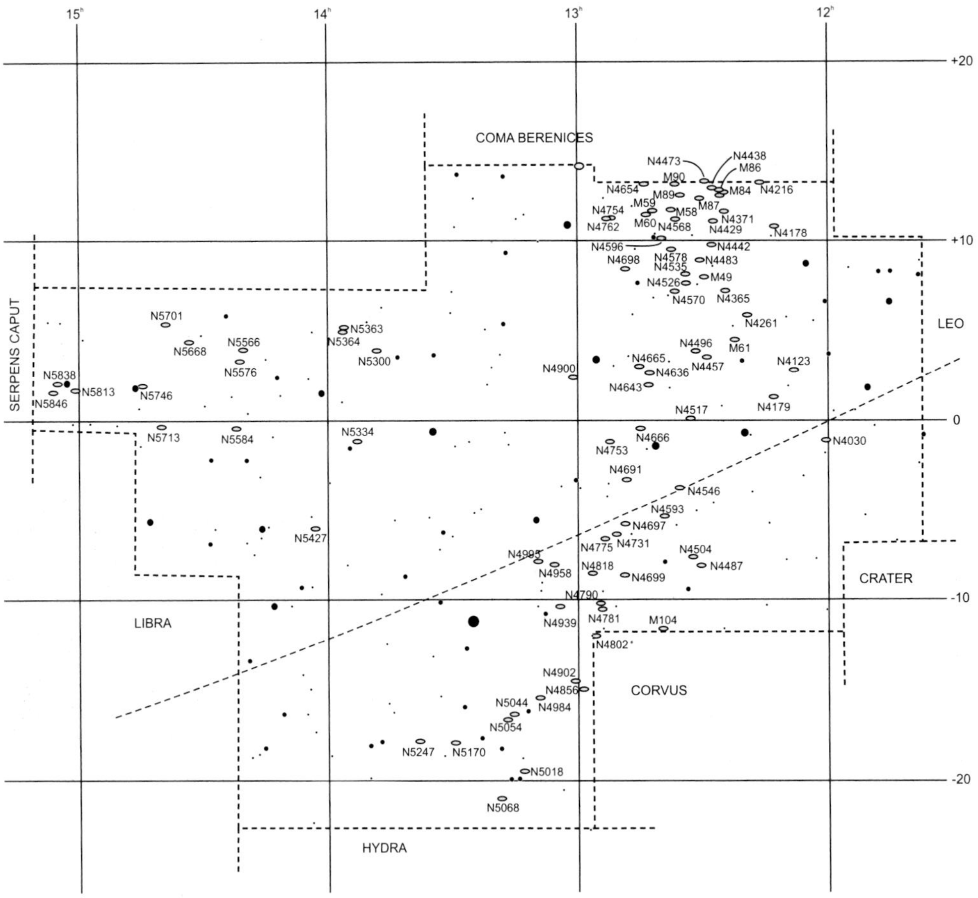

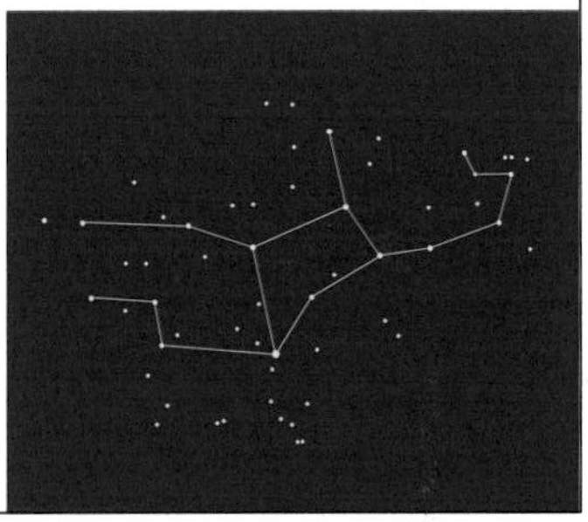

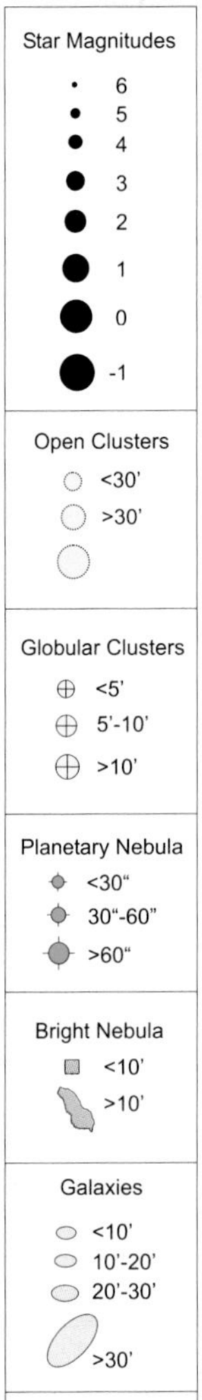

VIRGO

Constellation Facts:

Virgo; (VER-go)

Virgo, the Virgin.
The stars of Virgo lie along the celestial equator.
The constellation rises directly to the east, crosses the meridian halfway between the horizon and the zenith, and set near the western point of the horizon.
The constellation covers 1294 square degrees.

Constellation is visible from 67° N to 75° S. Partially visible from 67° N to 90° N.

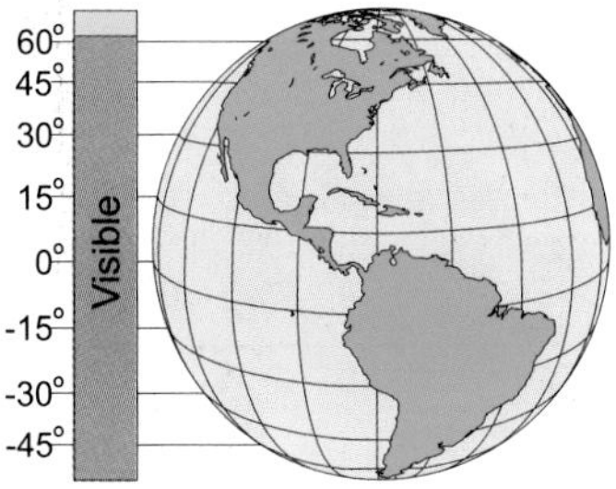

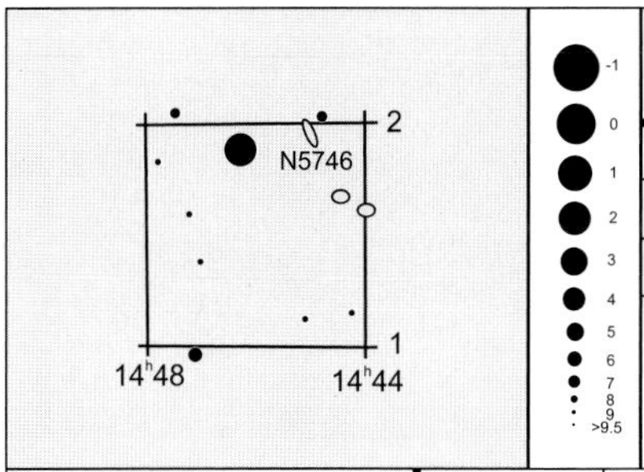

NGC 5746

RA:	14^h 44^m 58.1^s	Con:	Virgo
Dec:	01° 56' 46"	Type:	Spiral Galaxy
Size:	7.0' x 1.2'	Mag:	10.6

NGC 5746 appears edge-on to our line of site. Object displays visible dust lanes.

Telescope Aperture:	4" f/5	4" f/9	6" f/7	6" f/9	8" f/6.3	8" f/10	10" f/6.3	10" f/10	12" f/6.3	12" f/10
FOV(35mm film):	2.7° x 4.1°	1.50° x 2.26°	1.29° x 1.93°	1.0° x 1.50°	1.07° x 1.61°	0.68° x 1.02°	0.86° x 1.29°	0.54° x 0.81°	0.72° x 1.07°	0.45° x 0.68°

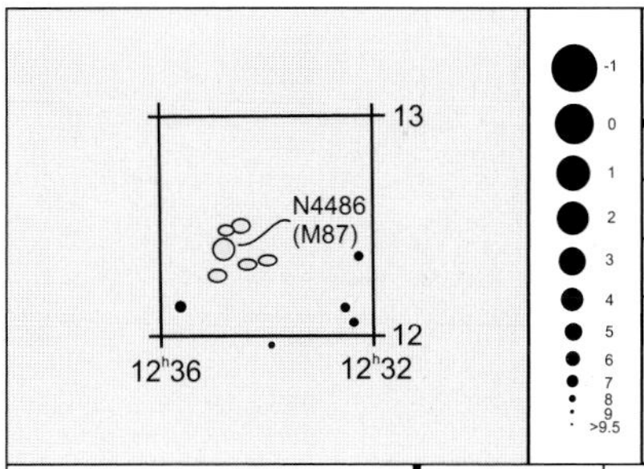

M87 (NGC 4486)

RA:	12^h 30^m 53.0^s	Con:	Virgo
Dec:	12° 23' 07"	Type:	Spiral Galaxy
Size:	8.3' x 6.6'	Mag:	9.6

M87 (NGC 4486) appears as a round galaxy with a bright central core.

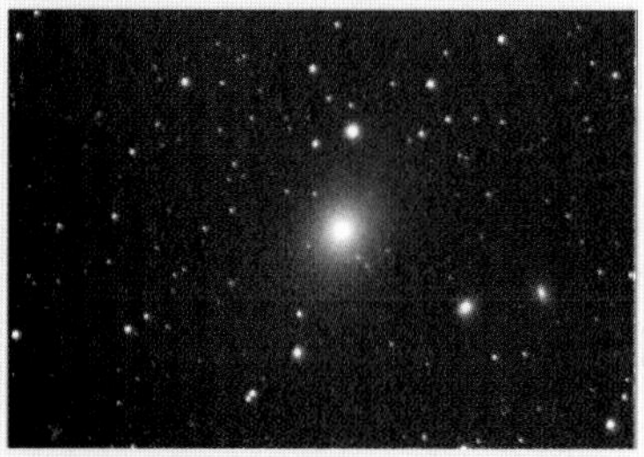

Telescope Aperture:	4" f/5	4" f/9	6" f/7	6" f/9	8" f/6.3	8" f/10	10" f/6.3	10" f/10	12" f/6.3	12" f/10
FOV(35mm film):	2.7° x 4.1°	1.50° x 2.26°	1.29° x 1.93°	1.0° x 1.50°	1.07° x 1.61°	0.68° x 1.02°	0.86° x 1.29°	0.54° x 0.81°	0.72° x 1.07°	0.45° x 0.68°

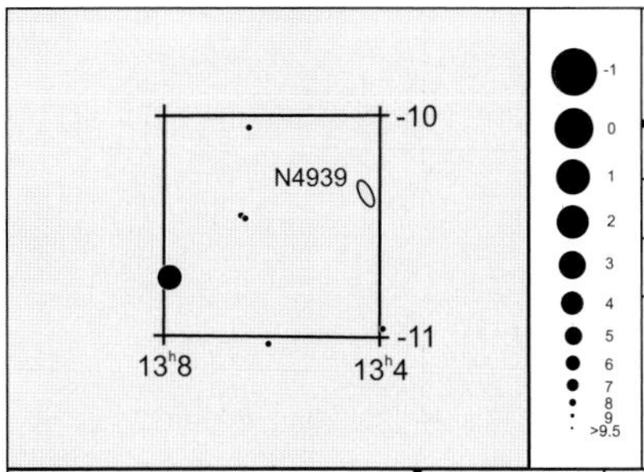

NGC 4939

RA:	13^h 04^m 15.7^s	Con:	Virgo
Dec:	-10° 20' 24"	Type:	Spiral Galaxy
Size:	5.3' x 2.0'	Mag:	11.0

NGC 4939 is a large galaxy who's true beauty is hidden by low surface brightness.

Telescope Aperture:	4" f/5	4" f/9	6" f/7	6" f/9	8" f/6.3	8" f/10	10" f/6.3	10" f/10	12" f/6.3	12" f/10
FOV(35mm film):	2.7° x 4.1°	1.50° x 2.26°	1.29° x 1.93°	1.0° x 1.50°	1.07° x 1.61°	0.68° x 1.02°	0.86° x 1.29°	0.54° x 0.81°	0.72° x 1.07°	0.45° x 0.68°

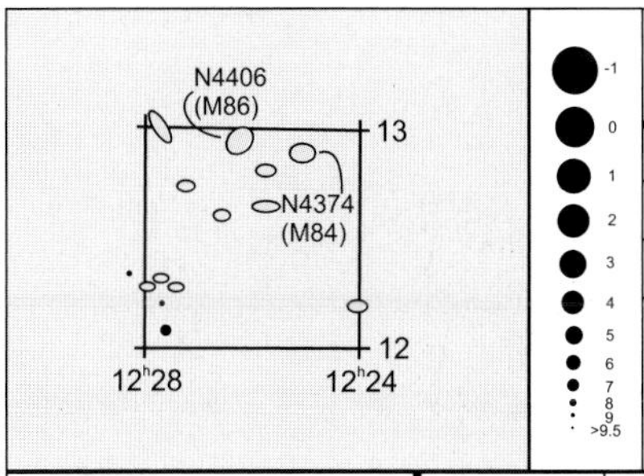

M84 (NGC 4374)

RA:	12^h 25^m 9.3^s	Con:	Virgo
Dec:	12° 52' 43"	Type:	Elliptical Galaxy
Size:	1.3' x 1.2'	Mag:	9.3

M84 (NGC 4374) is a bright elliptical galaxy. Object appears round with a bright core thru the eyepiece. Object is located in the northern reaches of the constellation.

Telescope Aperture:	4" f/5	4" f/9	6" f/7	6" f/9	8" f/6.3	8" f/10	10" f/6.3	10" f/10	12" f/6.3	12" f/10
FOV(35mm film):	2.7° x 4.1°	1.50° x 2.26°	1.29° x 1.93°	1.0° x 1.50°	1.07° x 1.61°	0.68° x 1.02°	0.86° x 1.29°	0.54° x 0.81°	0.72° x 1.07°	0.45° x 0.68°

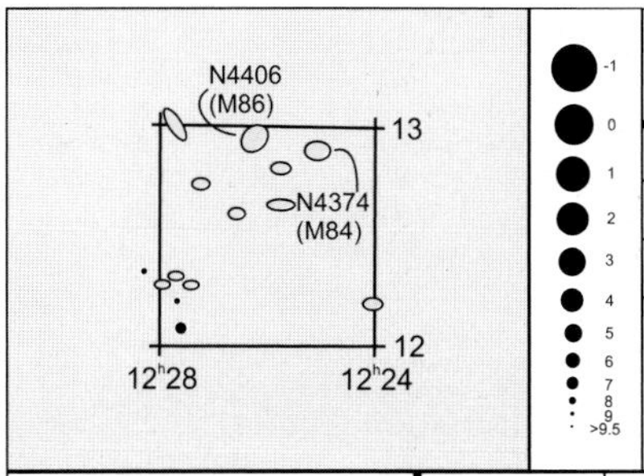

M86 (NGC 4406)

RA:	12^h 26^m 15.3^s	Con:	Virgo
Dec:	12° 56' 43"	Type:	Elliptical Galaxy
Size:	1.5' x 1.2'	Mag:	9.2

M86 (NGC 4406) is another elliptical galaxy appearing round with a bright core thru the eyepiece

Telescope Aperture:	4" f/5	4" f/9	6" f/7	6" f/9	8" f/6.3	8" f/10	10" f/6.3	10" f/10	12" f/6.3	12" f/10
FOV(35mm film):	2.7° x 4.1°	1.50° x 2.26°	1.29° x 1.93°	1.0° x 1.50°	1.07° x 1.61°	0.68° x 1.02°	0.86° x 1.29°	0.54° x 0.81°	0.72° x 1.07°	0.45° x 0.68°

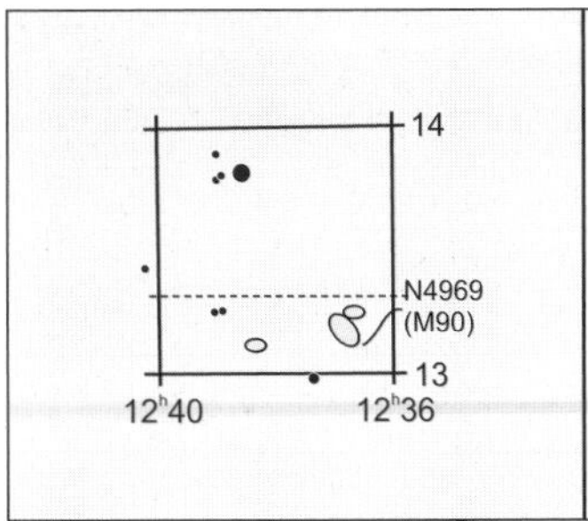

M90 (NGC 4569)

RA:	$12^h\ 36^m\ 51.4^s$	Con:	Virgo
Dec:	13° 09' 44"	Type:	Spiral Galaxy
Size:	11.0' x 3.0'	Mag:	9.5

M90 (NGC 4569) is a bright spiral galaxy that appears elongated with a bright core thru the eyepiece.

Telescope Aperture:	4" f/5	4" f/9	6" f/7	6" f/9	8" f/6.3	8" f/10	10" f/6.3	10" f/10	12" f/6.3	12" f/10
FOV(35mm film):	2.7° x 4.1°	1.50° x 2.26°	1.29° x 1.93°	1.0° x 1.50°	1.07° x 1.61°	0.68° x 1.02°	0.86° x 1.29°	0.54° x 0.81°	0.72° x 1.07°	0.45° x 0.68°

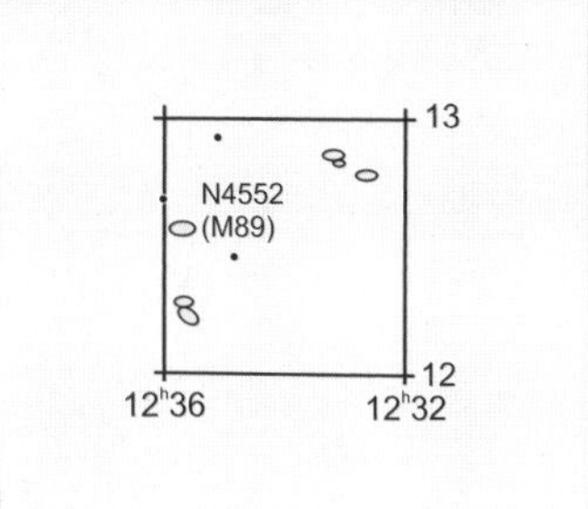

M89 (NGC 4552)

RA:	$12^h\ 35^m\ 45.4^s$	Con:	Virgo
Dec:	12° 32' 44"	Type:	Elliptical Galaxy
Size:	4.2' x 4.2'	Mag:	9.8

M89 (NGC 4552) is located 45' southwest of M90. Object is an elliptical galaxy that appears round with a bright core.

Telescope Aperture:	4" f/5	4" f/9	6" f/7	6" f/9	8" f/6.3	8" f/10	10" f/6.3	10" f/10	12" f/6.3	12" f/10
FOV(35mm film):	2.7° x 4.1°	1.50° x 2.26°	1.29° x 1.93°	1.0° x 1.50°	1.07° x 1.61°	0.68° x 1.02°	0.86° x 1.29°	0.54° x 0.81°	0.72° x 1.07°	0.45° x 0.68°

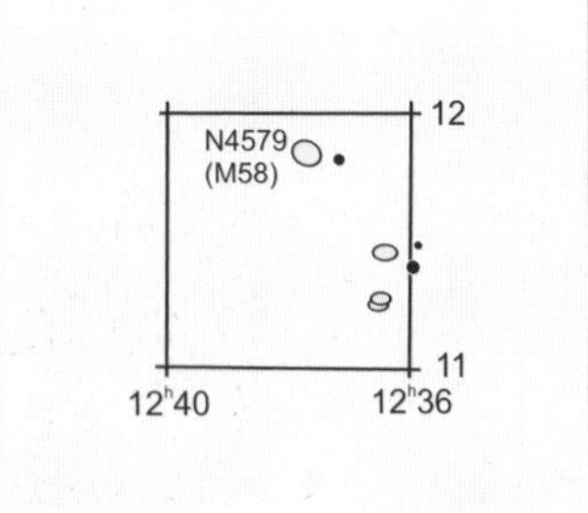

M58 (NGC 4579)

RA:	$12^h\ 37^m\ 45.4^s$	Con:	Virgo
Dec:	11° 48' 43"	Type:	Spiral Galaxy
Size:	5.0' x 4.0'	Mag:	9.8

M58 (NGC 4579) is a bright Sb-type spiral galaxy. Object is located 1° southeast of M89, and appears round with a bright core.

Telescope Aperture:	4" f/5	4" f/9	6" f/7	6" f/9	8" f/6.3	8" f/10	10" f/6.3	10" f/10	12" f/6.3	12" f/10
FOV(35mm film):	2.7° x 4.1°	1.50° x 2.26°	1.29° x 1.93°	1.0° x 1.50°	1.07° x 1.61°	0.68° x 1.02°	0.86° x 1.29°	0.54° x 0.81°	0.72° x 1.07°	0.45° x 0.68°

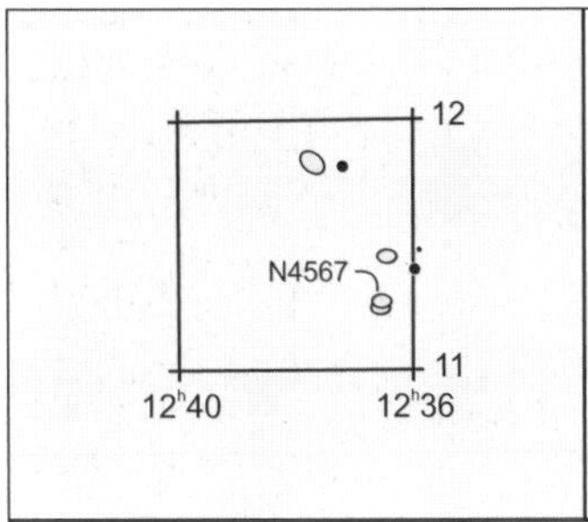

NGC 4567/68 "Siamese Twins"

RA:	$12^h\ 36^m\ 33.4^s$	Con:	Virgo
Dec:	11° 14' 43"	Type:	Spiral Galaxy
Size:	2.8' x 1.5'	Mag:	11.3

NGC 4567 known as the Siamese Twins is a bright spiral galaxy with a close companion. Object is located 30' southwest of M58.

Telescope Aperture:	4" f/5	4" f/9	6" f/7	6" f/9	8" f/6.3	8" f/10	10" f/6.3	10" f/10	12" f/6.3	12" f/10
FOV(35mm film):	2.7° x 4.1°	1.50° x 2.26°	1.29° x 1.93°	1.0° x 1.50°	1.07° x 1.61°	0.68° x 1.02°	0.86° x 1.29°	0.54° x 0.81°	0.72° x 1.07°	0.45° x 0.68°

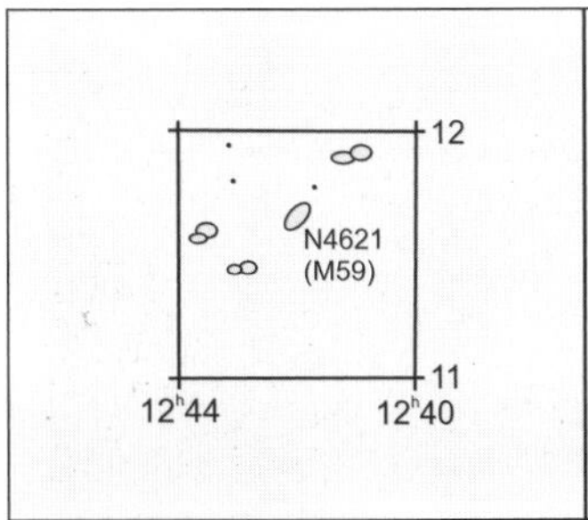

M59 (NGC 4621)

RA:	$12^h\ 42^m\ 3.4^s$	Con:	Virgo
Dec:	11° 38' 43"	Type:	Elliptical Galaxy
Size:	1.4' x 1.0'	Mag:	9.8

M59 (NGC 4621) is a bright elliptical galaxy located about 2° east of the Twins. Object appears as an elongated band with a bright core.

Telescope Aperture:	4" f/5	4" f/9	6" f/7	6" f/9	8" f/6.3	8" f/10	10" f/6.3	10" f/10	12" f/6.3	12" f/10
FOV(35mm film):	2.7° x 4.1°	1.50° x 2.26°	1.29° x 1.93°	1.0° x 1.50°	1.07° x 1.61°	0.68° x 1.02°	0.86° x 1.29°	0.54° x 0.81°	0.72° x 1.07°	0.45° x 0.68°

M60 (NGC 4649)

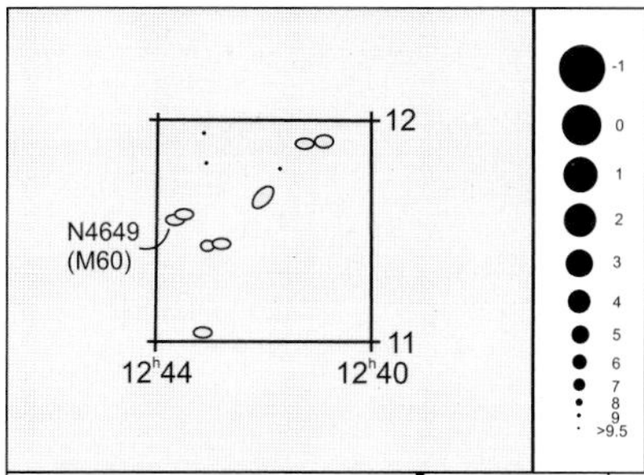

RA:	12h 43m 45.4s	Con:	Virgo
Dec:	11° 32' 43"	Type:	Elliptical Galaxy
Size:	3.0' x 2.0'	Mag:	8.8

M60 (NGC 4649) is an elliptical galaxy that has a small companion galaxy. Object appears roundish with a bright core.

Telescope Aperture:	4" f/5	4" f/9	6" f/7	6" f/9	8" f/6.3	8" f/10	10" f/6.3	10" f/10	12" f/6.3	12" f/10
FOV(35mm film):	2.7° x 4.1°	1.50° x 2.26°	1.29° x 1.93°	1.0° x 1.50°	1.07° x 1.61°	0.68° x 1.02°	0.86° x 1.29°	0.54° x 0.81°	0.72° x 1.07°	0.45° x 0.68°

M49 (NGC 4472)

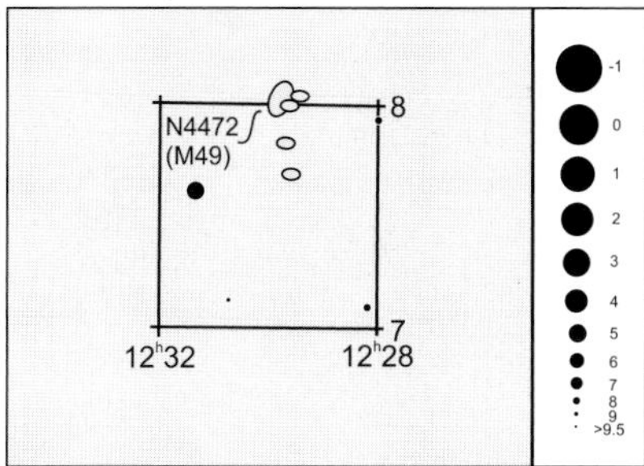

RA:	12h 29m 51.4s	Con:	Virgo
Dec:	07° 59' 42"	Type:	Elliptical Galaxy
Size:	2.2' x 1.8'	Mag:	8.4

M49 (NGC 4472) is found due south of the Virgo cluster center. Object is an elliptical galaxy that appears round with a bright core.

Telescope Aperture:	4" f/5	4" f/9	6" f/7	6" f/9	8" f/6.3	8" f/10	10" f/6.3	10" f/10	12" f/6.3	12" f/10
FOV(35mm film):	2.7° x 4.1°	1.50° x 2.26°	1.29° x 1.93°	1.0° x 1.50°	1.07° x 1.61°	0.68° x 1.02°	0.86° x 1.29°	0.54° x 0.81°	0.72° x 1.07°	0.45° x 0.68°

M61 (NGC 4303)

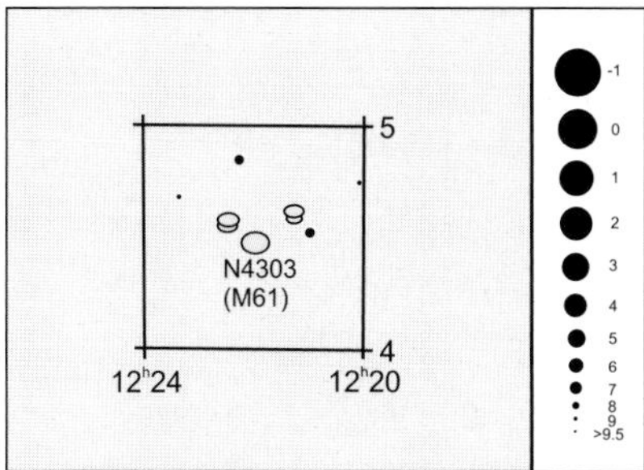

RA:	12h 21m 57.3s	Con:	Virgo
Dec:	04° 27' 41"	Type:	Spiral Galaxy
Size:	6.0' x 5.5'	Mag:	9.7

M61 (NGC 4303) is a face-on Sc-type spiral galaxy.

Telescope Aperture:	4" f/5	4" f/9	6" f/7	6" f/9	8" f/6.3	8" f/10	10" f/6.3	10" f/10	12" f/6.3	12" f/10
FOV(35mm film):	2.7° x 4.1°	1.50° x 2.26°	1.29° x 1.93°	1.0° x 1.50°	1.07° x 1.61°	0.68° x 1.02°	0.86° x 1.29°	0.54° x 0.81°	0.72° x 1.07°	0.45° x 0.68°

M104 (NGC 4594) "Sombrero Galaxy"

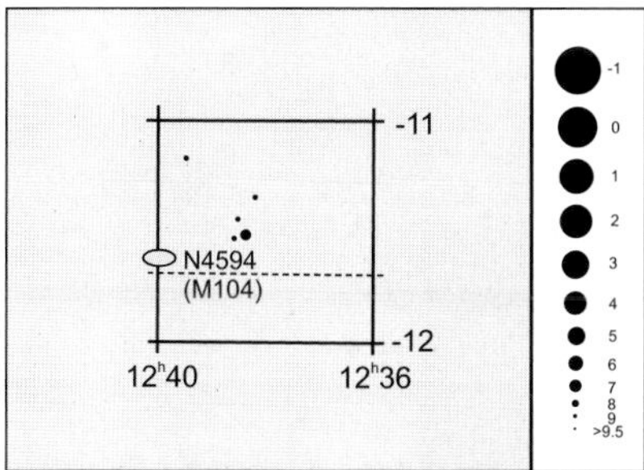

RA:	12h 40m 3.5s	Con:	Virgo
Dec:	-11° 37' 25"	Type:	Spiral Galaxy
Size:	8.0' x 5.0'	Mag:	8.3

M104 (NGC 4594) is known as the Sombrero galaxy. Object is edge-on to our line of sight, and is located on the southern border with Corvus. Object displays visible dust lanes.

Telescope Aperture:	4" f/5	4" f/9	6" f/7	6" f/9	8" f/6.3	8" f/10	10" f/6.3	10" f/10	12" f/6.3	12" f/10
FOV(35mm film):	2.7° x 4.1°	1.50° x 2.26°	1.29° x 1.93°	1.0° x 1.50°	1.07° x 1.61°	0.68° x 1.02°	0.86° x 1.29°	0.54° x 0.81°	0.72° x 1.07°	0.45° x 0.68°

Vul

Crosses Prime Meridian:
August thru September

SUMMER

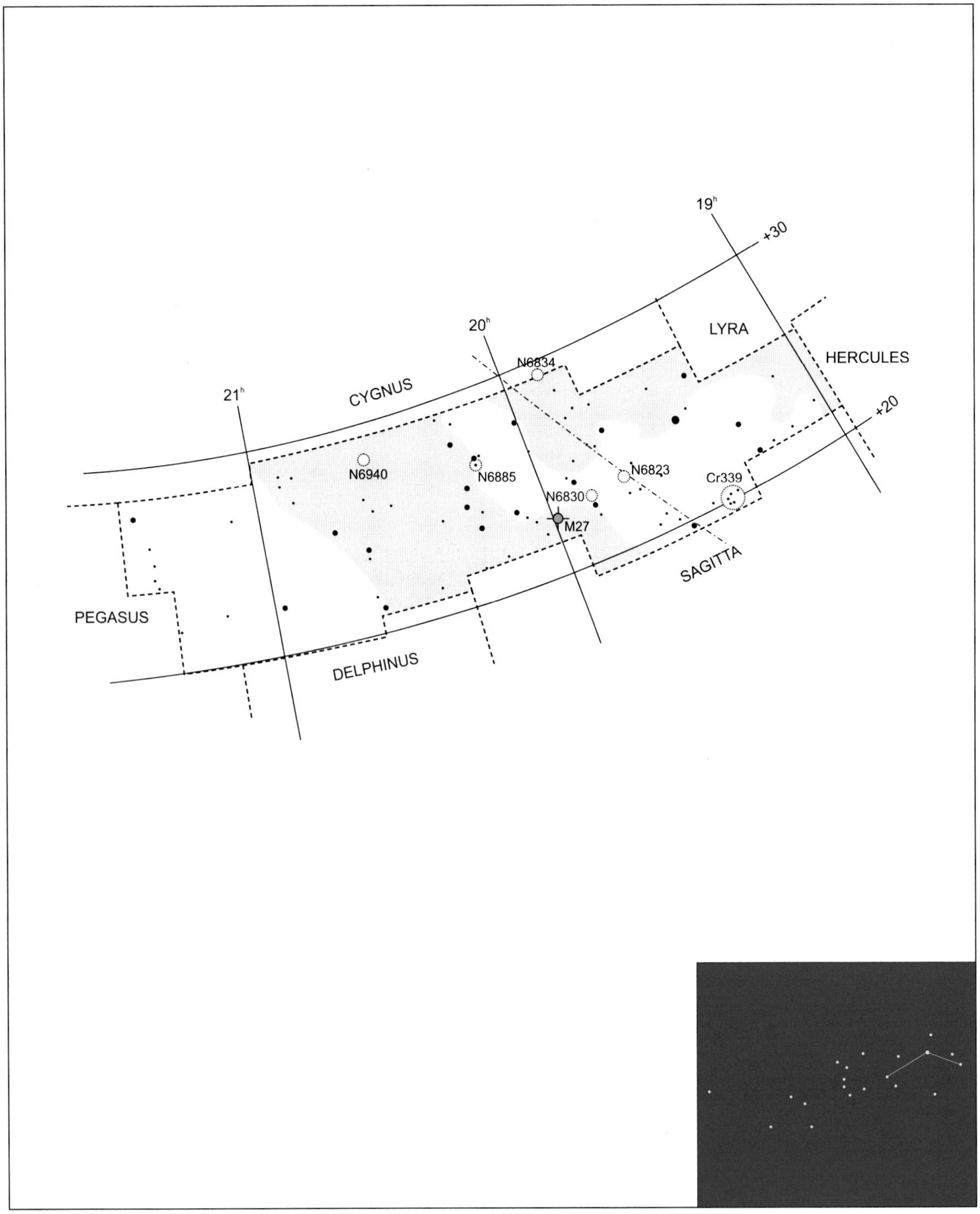

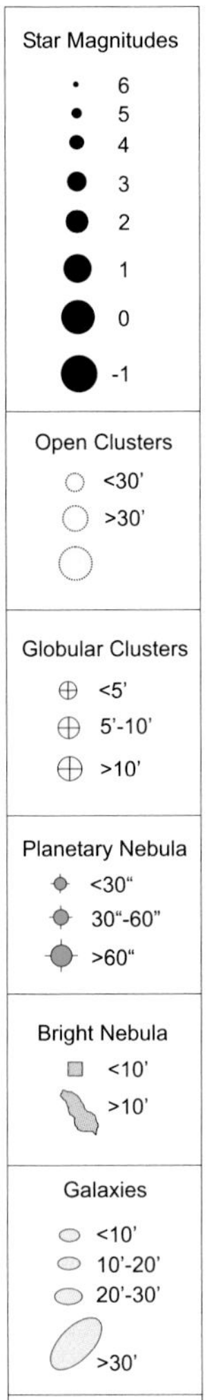

VULPECULA

Constellation Facts:

Vulpecula; (vul-PEK-u-luh)

Vulpecula, the Fox.
This constellation is in the Milky Way, south of Cygnus.
The constellation covers 268 square degrees.

Constellation is visible from 90° N to 61° S. Partially visible from 61° S to 90° S.

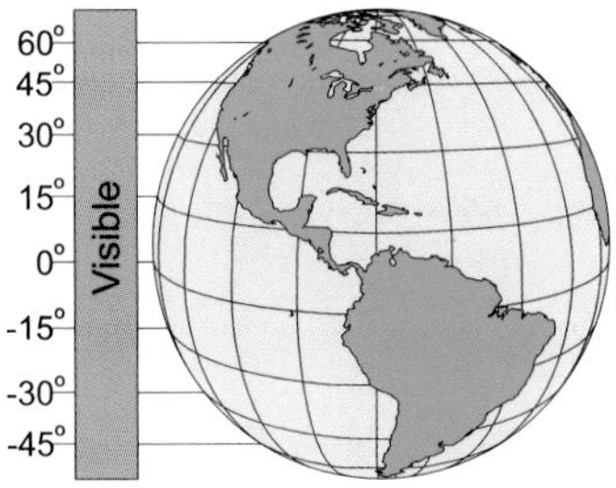

Sh2-88

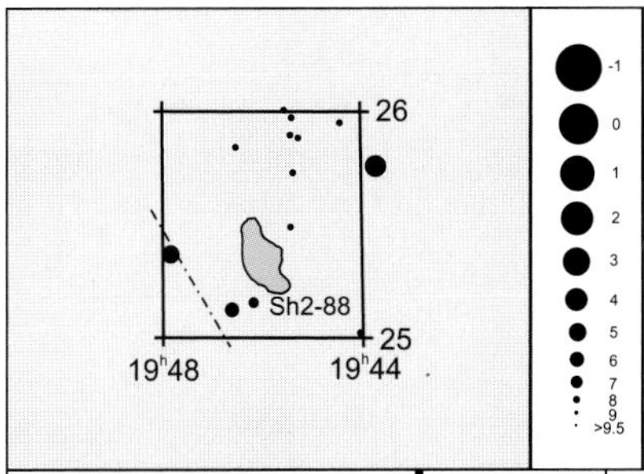

RA:	$19^h\ 46^m\ 0.0^s$	Con:	Vulpecula
Dec:	25° 20' 0.0"	Type:	Emission Nebula
Size:	18.0' x 6.0'	Mag:	

Sh2-88 is a bright emission nebula, with two bright knots.

Telescope Aperture:	4" f/5	4" f/9	6" f/7	6" f/9	8" f/6.3	8" f/10	10" f/6.3	10" f/10	12" f/6.3	12" f/10
FOV(35mm film):	2.7° x 4.1°	1.50° x 2.26°	1.29° x 1.93°	1.0° x 1.50°	1.07° x 1.61°	0.68° x 1.02°	0.86° x 1.29°	0.54° x 0.81°	0.72° x 1.07°	0.45° x 0.68°

NGC 6820

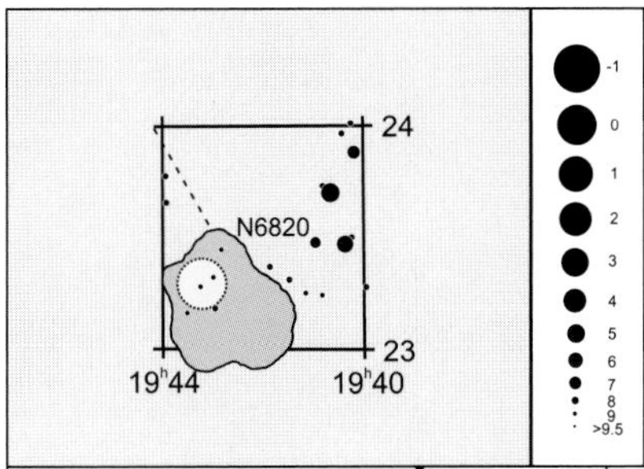

RA:	$19^h\ 43^m\ 10.5^s$	Con:	Vulpecula
Dec:	23° 17' 14"	Type:	Emission Nebula
Size:	40.0' x 30.0'	Mag:	

NGC 6820 is an emission nebula with low surface brightness. Object has a very mottled appearance.

Telescope Aperture:	4" f/5	4" f/9	6" f/7	6" f/9	8" f/6.3	8" f/10	10" f/6.3	10" f/10	12" f/6.3	12" f/10
FOV(35mm film):	2.7° x 4.1°	1.50° x 2.26°	1.29° x 1.93°	1.0° x 1.50°	1.07° x 1.61°	0.68° x 1.02°	0.86° x 1.29°	0.54° x 0.81°	0.72° x 1.07°	0.45° x 0.68°

M27 (NGC 6853) "Dumbell Nebula"

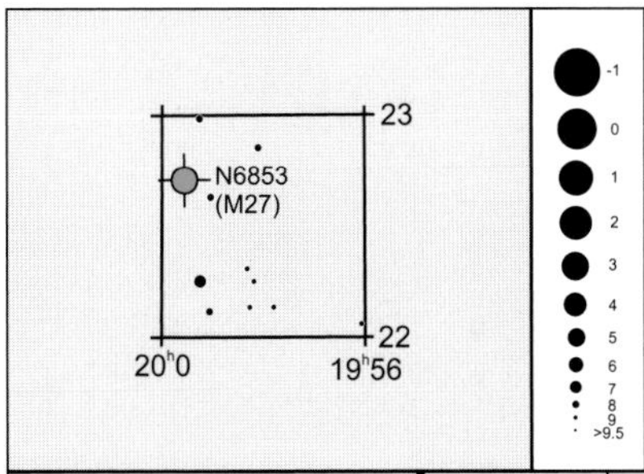

RA:	$19^h\ 59^m\ 40.6^s$	Con:	Vulpecula
Dec:	22° 43' 15"	Type:	Planetary Nebula
Size:	15.2'	Mag:	8.1

M27 (NGC 6853) is known as the Dumbell Nebula. Object is a bright irregular planetary nebula with a well defined double lobed structure.

Telescope Aperture:	4" f/5	4" f/9	6" f/7	6" f/9	8" f/6.3	8" f/10	10" f/6.3	10" f/10	12" f/6.3	12" f/10
FOV(35mm film):	2.7° x 4.1°	1.50° x 2.26°	1.29° x 1.93°	1.0° x 1.50°	1.07° x 1.61°	0.68° x 1.02°	0.86° x 1.29°	0.54° x 0.81°	0.72° x 1.07°	0.45° x 0.68°

Index

Index

Index

Index

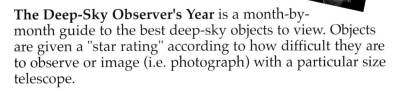

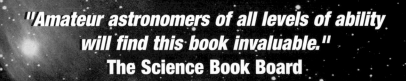